Y0-BCR-517

Introduction to

Broadband Communication Systems

Introduction to

Broadband Communication Systems

Cajetan M. Akujuobi • Matthew N. O. Sadiku

Prairie View A & M University
Texas, U.S.A.

Boca Raton London New York

Chapman & Hall/CRC is an imprint of the
Taylor & Francis Group, an **informa** business

Chapman & Hall/CRC
Taylor & Francis Group
6000 Broken Sound Parkway NW, Suite 300
Boca Raton, FL 33487-2742

Printed in the United States of America on acid-free paper
10 9 8 7 6 5 4 3 2 1

International Standard Book Number-13: 978-1-4200-6149-9 (Hardcover)

Library of Congress Cataloging-in-Publication Data

Akujuobi, Cajetan M.
Introduction to broadband communication systems / Cajetan M. Akujuobi and Matthew N.O. Sadiku.
p. cm.
Includes bibliographical references and index.
ISBN 978-1-4200-6149-9 (alk. paper)
1. Computer networks. 2. Broadband communication systems. I. Sadiku, Matthew N.O. II. Title.

TK5105.5.A3945 2007
621.382--dc22 2007020078

Visit the Taylor & Francis Web site at
http://www.taylorandfrancis.com

and the CRC Press Web site at
http://www.crcpress.com

Dedication

Dedicated to our families:
Caroline, Obinna, and Chijioke
Kikelomo, Motunrayo, Ann, and Joyce

Preface

Broadband communication is a type of telecommunication that supplies multiple channels of data in a single communications platform using some form of wave or frequency division multiplexing. In other words, broadband refers to the telecommunication in which a wide band of frequencies is available to transmit data. Broadband offers a new brand of services where data, voice, and video, commonly known as multimedia, can be delivered together as one packet. Some of the networks that are available for providing these types of services are asynchronous transfer mode (ATM), frame relay, and leased lines. These networks are instrumental in supplying customers with broadband services that have the potential of eventually overtaking the traditional dial-up Internet. Broadband communication systems distribute broadband services on the transmit end and also allow access to the services transmitted by the broadband provider on the receive end.

This book covers the typical broadband communication network systems. The question of why broadband networks are important in modern-day telecommunications is also covered. As evident from the outline, the book is divided into five parts. The book begins by providing the fundamental concepts of broadband communication systems. Part 1 discusses Internet-based networks—Internet, intranets, and extranets. Part 2 addresses networking technologies—X.25 and frame relay, fiber channel, synchronous optical network (SONET), virtual private network (VPN), integrated service digital network (ISDN), broadband ISDN (B-ISDN), and asynchronous transfer mode (ATM). Part 3 focuses on access networks—digital subscriber line (DSL), cable modem, and passive optical networks. Part 4 deals with wireless networks—wireless data services, personal communications service (PCS), and satellite communications. Part 5 covers network management and network security. The book concludes with networking testing, fault tolerance, and analysis.

The book is designed for a one- or two-semester course for senior-year undergraduate and beginning graduate engineering students. The prerequisite for taking the course is background knowledge of communication systems theory in general. The book can be used in giving short seminars on broadband communication systems. It may also serve as a reference for engineers developing broadband communication systems, managers involved in broadband communication systems planning, and researchers and instructors of broadband communication systems.

We would like to acknowledge the support we received from Dr. Milton Bryant, Dean of the College of Engineering, and Dr. John Attia, Head of the Department of Electrical and Computer Engineering, at Prairie View A&M University. Special thanks are due to the reviewers who went over the first and second drafts of the book. We extend our thanks to Shumon Alam, Research Associate at the Center of Excellence for Communication Systems Technology Research (CECSTR) at Prairie View A&M University, for his valuable contributions and help with drawing some of the figures. We also thank Jie Shen for his contributions. We thank our undergraduate and graduate students who we have taught Broadband and Advanced Broadband Communication Systems courses to over the years. Finally, we owe much to our wives (Caroline and Kikelomo) and our children (Obinna, Chijioke, Motunrayo, Ann, and Joyce). Without their constant encouragement and love, this book would never have been completed.

Send your comments and suggestions on this book to us at cmakujuobi@pvamu.edu or mnsadiku@pvamu.edu.

About the Authors

Cajetan M. Akujuobi received his O.N.D. from Institute of Management and Technology, Enugu, Nigeria in 1974, the B.S. degree from Southern University, Baton Rouge, LA., in 1980, the M.S. degree from Tuskegee University, AL., in 1983, all in electrical & electronics engineering, the Ph.D. degree from George Mason University, Fairfax, VA., in 1995, in electrical engineering, and the M.B.A. degree from Hampton University, Hampton, VA., in 1987.

He is a full Professor in the Department of Electrical Engineering and is the founding Director of the Analog Mixed Signal, DSP Solutions and High Speed (Broadband) Communication Programs at Prairie View A&M University, Prairie View, Texas. He is also the founding Director of the Center of Excellence for Communication Systems Technology Research (CECSTR). He has worked in such corporations as Texas Instruments, Advanced Hardware Architecture, Schlumberger, Data Race, Spectrum Engineering, Intelsat, and Bell Laboratories.

His research interests include signal/image processing and communication systems (broadband telecommunications) using such tools as wavelet and fractal transforms. His other research interests are in the areas of DSP solutions, analog mixed signal systems, and control-based communication systems. He was a participant and collaborative member of the ANSI T1E1.4 Working Group that had the technical responsibility of developing T1.413, Issue 2 ADSL standard.

He has published extensively and has written many technical reports. He was selected as one of the U.S. representatives for engineering educational and consultation mission to Asia in 1989. Prof. Akujuobi is a Senior Member of IEEE & ISA and a member of ASEE, SPIE, and Sigma Xi, the Scientific Research Society. He is one of the founding corporate members of the IEEE Standards Association (IEEE-SA), Industry Advisory Committee (IAC).

Matthew N. O. Sadiku received his B. Sc. degree in 1978 from Ahmadu Bello University, Zaria, Nigeria and his M.Sc. and Ph.D. degrees from Tennessee Technological University, Cookeville, TN in 1982 and 1984 respectively. From 1984 to 1988, he was an assistant professor at Florida Atlantic University, where he did graduate work in computer science. From 1988 to 2000, he was at Temple University, Philadelphia, PA, where he became a full professor. From 2000 to 2002, he was with Lucent/Avaya, Holmdel, NJ as a system engineer and with Boeing Satellite Systems as a senior scientist. He is presently a professor at Prairie View A&M University.

He is the author of over 150 professional papers and over 20 books including "Elements of Electromagnetics" (Oxford, 4th ed., 2007), "Fundamentals of Electric Circuits" (McGraw-Hill, 3rd. ed.,2007, with C. Alexander), "Numerical Techniques in Electromagnetics" (CRC, 2nd ed., 2001), and "Metropolitan Area Networks" (CRC Press, 1995). Some of his books have been translated into Korean, Chinese (and Chinese Long Form in Taiwan), Italian, Portuguese, and Spanish. He was the recipient of the 2000 McGraw-Hill/Jacob Millman Award for outstanding contributions in the field of electrical engineering.

His current research interests are in the areas of numerical modeling of electromagnetic systems and computer communication networks. He is a registered professional engineer and a senior member of the Institute of Electrical and Electronics Engineers (IEEE). He was the IEEE Region 2 Student Activities Committee Chairman. He was an associate editor for IEEE Transactions on Education.

Contents

1 Fundamental Concepts

You're either part of the solution or part of the problem.

—Eldridge Cleaver

Broadband communication is a type of telecommunication that supplies multiple channels of data in a single communications platform using some form of wave or frequency division multiplexing. In other words, broadband refers to telecommunication in which a wide band of frequencies is available to transmit data. In order to receive broadband services you must have broadband access, and this broadband access is the major issue in terms of allowing new applications to be used. Broadband offers a new brand of services where data, voice, and video, commonly known as multimedia, can be delivered together as one packet. Some of the networks that are available for providing these types of services are asynchronous transfer mode (ATM), frame relay, and leased lines. These networks are instrumental in supplying customers with broadband services that have the potential of eventually overtaking the traditional dial-up Internet. Broadband communication systems distribute broadband services on the transmit end and also allow access to the services transmitted by the broadband provider on the receive end.

In this chapter, the fundamental concepts of broadband communications will be discussed. The key components of broadband communication systems, communication network architecture, internetworking, and the various standards that influence them will be discussed.

1.1 COMPONENTS OF BROADBAND COMMUNICATION SYSTEMS

The components that affect the broadband telecommunication industry can be classified as the switched and dedicated components. The switched components, also known as circuit switching, or telephone calls, are connected on the public switch telephone network (PSTN). It is composed of three phases, which are

- call setup
- call transmission
- call termination

The dedicated components are the private lines for businesses that need 24-hour-a-day access. With the dedicated components, the user has permanent access with no dial-up.

The elements of data communication consist of the following components:

- transmitter
- receiver
- transmission medium
- communication equipment

Many components are classified as customer premise equipment (CPE). They are workstations, printers, servers, and mainframes. In order for the broadband networks to be able to communicate with each other, a common language called "protocol" must be established. There are different types of protocols that can be used on a broadband network depending on the classification and infrastructure of the network. The NetBIOS enhanced user interface (NetBEUI), the Internet work packet exchange/sequenced packet exchange (IPX/SPX), and the transmission control protocol/Internet protocol (TCP/IP) are three of the broadband network protocols in use today. The NetBEUI is nonroutable while the IPX/SPX and TCP/IP are routable.

DEFINITION: A set of rules and procedures that permit the orderly exchange of information within and across a network is called protocol.

There are various hardware components and media that make up the telecommunication network. These components are essential to the broadband telecommunication industry. The transmission of network signals depends on these components. Some of these are

- Fiber optics, used for high transmission rates, uses a modulated light beam to carry data signals. Fiber optics is manufactured in two versions, which are
  - single mode
  - multi mode
- A firewall is used to prevent unauthorized access to various Internet sites. It acts as an Internet police so that unauthorized users don't have access to private networks. Many companies have implemented firewalls to avoid private documents and information from leaking out into the public.
- A router, often called the "central offices" of the Internet, sends the messages from one system to another when there are many paths. The firewall and the filtering are contained in the router component.

1.2 COMMUNICATIONS NETWORK ARCHITECTURE

There are various basic broadband telecommunication network architectures. They are as follows:

- Local area network (LAN)
- Metropolitan area network (MAN)
- Wide area network (WAN)

These networks service various areas depending on the geographic size of the area. LAN was designed to connect computers, printers, and other devices within an office building and or college campus. It is often looked at as a cluster of network systems, which are capable of communicating. LAN can be connected in three different topologies: bus, ring, and star. Metropolitan access network was designed to aid a city or a large business facility with different buildings. MAN is just a larger version of LAN and it stands between LAN and WAN. MAN is often connected through a fiber-distributed data interface. Wide access network helps various cities and states communicate with each other. WAN was designed to service wide separate geographic areas. WAN is used to help countries stay connected to other countries across the world. Without WAN, much of the global trading and commerce would not be possible.

Figure 1.1 shows a typical broadband communication network architecture. It consists of the central office (CO) and the residential and business (user) sections. The digital subscriber line access multiplexer (DSLAM) that handles all the traffic generated from the user (residential or business) and the data center make up part of the CO. The DSLAM takes many customer connections and puts them on a single high-speed connection to the Internet. Figure 1.2 and Figure 1.3 describe the individual sections of the typical broadband network. Figure 1.2 shows a diagram of the DSLAM and switch together in the CO for access to the public switched telephone network (PSTN) and Internet.

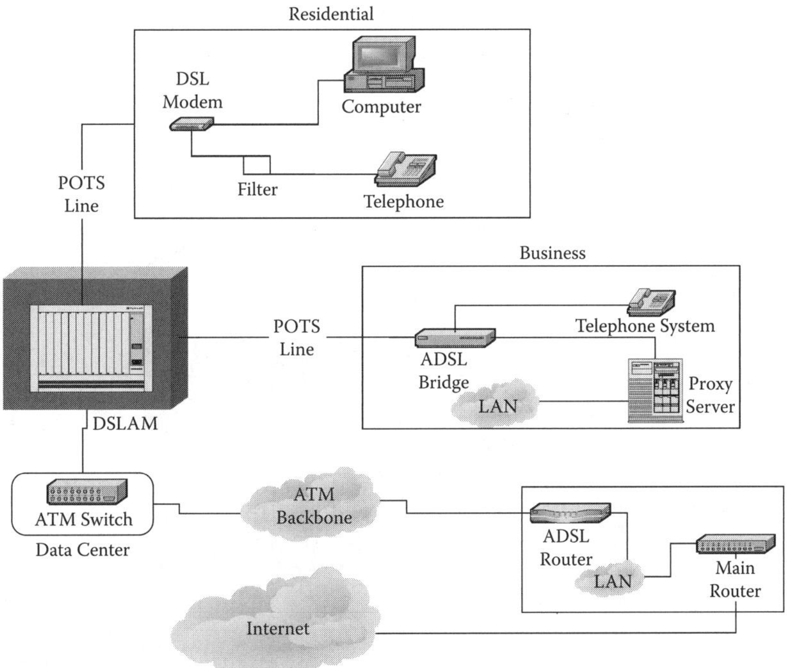

FIGURE 1.1 Typical broadband network architecture.

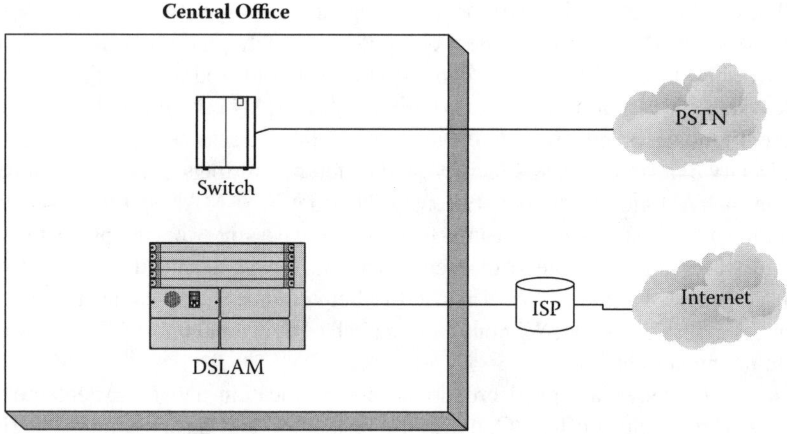

FIGURE 1.2 Diagram of the DSLAM and switch together in the CO for access to the PSTN and Internet.

The asynchronous transfer mode (ATM) switch in Figure 1.2 is a local exchange switch that allows a connection to the PSTN for telephone access. The Internet service provider (ISP) is the company that provides the Internet services that the customer pays for. The CO is connected to two types of customers—residential and business. Each of these customers will normally have different equipment that allows access to the network provided to them. Figure 1.3 displays a diagram of the typical setup of an office network used by a business customer.

The proxy server in Figure 1.3 is a server that acts as an intelligent access point between the business user and the Internet in order to provide security, administrative

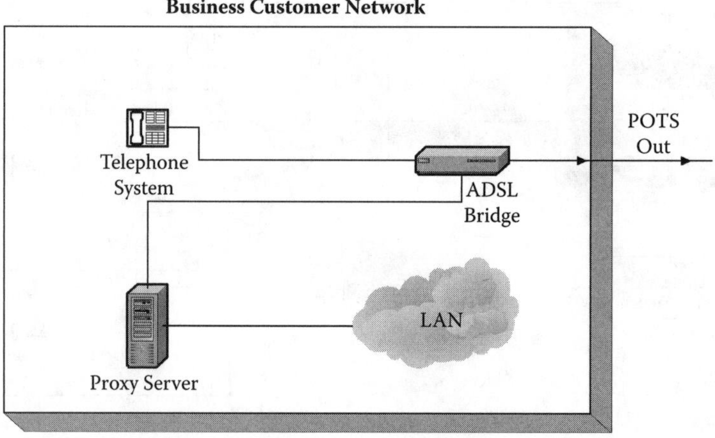

FIGURE 1.3 Typical office digital subscriber line (DSL) network setup for a business customer.

control, and caching service for the LAN in the business. The proxy server as a whole separates the internal network of the business from the outside network. The ADSL bridge in Figure 1.3 is a connection device that connects the LAN of the reference business user to another LAN that encompasses the same type of network protocol. The bridge is intelligent enough to decipher whether to "pass" information sent by a user that is intended for someone inside the same network LAN or whether to "forward" the information sent by the user that is intended for someone outside the reference network to an interconnected LAN. The ADSL bridge also separates telephone usage from Internet usage by the user.

In the residential network, the user needs a DSL modem for high-speed Internet access, which is obtained by using an Ethernet adapter for connecting the DSL to the computer. Ethernet cable, also known as CAT 5 is used to connect the DSL to the computer. A filter for the phone line in the house is also needed so that certain signals cannot be passed through to avoid damaging the phone system. A cable modem could also be used instead of DSL, but the configuration would be slightly different and will be discussed in the next section.

1.3 CABLE BROADBAND DATA NETWORK ARCHITECTURE

The basic broadband cable network architecture as shown in Figure 1.4 consists of three distinct blocks: regional cable headend, distribution hub, and fiber node (or local distributor). From the distribution hub, coax cable is usually used to connect customer premises. The local hub can work as stand-alone for data services or incorporate TV signals for cable TV operations.

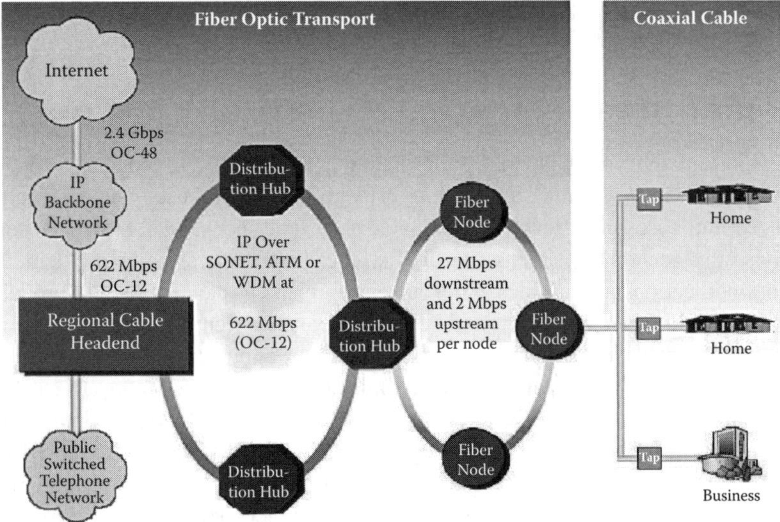

FIGURE 1.4 Complete cable data architecture from Kinetic Strategies, Inc.

1.4 THE IMPORTANCE OF BROADBAND NETWORK ARCHITECTURES

Broadband network architecture is not a luxury but rather a necessity for various reasons, in this day and age. On September 11, 2001 during the terrorism attack in New York, there was an unbelievable number of families trying to get in touch with their loved ones, but the phone lines were overloaded by the number of calls. The cable broadband networks played a vital role in enabling millions of Americans to contact their friends and co-workers.

While broadband network architectures can help connect more people in times of crisis, the main driving factors are economics and revenues. Almost every use for broadband network architecture is geared toward generating profit, and the telecommunications industry today is desperately seeking a gigantic boost because it may not be in good shape. Broadband networks are still in their adolescence and maturation is slow and steady. The Organization of Economic Cooperation and Development reports that there are 500 million subscriptions to standard telephone and about 22 million subscriptions to broadband Internet access; this shows how much broadband Internet access architecture can potentially grow.

The growth of applications in education, health care, e-government, and business offers substantial productivity growth for the broadband sector. The authors of a report in In-Stat and MDR, "Challenges and Opportunities in Broadband Networks," note that about 90% of K-12 schools and 95% of higher learning institutions employ broadband Internet access in the United States. The addition of broadband allows teachers to use Internet content in their lessons, and without this type of access, students would have to wait for the content to download, which could be frustrating to the students and cause them to lose interest. The ability to incorporate real-time voice and video for the students by the teachers is also enhancing the learning experience. Distance learning applications are also being used to teach students by online courses using broadband communication systems network architectures.

1.5 THE FUTURE OF BROADBAND TELECOMMUNICATIONS

One important application is videoconferencing in the health care industry, which will allow specialists to pay more attention to critical patients and use videoconferencing in less serious cases, in order to save valuable time. The specialist doctors will be able to advise the general physicians on cases where distance is a major problem. With telecommunications in shambles at the current time, the implementation of broadband networks would create more jobs and boost the economy significantly. Since the need for integrating data, voice, and video is needed for new applications that have been proposed, broadband networks are the only means of achieving these innovative ideas.

The future of broadband is in wireless telecommunications because of the flexibility and creativeness of the technology. Applications that are on the horizon are voice over IP (VOIP), which will use data streaming in Web access through wireless communication, interactive data (email), which will incorporate interactivity among parties involved, and video on demand, which will enable users to look at any movie desired by just uploading the movie from the provider.

1.6 INTERNETWORKING

Internetworking consists of individual networks connected together by network connection devices. An enterprise internetwork consists of a LAN and WAN connected to a network interconnection device. The types of devices that can be used in the interconnection of broadband networks are as follows:

- bridges
- WAN switches
- routers
- gateways

Most of the internetworks use telecommunication facilities with data transferring speeds of 19.2 kbps and 56 kbps. Others include digital T1 (1.544 Mbps) and T3 (44.736 Mbps) transfer speeds. The following are the wide area networking facilities used in computer networks:

- fiber channel connectivity
- frame relay access
- switched multimegabit data service (SMDS) access
- narrowband ISDN (N-ISDN) access
- broadband ISDN (B-ISDN) access
- X.25 packet-switched public data network circuits
- synchronous optical network (SONET) facilities
- distributed queue dual bus (DQDB)/MAN access

A signaling mechanism is also used in internetworking broadband system networks. Internetworking or intersystem signaling of a network uses signaling to communicate between networks not users. DTMF is an example of an internetworking protocol that connects networks and switches. Multi-frequency (MF) signaling is one of the common modes used in internetworking. MF is the process of using tone pulses in the frequency range to transmit digits by combining two of six frequencies. Each combination of digits represents a single digit. MF also utilizes KP and ST tones for use in transmitting data. The disadvantage of MF is the fact that MF cannot support the broadband applications that the business world has deemed the future of broadband. Another standard protocol, ATM is the backbone of signaling protocols designed to interconnect networks via the asynchronous transmission line. ATM is a great way to interconnect networks, as there are several advantages:

- granular control and network flexibility
- dynamic bandwidth allocation
- service transparency
- scalability

In addition to the many advantages of ATM, there are also several disadvantages:

- ATM lacks mature congestion control
- ATM can suffer from variable cell delays

There are also other issues with ATM but they are more business oriented.

1.7 STANDARDS

Standards play an important role in broadband communication systems. There are different standards organizations that help in the development of the different standards that govern the various products available in the world. These highly complex technical standards are usually set by committees formed by industry consortia and/or government organizations. The International Standards Organization (ISO) and the Institute of Electrical and Electronics Engineers (IEEE) set the standards for various network systems. Some of the principal communication standards bodies are the United States ANSI Committee T1 on Telecommunications and the Telecommunications Industry Association (TIA); the Japanese Telecommunications Technology Committee (TTC); the European Telecommunications Standards Institute (ETSI); and the International Telecommunications Union (ITU).

The standards take years to develop and ratify and it is a gradual process (see Figure 1.5). For example, in the development of the ANSI T1.413-1995 (Network and Customer Installation Interfaces - Asymmetric Digital Subscriber Line (ADSL) Metallic Interface) standard, some of the work used, especially some of the normative requirements, dated as far back as 1978 and progressed until 1995 when it was finally approved. There were 53 different organizations represented on the subcommittee. Other standards such as ISO/IEC IS 13818-1 (1994), MPEG-2 (systems); ISO/IEC IS 13818-2 (1994), MPEG-2 (video); and ATSC A/52 (1995), (digital audio compression) went through the same process. In the case of the integrated service digital network (ISDN), the concept has been around for over 20 years; however, it has not grown as quickly as expected.

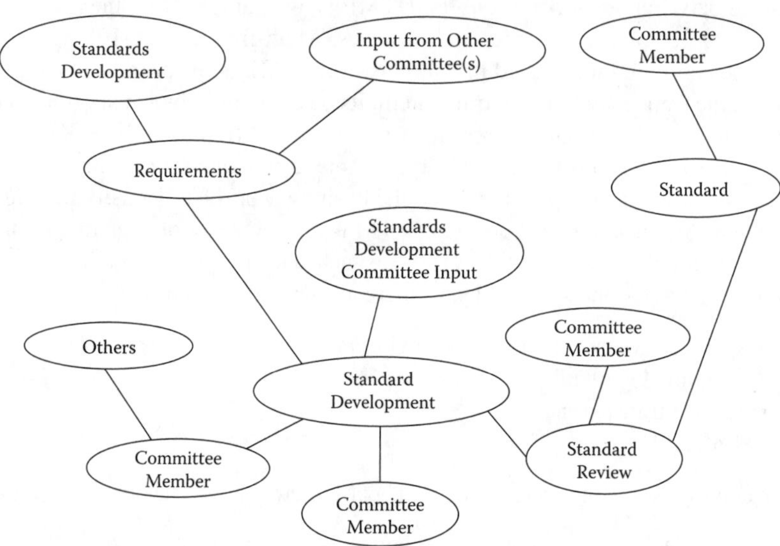

FIGURE 1.5 Flow chart of a typical standards development and approval process.

In 1984, the first red book standard was published by the International Consultative Committee on Telecommunications and Telegraphy (CCITT) now called ITU. It was only in 1992 that standards to permit ISDN products from different vendors became available in the United States. As a result of that, it then became possible to have a coast-to-coast ISDN connection as late as 1993 in North America. Different countries and regions often have their own standards. ANSI and TTC are different standard bodies from the United States and Japan respectively, while ETSI and CITEL (the Inter-American Telecommunications Commission) are European and Latin-American standardization regional bodies respectively. Competing standards within a geographic area are not unheard of; for example, TSACC (the Telecommunications Standards Advisory Council of Canada) and ANSI (American National Standards Institute) are competing standards within the same North American geographic area. It is not unusual for standards to emerge that specify contradictory protocols for the same function; for example, in ISDN, the ANSI uses the T1 protocol (23B + D) while ETSI uses the E1 (30B + D) protocol for the ISDN primary rate interface (PRI). However, when it comes to global standardization, ITU may come in to coordinate the standardization of such issues.

SUMMARY

1. This chapter has introduced us to the fundamental concepts of broadband communication systems.
2. There are many components of a successful networking system.
3. Many devices such as the routers, bridges, switches, and gateways make up a system.
4. There are various ways to connect a broadband system network such as LAN, MAN, and WAN.
5. These networks service various geographic areas depending on the size of the area.
6. Finally, these connections are governed by various organizations that make sure they agree with standard procedures.
7. The International Standards Organization (ISO), ANSI, ITU, IEEE and others set the standards for various network systems.
8. The devices have been constantly improved to increase faster data transferring speeds and error-free transferring.
9. As broadband continues to grow, it will only get better and as broadband is further integrated into our lives, the need for more speed will always put pressure on service providers to come up with a cost-effective way of meeting this need.
10. Broadband systems have taken telecommunications to another level in terms of speed and creative bandwidth management at a cost affordable to potential customers.
11. It has become the access technology application of the future that can enrich the lives of people.
12. As we move towards the next generations of broadband phases of 3G and 4G, wireless broadband will be the leading technology used in the new networks.

REVIEW QUESTIONS

1.1 Broadband offers a new brand of services where data, voice, and video, commonly known as multimedia, can be delivered together as one packet.
   a. True
   b. False

1.2 The components that affect the broadband telecommunication industry can be classified as:
   a. Switched and setup components
   b. Dedicated and call transmission components
   c. Switched and dedicated components
   d. Switched, dedicated, and termination components

1.3 Workstations, printers, servers, and mainframes are classified as:
   a. CPE
   b. CO
   c. TCP/IP
   d. NetBUI

1.4 The basic broadband telecommunication network architectures are LAN, MAN, and WAN.
   a. True
   b. False

1.5 DSL is not a high-speed access technology that works over ordinary telephone lines.
   a. True
   b. False

1.6 What is the cable modem bandwidth range for the downward transmitter?
   a. 50–100 MHz
   b. 80–500 MHz
   c. 500–750 MHz
   d. 70–1000 MHz

1.7 Voice POTS frequency ranges from 0–10 KHz.
   a. True
   b. False

1.8 The transferring speed for a T1 network.
   a. 1.544 Mbps
   b. 44.736 Mbps
   c. 56 Kbps
   d. 19.2 Kbps

1.9 Standards play an important role in broadband communication systems.
   a. True
   b. False

1.10 ANSI T1.413 is an ADSL-based standard.
   a. True
   b. False

Answer: 1.1a, 1.2c, 1.3a, 1.4a, 1.5b, 1.6c, 1.7b, 1.8a, 1.9a, 1.10a.

PROBLEMS

1.1 Define broadband. What is significant about broadband systems?
1.2 Name and describe three basic broadband telecommunication network architectures.
1.3 Describe DSL architectures and how they differ from cable architectures.
1.4 Describe cable architectures and how they differ from DSL architectures.
1.5 What are the key areas of commonality between the DSL and cable architectures?
1.6 What is the importance of broadband network architectures?
1.7 Define what a protocol is in broadband network systems and why it is important.
1.8 Is internetworking a vital aspect of broadband systems and if so, why?
1.9 What are the types of devices that can be used in the interconnection of broadband networks?
1.10 Name the types of WAN facilities that can be used in computer networks?
1.11 Why are standards important in broadband systems?
1.12 Describe the typical standards development and approval process.
1.13 Name the components that can be classified as CPE in broadband communication systems. Why are these components classified as CPE?
1.14 Name the components that can be classified as CO equipment in broadband communication systems. Why are these components classified as CO equipment?
1.15 Name the elements of data communication systems and describe the functions of each.
1.16 Why is ATM a great way to interconnect broadband networks?
1.17 What are the advantages of ATM and why?
1.18 What are the disadvantages of ATM and why?
1.19 Name the different standards-making organizations that may influence broadband communication systems.
1.20 Why is wireless broadband technology a leading technology used in the new networks?

Part 1

Internet-Based Networks

2 Internet

The only thing you take with you when you're gone is what you leave behind.

—John Allston

The Internet, the so-called "information superhighway," is being used by an increasing number of people worldwide. As a result, the Internet has been growing exponentially with the number of machines connected to the network and the amount of network traffic roughly doubling each year. The Internet today is fundamentally changing our social, political, and economic structures, and in many ways obviating geographic boundaries.

In this chapter, we will consider the Internet protocol suite and the next generation Internet (IPv6). We will also consider Internet applications, Internet security, and voice over IP networks.

2.1 INTERNET PROTOCOL SUITE

The Internet is a combination of networks, including the Arpanet, NSFnet, regional networks such as NYsernet, local networks at a number of universities and research institutions, and a number of military networks. A portion of the Internet is illustrated in Figure 2.1.

The **Internet** is a global network of computer networks (or wide area networks) that exchange information via telephone, cable television, wireless networks, and satellite communication technologies.

Each network on the Internet contains anywhere from two to thousands of addressable devices or nodes (computers) connected by communication channels. All computers do not speak the same language, but if they are going to be networked they must share a common set of rules known as ***protocols***. That is where the two most critical protocols, Transmission Control Protocol/Internetworking Protocol (**TCP/IP**), come in. Perhaps the most accurate name for the set of protocols is the *Internet protocol suite*. (TCP and IP are only two of the protocols in this suite). TCP/IP is an agreed upon standard for computer communication over Internet. The protocols are implemented in **software** that runs on each node.

The **TCP/IP** is a layered set of protocols developed to allow computers to share resources across a network.

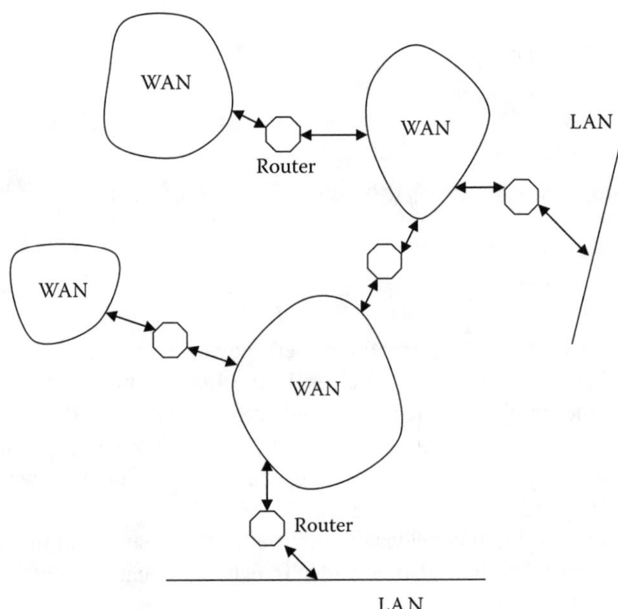

FIGURE 2.1 A piece of the Internet.

Figure 2.2 shows the Internet protocol architecture. The figure is by no means exhaustive, but shows the major protocols and application components common to most commercial TCP/IP software packages and their relationship. As a layered set of protocols, Internet applications generally use four layers:

- *Application layer:* This is where application programs that use the Internet reside. It is the layer with which end users normally interact. Some application-level protocols in most TCP/IP implementations include FTP, TELNET, and SMTP. For example, FTP (file transfer protocol) allows a user

Application Layer	TELNET, FTP, Finger, Http, Gopher, SMTP, etc.	DNS, RIP, SNMP, etc.	
Transport Layer	TCP	UDP	
Internet Layer	IP		ARP
Network Layer	Ethernet, Token ring, X.25, FDDI, ISDN, SMDS, DWDM, Frame Relay, ATM, SONET/SDH, Wireless, xDSL, etc.		

FIGURE 2.2 Abbreviated Internet protocol suite.

to transfer files to and from computers that are connected to the Internet. Security is handled by requiring the user to specify a user name and password for the other computer. TELNET (network terminal protocol for remote login) allows a user to log on to and use other computers that are connected to the Internet regardless of their location. SMTP (simple mail transfer service; for computer mail) allows a user to send messages to users on other computers. Originally, people tended to use only one or two specific computers. They would maintain "mail files" on those machines. The computer mail system is simply a way for you to add a message to another user's mail file.

- *Internet Layer:* This handles addressing and routing of the data. IP (internet protocol) provides the basic service of getting datagrams to their destination. ARP (address resolution protocol) figures out the unique address of devices on the network from their IP addresses. The IP can be described as the common thread that holds the entire Internet together. It is responsible for moving datagrams from one host to another, using various techniques (or "routing" algorithms). Prior to transmitting data, the network layer might subdivide or fragment it into smaller packets for ease of transmission. When all the pieces finally reach the destination, they are reassembled by the network layer into the original datagram. Thus the Internet layer is responsible for breaking up large messages and reassembling them at the destination.
- *Network layer:* This is responsible for routing datagrams from one host to another. It contains the IP as well as several routing protocols that determine the routes of the datagrams. The network layer involves every host and router in the network. It also supervises addressing and congestion control. Protocols at this layer are needed to manage a specific physical medium, such as Ethernet or a point-to-point line.

Although each of these layers provides unique and valuable services, the IP is perhaps the most important to the overall operation of the Internet in general because it is responsible for getting data from one host to another.

TCP/IP is built on connectionless technology. IP provides a *connectionless, unreliable, best-effort* packet delivery service. Information is transferred as a sequence of datagrams. Those datagrams are treated by the network as completely separate. For example, suppose we want to transfer a 15,000 octet file. Most networks cannot handle a 15,000 octet datagram, so the protocols will break this up into something like 30,500-octet datagrams. Each of these datagrams will be sent to the other end. At that point, they will be put back together into the 15,000-octet file. However, while those datagrams are in transit, the network does not know that there is any connection between them. It is perfectly possible that datagram 27 will actually arrive before datagram 19. It is also possible that somewhere in the network, an error will occur, and some datagrams do not get through at all. In that case, that datagram has to be sent again.

For example, when TCP/IP is used on top of X.25, the X.25 interface breaks the datagrams up into 128-byte packets. This is invisible to IP, because the packets are put back together into a single datagram at the other end before being processed by TCP/IP. So in this case, one IP datagram would be carried by several packets.

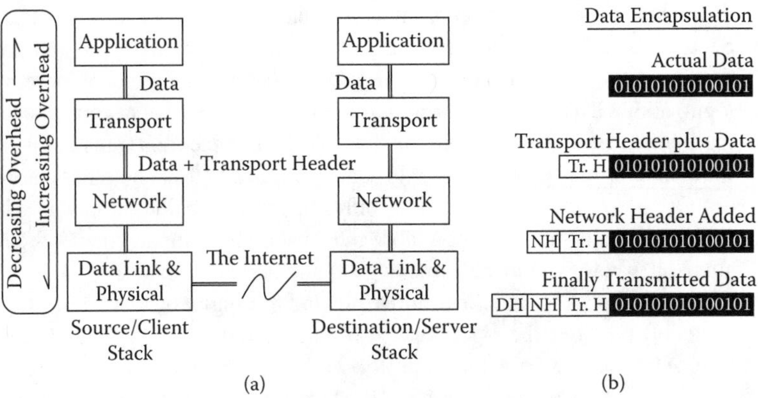

FIGURE 2.3 How data travels through the TCP/IP stack.

As shown in Figure 2.3, each layer of the protocol stack adds a header containing layer-specific information to the data packet. A header for the network layer might include information such as source and destination addresses. The process of appending data with headers is called *encapsulation*. Figure 2.3 shows how data are encapsulated by various headers. The reverse occurs during *decapsulation:* the layers of the receiving stack extract layer-specific information and process the encapsulated data accordingly. It is interesting to note that the process of encapsulation increases the overhead involved in transmitting data.

2.1.1 TCP Level

- TCP puts a header at the front of each datagram. This header actually contains at least 20 octets, but the most important ones are a source and destination "port number" and a "sequence number." Figure 2.4 shows the datagram format. The port numbers are used to keep track of different conversations.

Bit 0 ... 31

Source Port (16)			Destination Port (16)
Sequence Number (32)			
Acknowledgement Number (32)			
Data Offset (4)	Reserved (6)	Flags (6)	Window (16)
Checksum (16)			Urgent Pointer (16)
Options (variable)			Padding (variable)

FIGURE 2.4 TCP header format (20 bytes).

- *Source address*: This defines the application program in the source computer.
- *Destination address*: This defines the application program in the destination computer.
- *Sequence number*: Each datagram has a sequence number that is used at the other end to make sure that the datagrams are in the right order and that no datagrams are missing.
- *Window size*: This field is used to control how much data can be in transit at any one time. Each end indicates how much new data it is currently prepared to absorb by putting the number of octets in its "window" field. As the computer receives data, the amount of space left in its window decreases. When it goes to zero, the sender has to stop. As the receiver processes the data, it increases its window, indicating that it is ready to accept more data. Often the same datagram can be used to acknowledge receipt of a set of data and to give permission for additional new data (by an updated window).
- *Checksum*: This is a number that is computed by adding up all the octets in the datagram. The result is put in the header. TCP at the other end computes the checksum again. If they disagree, then something bad happened to the datagram in transmission, and it is thrown away.
- *Urgent pointer*: This field allows one end to tell the other to skip ahead in its processing of a particular octet.

Other items in the header are generally involved with managing the connection. In order to make sure the datagram has arrived at its destination, the recipient has to send back an "acknowledgement." This is a datagram whose "acknowledgement number" field is filled in. For example, sending a packet with an acknowledgement of 1500 indicates that you have received all the data up to octet number 1500. If the sender does not get an acknowledgement within a reasonable amount of time, it sends the data again.

2.1.2 IP Level

IP is the standard that defines the manner in which the network layers of two hosts interact. All IP packets or datagrams consist of a header part and a text part (payload). The payload has a maximum size limit of 65,536 bytes per packet. The IP header consists of a 20-byte fixed part plus a variable part. Its size is optimized to maximize the packet processing rate without using excessive resources. The header is illustrated in Figure 2.5. The main things in this header are the source and destination Internet address (32-bit addresses, like 128.6.4.194), the protocol number, and the header checksum.

- *Version*: The header begins with a 4-bit version field that keeps track of the version of the IP protocol to which the datagram belongs. This field helps smooth the transition from one version of IP to another, which can take months or even years.
- *Time to live*: The time to live (TTL) is used to limit the life of the packet on the network. This is to take care of a situation in which an IP packet gets

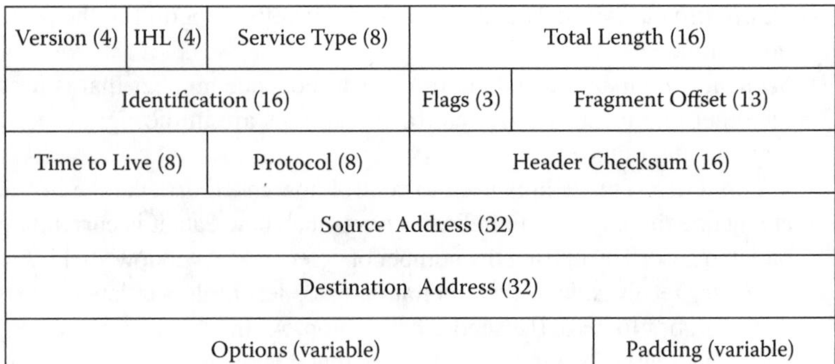

FIGURE 2.5 IP header format (20 bytes).

caught in the system and becomes undeliverable. The TTL field maintains a counter that is normally initialized to thirty count and is decremented each time the packet arrives at a routing step. If the counter reaches zero, the packet is discarded.

- *Protocol*: The protocol number tells IP at the other end to send the datagram to TCP. Although most IP traffic uses TCP, there are other protocols (UDP, ICMP, etc.) that can use IP, so you have to tell IP which protocol to send the datagram to.
- *Header checksum*: The checksum allows IP at the other end to verify that the header was not damaged in transit. Note that TCP and IP have separate checksums. IP needs to be able to verify that the header did not get damaged in transit, or it could send a message to the wrong place.
- *Source address*: The source address designates the originating node's interface to the network.
- *Destination address*: The destination address specifies the interface for an intended recipient or multiple recipients (for multicasting).

TCP sends datagrams to IP with the Internet address of the computer at the other end. IP's job is simply to find a route for the datagram and get it to the other end. In order to allow gateways or other intermediate systems to forward the datagram, it adds its own header, as shown in Figure 2.5.

2.1.3 Addresses and Addressing Scheme

For IP to work, every computer must have its own number to identify itself. This number is called the IP address. You can think of an IP address as similar to your telephone number or postal address. All IP addresses on a particular LAN must start with the same numbers. In addition, every host or router on the Internet has an address that uniquely identifies it and also denotes the network on which it resides. No two machines can have the same IP address. To avoid addressing conflicts, the network numbers have been assigned by the INTERNIC (formerly known simply as NIC). (INTERNIC is the registration center in North America; RIPNIC is the registration center for Europe; and APNIC is for Asia and Pacific countries.)

The network part of the IP address is common for all machines on a local network. It is similar to a postal zip code that is used by a post office to route letters to a general area. The rest of the address on the letter (i.e., the street and house number) are relevant only within that area. It is only used by the local post office to deliver the letter to its final destination. The host part of the IP address performs this same function. There are five types of IP addresses:

- *Class A* format: 126 networks with 16 million hosts each; an IP address in this class starts with a number between 0 and 127.
- *Class B* format: 16,382 networks with up to 64K hosts each; an IP address in this class starts with a number between 128 and 191.
- *Class C* format: 2 million networks with 254 hosts each; an IP address in this class starts with a number between 192 and 223.
- *Class D* format: Used for multicasting, in which a datagram is directed to multiple hosts.
- *Class E* format: Reserved for future use.

Thus, blocks of IP addresses are assigned to individuals or organizations according to one of three categories—Class A, Class B, or Class C. The IP address formats for the three classes are shown in Figure 2.6.

Conventionally, IP addresses are usually written as four integers separated by dots; each integer corresponds to 8 bits. For example, the binary address 10000000 00001011 00000110 00011110 is written in decimal form as 128.11.6.30

Thus, IP addresses are usually written as a sequence of four numbers separated by three dots such as NNN.NNN.HHH.HHH, where N stands for octets that identify network and H denotes octets that specify the host. Each number can be between 0 and 255 except the last number must be between 1 and 254. Inside your computer, an IP address is stored as a 32 bit (4 byte) integer. Dotted decimal notation is just an easy way for humans to write an IP address without having to know the binary numbers that computers work with. Some examples of valid and invalid IP addresses are shown in Table 2.1.

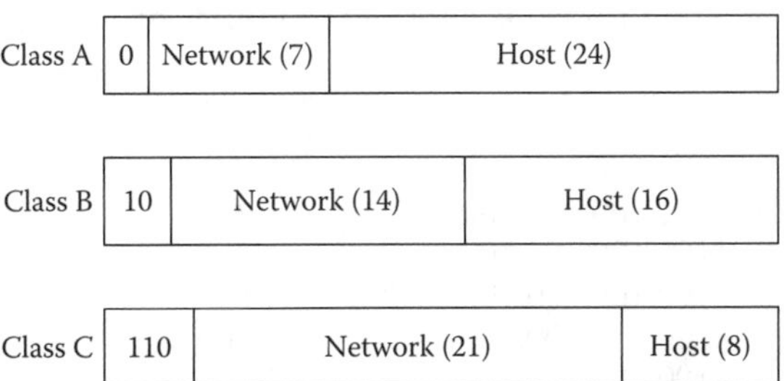

FIGURE 2.6 IP address formats.

TABLE 2.1
Typical Examples of IP Numbers

IP Number	Valid?
200.100.50.25	Yes
1.1.1.1	Yes
1.2.3.0	No
1.1.2.255	No
1.1.255.1	Yes
1.1.700.1	No
6.6.6.6.6.6	No

Because personal computers are not as smart as postmasters they need a fool-proof mathematical method of working out if they should attempt to deliver a packet locally (because the address is on the local LAN) or give it to the router (because the address is not on the local LAN). This is where the *netmask* comes in.

In dotted decimal notation, the most common netmask is 255.255.255.0. Without getting bogged down in mathematics, this netmask says, "If a computer has the first *three* IP numbers the same as mine, then it must be on the same LAN as I am and packets should be sent directly to that computer and not to the router." On the other hand, a netmask of 255.255.0.0 says, "If a computer has the first *two* IP numbers the same as mine, then it must be on the same LAN as I am and packets should be sent directly to that computer and not to the router." More examples are presented in Table 2.2.

We all know that computers like to work with numbers, and humans prefer *names.* With this in mind, the designers of the Internet have set up a system to give names to computers on the Internet. A *DNS Server* is a computer somewhere that can change a *hostname* into an *IP address*, and vice versa. It holds a database similar to a telephone book. It also knows the address of other DNS servers it can query if it does not have an entry for address you are looking for. For example, when you ask your

TABLE 2.2
Checking Whether a Packet is Sent to the Router or Delivered Directly

My Netmask	My IP Address	Recipient's IP Address	Delivery Method
255.255.255.0	100.100.100.100	100.100.100.1	Local
255.255.255.0	100.100.100.100	100.100.100.2	Local
255.255.255.0	100.100.100.100	100.100.99.1	Give to router
255.255.255.0	100.100.100.100	100.99.100.1	Give to router
255.255.255.0	100.100.100.100	99.98.97.100	Give to router
255.0.0.0	100.100.100.100	100.99.100.1	Local
255.255.0.0	100.100.100.100	100.100.99.1	Local

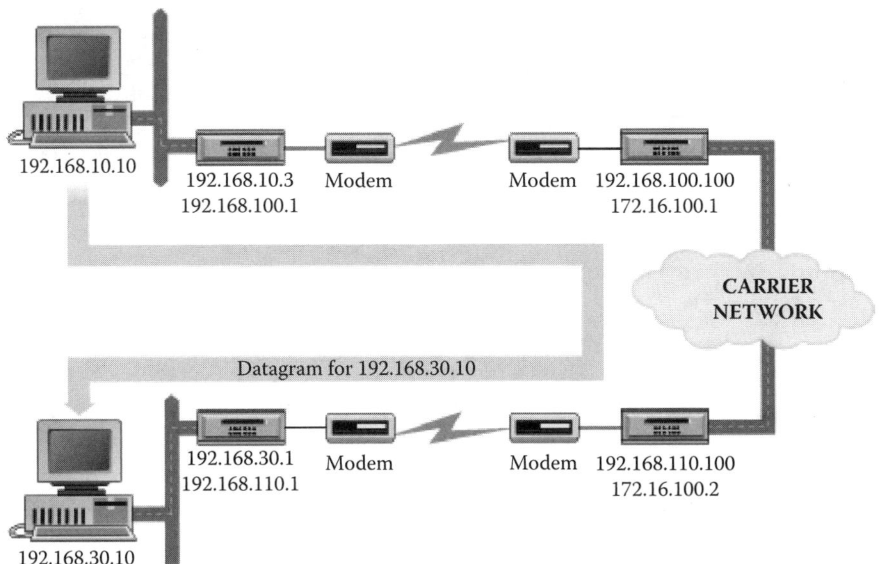

FIGURE 2.7 A typical multi-hop network path.

web browser to connect to www.yahoo.com, your computer asks a DNS server to look up the IP address of www.yahoo.com. The DNS server will tell your computer that it is 204.71.200.68. Your computer then uses that IP address to contact yahoo.

IP handles complex networks the same way it handles small networks: one hop at a time. Eventually, the datagrams will get through. This is illustrated in Figure 2.7, which shows different network segments in between the sending and destination systems.

Example 2.1

A class B address has the binary format 10NNNNNN.NNNNNNNN.HHHHHHHH.HHHHHHHH where N stands for network ID and H denotes host ID. There are 16,384 (64 × 256) class B addresses and each address can accommodate 65,434 (2^{16}–2) host addresses, not counting those with all 0s and all 255s in the least significant two bytes. For class B, the most significant byte of the network ID is between 128 and 191. Therefore, an IP address such as 142.381.7.23 is a class B address.

2.2 IPV6: NEXT-GENERATION INTERNET

Most of today's Internet uses Internet Protocol Version 4 (IPv4), which is now nearly twenty-five years old. Due to the phenomenal growth of the Internet, the rapid increase in palm computers, and the profusion of smart cellular phones and PDAs, the demand for IP addresses has outnumbered the limited supply provided by IPv4. In response to this shortcoming of IPv4, the Internet Engineering Task Force (IETF) approved IPv6 in 1997. IPv4 will be replaced by Internet Protocol Version 6 (IPv6),

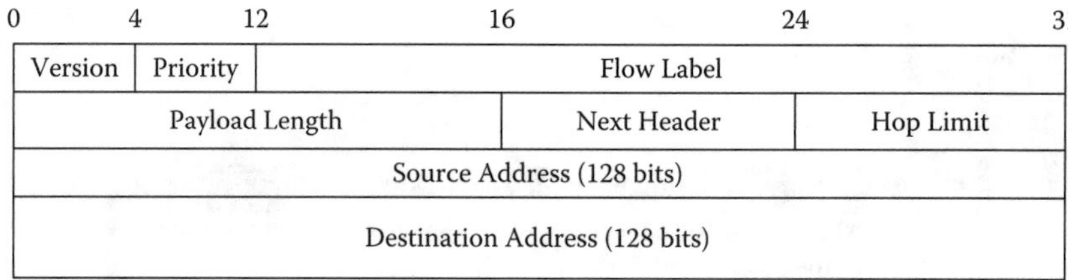

FIGURE 2.8 IPv6 header format.

which is sometimes called the Next-Generation Internet Protocol (or IPng). IPv6 adds many improvements and fixes a number of problems in IPv4, such as the limited number of available IPv4 addresses.

With only a 32-bit address field, IPv4 can assign only 2^{32} different addresses, i.e., 4.29 billion IP addresses, which is inadequate in view of rapid proliferation of networks and the two-level structure of the IP addresses (network number and host number). To solve the problem of severe IP address shortage, IPv6 uses 128-bit addresses instead of the 32-bit addresses of IPv4. That means, IPv6 can have as many as 2^{128} IP addresses, which is roughly 3.4×10^{38} or about 340 billion billion billion billion unique addresses.

The IPv6 packet consists of the IPv6 header, routing header, fragment header, the authentication header, TCP header, and application data. The IPv6 packet header is of fixed length, whereas the IPv4 header is of variable length. The IPv6 header consists of 40 bytes as shown in Figure 2.8. It consists of the following fields:

- *Version* (4 bits): This is the IP version number, which is 6.
- *Priority* (4 bits): This field enables a source to identify the priority of each packet relative to other packets from the same source.
- *Flow label* (24 bits): The source assigns the flow label to all packets that are part of the same flow. A flow may be a single TCP connection or a multiple of TCP connections.
- *Payload length* (16 bits): This field specifies the length of the remaining part of the packet following the header.
- *Next header* (8 bits): This identifies the type of header immediately following the header.
- *Hop limit* (8 bits): This is to set some desired maximum value at the source and the field denotes the remaining number of hops allowed for the packet. It is decremented by 1 at each node the packet passes and the packet is discarded when the hop limit becomes zero.
- *Source address* (128): The address of the source of the packet.
- *Destination address* (128 bits): The address of the recipient of the packet.

An IPv6 address consists of 16 bytes, making it 128 bits long. The address is usually written as eight 16-bit pieces, separately by colons. Each piece is represented by four hexadecimal digits. Two typical examples are FDCE:BC96:4510:7252:CA84:3567:3789:AFFF and 1401:0000:0000:0000:0004:0060:300A:521B.

These addresses are large and not easy to write down or remember. There are ways to compress them, especially when an address has many leading zeros. Address 1 above does not have any leading zeros in any field and cannot be compressed, but address 2 has some leading zeros and can be compressed as follows: 1401:0:0:0:4:60:300A:521B. Note that at least one digit must be present in every field.

There are three types of IPv6 addresses:

1. Unicast is used to identify a single interface.
2. Anycast identifies a set of interfaces. A source may use an anycast address to contact any node from a group of nodes.
3. Multicast identifies a set of interfaces. A packet with multicast address is delivered to all members of the group.

IPv6 offers several advantages over IPv4. First, IPv6 has more addresses. IPv4's 32-bit addressing scheme can only support a theoretical maximum of 4.29 billion IP addresses, whereas IPv6 can support $2^{1,033}$ useful IP addresses. Second, although IPv4 supports extensibility, it limits to only 40 bytes the amount of data than can be used to describe additional functionality provided by the extensions. IPv6 allows considerably longer option descriptions. Third, since IPv6 has a large number of addresses, each device on the Internet can have its own address and this would eliminate the need for manual device configuration or DHCP (dynamic host configuration protocol) servers, which are used to manage and assign addresses within IPv4 networks. Fourth, IPv6 has support for resource allocation. A mechanism known as *flow label* has been added to enable the source to request special handling of the packet. This mechanism can be used to support traffic such as real-time video and audio.

IPv6 is expected to gradually replace IPv4, with the two coexisting for a number of years during a transition period. IPv6 may be most widely deployed in mobile phones, PDAs, and other wireless terminals in the future.

2.3 QUALITY OF SERVICE

This section describes the quality of service (QoS) provided by the Internet. A large portion of Internet users access network resources through web browsers. The QoS experienced by Internet users has gained importance in recent years. The Internet offers a point-to-point delivery service, which is based on "best effort" delivery, which means that data will be delivered as soon as possible, but with no guarantees as to timeliness or actual delivery and no special resources allocated for any of the packets. Under a best-effort scheme, the Internet treats all packets fairly. The original TCP/IP protocol suite was built on the idea of fair and equitable service for everyone. This is adequate for traditional data applications like FTP and Telnet, but inadequate for applications requiring timeliness. The needs of users have changed. QoS guarantees are important if the network capacity is limited, especially for real-time multimedia applications such as voice over IP and IP-TV; such applications often require fixed bit rate and may be delay sensitive.

Since QoS means different things to different people, it is difficult to define QoS to satisfy everybody. However, QoS may be regarded as a standard specifying that data will be delivered within a particular time after transmission.

Quality of service (QoS) refers to the capability of a network to provide better service to selected network traffic over LAN, MAN, and WAN technologies.

QoS is a set of quality requirements on the collective behavior of one or more objects. It may also be regarded as the capability to differentiate between service types so that users can treat classes of traffic differently. For example, QoS parameters that describe the speed and reliability of data transmission include throughput, transit delay, and error rate. QoS is a multidimensional performance objective function including parameters such as delay, throughput, call blocking probability, packet dropping probability, packet loss rate, and delay jitter. The primary goal of QoS is to provide parameters, including dedicated bandwidth, jitter, and delay.

Do we even need to provide QoS? This is a hotly debated issue. Some have the opinion that fibers and wavelength-division multiplexing (WDM) will make bandwidth so abundant and cheap that QoS will be automatically delivered. Others believe that no matter how much bandwidth the networks can provide, a new application will be needed to provide QoS. The debate is beyond the scope of this book. Here we simply note that even if the bandwidth becomes abundant and cheap, it is not going to happen soon. So for now, some simple mechanisms will be needed in order to provide QoS on the Internet.

2.4 APPLICATIONS AND SERVICES

Some of the common applications of the Internet include electronic mail (email), file transfer protocol (FTP), Telnet, and the World Wide Web (WWW). These applications are briefly discussed as follows.

- *Electronic mail*: This is perhaps the most prevalent application of the Internet. In fact, electronic mail was the first application developed for the TCP/IP suites in the 1960s. People have always wanted to correspond with each other in the fastest way possible. Electronic mail (or email) allows people to write back and forth without worrying about the cost. In addition to regular letter writing, email allows you to send documents and electronic greeting cards.
- *File Transfer Protocol (FTP):* This is a standard Internet protocol that uses TCP/IP protocols. The objectives of FTP are to promote sharing of file data reliably and efficiently. FTP provides the simplest way to exchange files between computers on the Internet. As a user, you can use FTP with a simple command line interface to copy, delete, and move files to a server.
- *Telnet:* This is the terminal-handler portion of the TCP/IP protocol suite. Telnet handles the remote login to another Internet host. In other words, Telnet provides a way to access someone else's computer, assuming that you have been permitted to do so. FTP allows you to request specific files from

remote computers, but not to actually be logged on as a user of that computer. With Telnet, you can log on as a regular user with whatever privileges you may have been granted to the specific application and data on that computer. Telnet operates in a client/server environment in which one host (the computer you are using) negotiates opening a session on another computer.

- *World Wide Web (WWW):* The WWW incorporates all of the Internet services. It allows one to retrieve documents, view images, listen to sound files, speak and hear voice, and view programs that run on practically any software in the world. When you log onto the Internet using Netscape or Microsoft's Internet Explorer or some other browser, you are viewing documents on the World Wide Web. The current foundation on which the WWW functions is the programming language called HTML (HyperText Markup Language). HTML is the coded format language used for creating hypertext documents on the World Wide Web and controlling how Web pages appear. Hypertext allows Web pages to contain links, which are areas on a page on which you can click your mouse button to retrieve another document onto your computer. The World Wide Web Consortium (W3C at www.w3.org) was created in 1994 and consists of over 400 organizations that jointly develop interoperable technologies to lead the Web to its full potential.

In addition to these common Internet applications, the Internet provides the following services:

- Educational services such as research using search engines, lifelong learning, and virtual library
- Health care
- Global e-commerce and e-business
- Electronic journals
- Internet relay chat (IRC)
- Network news transfer protocols for news distribution
- Virtual private networks
- Interconnecting networks
- Audio and video players
- Video- and audioconferencing
- Voice over IP networks

There are several emerging applications that can potentially alter the way people perform their routine daily activities. It is needless to say that such applications will create new demands on network resources. One of these will be discussed next.

2.5 VOICE OVER IP

The Public Switched Telephone Network (PSTN) has served voice traffic well over the last 100 years, but its success has been paralleled by the rise of separate networks to support data traffic. Since data traffic is growing much faster than telephone traffic, a lot of interest has been generated in transporting voice over data networks. Voice over IP (VoIP) represents a major step toward the integration of voice and data networks.

> **Voice over IP** (VoIP) or Internet telephony is the ability to make telephone calls (or do whatever we can do today with the PSTN) over IP networks with a suitable quality of service (QoS) and a much superior cost/benefit.

It consists of hardware and software that enable data packet networks to transport real-time voice traffic.

The nifty idea of voice signal traveling over the Internet, rather than the regular telephone wires, became a reality in 1995 when Vocaltec Inc. introduced its Internet phone software. Since then, Internet telephony has advanced rapidly. Internet telephony is attracting more and more users because it offers tremendous cost savings relative to the PSTN.

2.5.1 VoIP Network

VoIP technology enables the real-time transmission of voice signals as packetized data over IP networks that employ the transmission control protocol (TCP), real-time transport protocol (RTP), user datagram protocol (UDP), and Internet protocol (IP) suite. Figure 2.9 illustrates IP network protocols that are currently being used to implement VoIP.

In VoIP systems, analog voice signals are digitized and transmitted as a stream of packets over a digital data network. IP networks allow each packet to independently find the most efficient path to the intended destination, thereby best using the network resources at any given instant. The packets associated with a single source may thus take many different paths to the destination in traversing the network, arriving with different end-to-end delays, arriving out of sequence, or possibly not arriving at all. At the destination, however, the packets are re-assembled and converted back into the original voice signal. VoIP technology insures proper reconstruction of the voice signals, compensating for echoes made audible due to the end-to-end delay, for jitter, and for dropped packets.

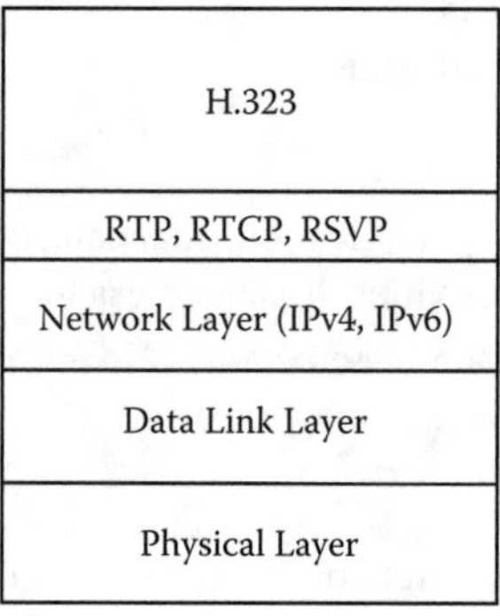

FIGURE 2.9 VoIP protocol structure.

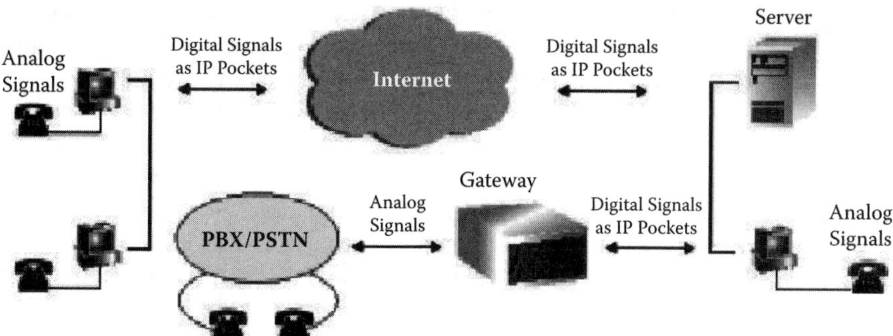

FIGURE 2.10 VoIP infrastructure.

A typical VoIP network consists of a switched LAN infrastructure with WAN links between its main campuses. The network generally requires at least two VoIP gateways. VoIP gateways provide the bridge between the local PSTN and the IP network for both the originating and terminating sides of a call. To support standard telephony signaling, the gateways must employ a common protocol such as the H.323 or MGCP or a proprietary protocol. A typical VoIP infrastructure is shown in Figure 2.10.

In order to provide a voice service with a reasonable measure of quality and deal with network congestion and delay by making better use of bandwidth, there are many techniques that are employed. These techniques include prioritization, fragmentation, jitter buffering, voice compression, silence suppression, and echo cancellation. This is where the various protocols, such as H.323 come in.

- *Prioritization* techniques are related to QoS and are designed to guarantee throughput for certain traffic on the network. This can ensure that voice traffic on a data network is given high priority. Prioritization can be based on location, protocol, or application type. Protocols used to ensure this QoS are RTP (real time protocol) and RSVP (resource reservation protocol).
- *Fragmentation* divides packets into smaller fragments so that their priority can be ensured. This can help reduce the overall delay of voice delivery. However, on IP based networks, this can create extra overhead because of the large size of IP headers (20 bytes). So although necessary, fragmentation alone cannot ensure the reliable delivery of real-time voice applications.
- *Compression* is also necessary. Video signal coming from cameras or tape players is first digitized into a 135-Mbps feed. Since this is a rather high bandwidth, it must first be compressed to a more manageable bit rate. The decompressor reverses the process. Various codecs (coder/decoder) standards have been implemented.
- *Jitter buffering* contributes to latency. Jitter is variation in delay. It is due to the speed variation between quickly and slowly traveling packets. As a result of jitter, voice transmissions will sound unnatural. The jitter buffer stores packets, allowing most of the slower packets to catch up. The less

control in routing, the more jitter that results, and more jitter means a longer jitter buffer. But a longer jitter buffer introduces more latency. Too short a jitter buffer loses too many packets, causing voice quality to tumble.

- *Silence compression* is necessary because in human telephone conversations, generally only about 50% of the full duplex bandwidth is used at any given time. This means that between 50 and 60% of the available bandwidth is not being used. Silence suppression techniques take advantage of this by detecting when there is a gap and then suppresses the transmission of these silences. This can result in more bandwidth being available for other transmissions.
- *Echo* becomes a problem that can detract from the quality of the conversation when the total end-to-end delay of a voice transmission is greater than 50 ms. An echo cancellation unit solves this problem by performing echo cancellation on the signals. ITU G.165 or G.168 provides the standards and requirements for echo cancellation.

2.5.2 Applications and Benefits

Benefits of VoIP will generally include:

- *Cost reduction* (low-cost phone calls): The most common reason to use IP telephony is to save money. IP telephony can yield big cost savings to both corporations and consumers. It is more efficient than the plain old telephone service (POTS) and is poised to undergo huge growth.
- *Convergence of data/voice networks* (simplification): An integrated infrastructure that supports all forms of traffic allows more standardization and reduces the total equipment cost.
- *Consolidation* (centralized management): Since people are among the major significant cost elements in a network, combining operations will consolidate accounting systems and be beneficial. Universal use of IP protocols holds out promise of both reduced complexity and more flexibility.

IP telephony promises better efficiencies in the transport of voice and data, and, as a result, lower telecommunications costs to end users. The real promise of IP telephony, however, will be realized when its advanced services begin to surpass the capabilities of the PSTN. Such future services include:

- Multimedia conferencing: allowing multiple users to communicate by voice and video.
- Multicasting: disseminating voice, video, and data to a large targeted group.
- Collaborative workgroup applications: facilitating verbal and visual interaction of multiple users while sharing access to common data and applications.
- Call center applications: using IP telephony to enhance today's Web-based ecommerce through interaction with live service representatives.
- Unified messaging: consolidating email, pager, voicemail, and fax services into a single IP-based service.
- IP call waiting: expanding the number and type of simultaneous incoming calls beyond the two-voice calls that the PSTN currently supports.

VoIP suffers from QoS problems, which can single-handedly limit the adoption of the technology. When the network is congested, the quality of a VoIP call is quite bad, partly because VoIP bandwidth is not dedicated. The IP infrastructure was designed to send data in non-real time, not for real-time communications. This is a problem that many IP vendors are currently working on. As IP telephony evolves, it will be able to match all the features of voice communications currently supported by the PSTN.

2.6 INTERNET SECURITY

The Internet has become an irreplaceable tool for most businesses. It is the primary source of market news, a medium of direct commerce, and an essential tie to partners and suppliers. But when a private network connects to the Internet, it opens physical links to numerous unknown networks and unknown users. While this interconnectedness brings exciting opportunities for information sharing, it also brings threats to information not meant for sharing. As the number of networks, hosts, and users on the Internet has increased, so too have security incidents. Internet security risks should never be taken lightly, but they can all be managed and minimized—just like other security risks in business. The issue of Internet security is so broad that we cannot do it justice here. (See Chapter 17 for more on network security.) Here are common Internet risks and how they can be avoided:

- *Hackers*: These are unwelcome visitors that tamper with the information on the Internet. If your electronic communications are valuable, interception by clever and greedy hackers cannot be ruled out. There are routine and simple security measures to protect your Internet traffic against the junior-grade hackers. Some hacker masterminds can find new ways to break into computers, but such people are rare.
- *Industrial espionage*: If your competitors are big and wealthy, and your company information could be valuable to them, you may be at risk using the Internet. It is easy to stop this sort of spying. You scramble (encrypt) your messages, and do it in such a way that even well-financed spies will find the messages practically impossible to read.
- *Viruses*: As viruses increase in number and impact, content validation of emails and other information off the Internet is becoming an increasingly critical component of security management. Internet viruses are no more menacing than viruses from CD ROMs or floppy disks. This is one of the easiest risks to manage. One needs anti-virus software to scan files passing through the gateway. This approach, known as *content security*, is an important component of every security policy.
- *Network packet sniffers*: A packet sniffer is a software application that uses a network adapter card in promiscuous mode (a mode in which the network adapter card sends all packets received on the physical network wire to an application for processing) to capture all network packets that are sent across a local area network. Because several network applications distribute network packets in clear text, a packet sniffer can provide

its user with meaningful and often sensitive information, such as user account names and passwords. In a worst-case scenario, an attacker gains access to a system-level user account, which the attacker uses to create a new account that can be used at any time as a back door to get into your network and its resources. Packet sniffers provide information about the topology of your network that many attackers find useful. This information, such as what computers run which services, how many computers are on your network, which computers have access to others, and so on, can be deduced from the information contained within the packets that are distributed across your network as part of necessary daily operations. In addition, a network packet sniffer can be modified to interject new information or change existing information in a packet. By doing so, the attacker can cause network connections to shut down prematurely, as well as change critical information within the packet. The effects of such attacks can be difficult to detect and very costly to correct.

- *IP Spoofing*: This is the use of an internal address to mask the true source of an external transmission. IP spoofing can yield access to user accounts and passwords. For example, an attacker can send e-mail messages to business partners that appear to have originated from someone within your organization. Such attacks are easier when an attacker has a user account and password.
- *Firewall*: A firewall is a secured gateway located between the router and your network to restrict access to your internal systems. A secure gateway forces all network connections to pass through the gateway where they can be examined, evaluated, and then given restricted access to and from selected systems. A firewall is a system that enforces a policy controlling access between multiple networks. It operates at the network layer and inspects packets as they are received. In other words, a firewall's job is to keep intruders out of your network, while still letting you do your job.
- *Password Attacks*: Passwords have long been the front line of defense in protecting information systems and networks. Unfortunately, they are usually the first thing a hacker will try to "break" to gain access to your system, and even well-composed passwords are vulnerable to being intercepted and "stolen" by today's more sophisticated system attackers. Password attack can provide access to accounts that can be used to modify critical network files and services. An example that compromises your network's integrity is an attacker modifying the routing tables for your network. By doing so, the attacker ensures that all network packets are routed to him or her before they are transmitted to their final destination. In such a case, an attacker can monitor all network traffic, effectively becoming a middle man.
- *Encryption*: This is the transformation of data into a form unreadable by anyone without a secret decryption key. Its purpose is to ensure privacy by keeping the information hidden from anyone for whom it is not intended, even those who can see the encrypted data. For example, one may wish to encrypt files on a hard disk to prevent an intruder from reading them.

- *Authentication*: This is a process whereby the receiver of a digital message can be confident of the identity of the sender and/or the integrity of the message. Authentication protocols can be based on either conventional secret-key cryptosystems like DES or on public-key systems like RSA, which uses digital signatures.

These are some of the means available to combat the hacker.

2.7 FLOW CONTROL

Networks cannot afford to accept all the traffic that is offered without control. There must be rules that govern the acceptance of traffic from outside and coordinate the flow inside the network. These rules are referred to as flow control procedure.

> **Flow control** is a mechanism that allows a destination to regulate the flow of packets sent by a source.

It may also be regarded as the process of adjusting the flow of data from one device to another to ensure that the receiving device can handle all of the incoming data. This is particularly important where the sending device is capable of sending data much faster than the receiving device can receive it. In a network, the goal of an effective flow control scheme is to maximize the throughput of every virtual circuit with as small a delay as possible.

There are many flow control mechanisms. A common one is called *xon-xoff*. In this case, the receiving device sends an *xoff* message to the sending device when its buffer is full. This will help the sending device to stop sending data. When the receiving device is ready to receive more data, it sends an *xon* signal. Other mechanisms include: *credit* (or *tokens*), which give permission for message flow; a *rate* at which a given flow may proceed; a *stop-and-go* procedure, which turns a flow on and off according to some criteria; and the introduction of *delay*, so as to slow down the flow. The rate-based and credit-based flow controls are used in ATM (asynchronous transfer mode) networks.

Besides these protocols, there are two popular methods for controlling the flow of data.

- *Stop-and-wait flow control*: In this simplest form of flow control, a source device transmits a frame. After reception, the destination device indicates its willingness to accept another frame by returning an acknowledgement (sequence number of last correctly received frame) for the frame just received. The source must wait until it receives the acknowledgement (ACK) before sending the next frame. Rather than sending all the frames at once, this protocol sends one, waits for an ACK, sends another frame, waits for an ACK, and so on. The destination can thus stop flow of data by simply withholding the ACK. The source must wait until it receives the ACK before sending the next frame. This method fails if the message transmitted is very large.
- *Sliding window flow control*: The basic idea of sliding window protocol is that both sender and receiver keep a "window"' of acknowledgment.

The window is the maximum amount of data that can be sent without waiting for an ACK. With the *sliding window* method, the receiving device can send a single acknowledgement message for multiple packets of data sent in one window. Within that acknowledgement message is information about the receiving device's buffer size, which tells the sending device to increase or decrease the number of packets in the next transmission (this is where the "sliding" in the name comes in). If the receiving device processes the packets at a slower rate than the sending device is transmitting them, it will tell the sending device to decrease the number of packets or temporarily cease transmission altogether, in order to free up room in the buffer. If the receiving entity can process the packets faster than the sending device is transmitting them, it will tell the sending device to increase the number of packets in the next window as the application's buffer can handle more data. The sender keeps the value of expected acknowledgment while the receiver keeps the value of expected receiving frame. When it receives an acknowledgment from the receiver, the sender advances the window. When it receives the expected frame, the receiver advances the window.

In the Internet protocol suite, IP is a best-effort service and TCP is a reliable service. IP provides basic packet forwarding while TCP implements flow controls, acknowledgements, and retransmissions of lost or corrupted packets. TCP manages limited network bandwidth by performing flow control. TCP uses a sliding window algorithm. The Internet allows packets to be discarded if network resources are not available, and demands that protocols make provisions for retransmission.

SUMMARY

1. The TCP/IP has become the foundation on which the Internet and the World Wide Web (WWW) are based.
2. Approved by the Internet Engineering Task Force (IETF), IPv6, as the next generation protocol, can solve the IP addressing space depletion problem of IPv4.
3. Internet QoS will have a significant economic impact as the Internet evolves from a best-effort connection engine into a universal service-delivery medium for large volumes of real-time voice and data.
4. Common applications of the Internet include electronic mail (email), file transfer protocol (FTP), Telnet, and the World Wide Web (WWW).
5. Voice-over-IP technology allows voice calls originated and terminated at standard telephones to be conveyed over IP networks.
6. Internet security is concerned with how information can be protected. Due to the fact that software is constantly evolving, security is an endless process.
7. Flow control is the set of mechanisms for assuring that a transmitting entity does not overwhelm a receiving entity with data.

REVIEW QUESTIONS

2.1 The IP in TCP/IP stands for:
   a. Internet provider
   b. Internetworking program
   c. Internet protocol
   d. International protocol

2.2 Which OSI layer corresponds to Telnet?
   a. Application layer
   b. Presentation layer
   c. Session layer
   d. Data link layer
   e. Physical layer

2.3 Which TCP/IP layer corresponds to the top three OSI model layers?
   a. Application layer
   b. Transport layer
   c. Network layer
   d. Internet layer

2.4 TCP and UDP are the two common transport layer protocols.
   a. True
   b. False

2.5 Information sent using TCP/IP is divided up into pieces called:
   a. Packets
   b. Frames
   c. Cells
   d. Datagrams

2.6 Which of the following are not valid IP addresses?
   a. 45.6.9.36.2
   b. 215.146.31.9
   c. 13.214.41.210
   d. 21.401.42.1
   e. 49.25.32.22

2.7 Which of the following is a class C host address?
   a. 20.4.5.9
   b. 220.1.4.1
   c. 120.1.5.3
   d. 125.1.2.3

2.8 IPv6 was approved by:
   a. IEEE
   b. ANSI
   c. ITU
   d. IETF

2.9 Which Internet protocol makes it possible to move files across the Internet?
   a. email
   b. Telnet

c. FTP
d. HTTP

2.10 Which of the Internet protocols provides a terminal session on a remote computer?
a. TCP/IP
b. FTP
c. SNMP
d. SMTP
e. TELNET

Answer: 2.1c, 2.2a, 2.3a, 2.4a, 2.5d, 2.6a,d, 2.7b, 2.8d, 2.9c, 2.10e

PROBLEMS

2.1 It is called TCP/IP. Mention three other protocols in the suite and what they are for.
2.2 What is the difference between TCP and IP in TCP/IP?
2.3 What is the difference between TCP and UDP if they both operate at the transport layer?
2.4 What are the layers of the TCP/IP model and how are they related to the OSI model?
2.5 Show how Telnet uses TCP/IP protocols to communicate between hosts A and B.
2.6 Describe SMTP.
2.7 What is the purpose of address resolution protocol (ARP)?
2.8 How do we prevent a datagram from wandering the Internet forever?
2.9 What field in the TCP header is responsible for flow control?
2.10 How is the IP address constructed? Give two examples of IP addresses.
2.11 Determine how many Class C IP addresses there are. Give three examples of such addresses.
2.12 Using the Web, find the Request for Comment (RFC) for the Transmission Control Protocol (TCP).
2.13 What is the class of each of the following IP addresses?

a.	10011001	10001111	01111000	10011101
b.	11011001	10001111	01111000	10011101
c.	01111001	10001111	01111000	10011101
d.	11101001	10001111	01111000	10011101
e.	11110001	10001111	01111000	10011101

Netmask	Subnetwork Number	Next Hop
255.255.255.0	157.84.134.0	R1
255.255. 110.1	157.84.134.182	R2
255.172.45.6	154.9.67.132	R3

2.14 What advantages does IPv6 have over IPv4?
2.15 Describe three common applications of the Internet.
2.16 Most VoIP applications are considered to be real-time activities. Mention three of these applications.
2.17 How is QoS a major problem with VoIP?
2.18 Mention two *common network attacks used to undermine network security.*
2.19 What is a firewall?
2.20 Discuss two flow control mechanisms.

3 Intranet and Extranet

> Do not worry about your difficulties in mathematics. I can assure you mine are still greater.
>
> —Albert Einstein

As WWW-based Internet spread everywhere, the IP technology was adopted in enterprises to set up their private or semiprivate networks similar to the Internet. The Internet-like network inside an enterprise is called Intranet. When the enterprise shares some of its Intranet resources with its partners, the expanded network is called Extranet.

Toward the end of 1995 the term "Intranet" was coined to describe the uptake of standard Internet tools and services, internally within corporations. One of the best reasons why Intranet rose so quickly is that it leverages existing technologies based on TCP/IP. The cost to deploy Intranet is thus comparatively cheap. HTTP is the most important application protocol above TCP/IP and is the foundation of Intranet. Most information is formatted in HTTP hypertext and is displayed to users in browsers. The consistent interface and intuitive operations of browser-based applications require little or no need to train the users. There are three discernable stages in the development of an Intranet: first, information publishing; second, interactive application; and third, integration with suppliers and customers—the growth of Extranet.

With Intranet, enterprises connect their internal resources on a common platform, which reduces the operational cost. Similarly, enterprises would like to cut cost by moving more and more business online. The business-to-business communication over the Internet is implemented by developing Extranets. Extranet could be understood as a part of one or more organization's Intranet that has been expanded onto the Internet. Security is a critical issue in Extranet. Users who want to access a resource from other organizations' Extranet must provide a password. Communication on Extranet must be encrypted.

3.1 INTRANET OVERVIEW

An Intranet is a privately maintained, secure computer network based on the communication standards of the Internet. It is a smaller version of the Internet that only the internal members of an organization can see. At the same time it is available remotely over the Internet allowing access from any physical location. Organizations can create, within their walls, a manageable, secure version of the World Wide Web. These internal Webs are growing from an explosion in the use and understanding of Internet technology.

Intranet is an IP-based system, which means the whole set of higher-level protocols above TCP/IP could be applied in Intranet without modification. Some most famous examples include: HTTP (hyper text transfer protocol, used for web access), SMTP (simple mail transfer protocol, used for email service), and FTP (file transfer protocol, used for remote file access). These protocols were originally developed in the context of global Internet. But since the Intranet can be considered as a small Internet, all these protocols and their applications are naturally ready for use in Intranet.

Maybe the first feeling for Intranet users is the convenience of information distribution. All information is available online and is up-to-date. There is no need to print and distribute hard copies of information to all employees. A lot of information displayed on web pages is generated dynamically based on the content of various enterprise databases. Any change of the database caused by anybody is reflected on the web page immediately. This dynamic manner of information publication significantly increased the work efficiency of an enterprise. Clearly, the value of information is dependent upon its accuracy and currency.

The benefits and implications of an Intranet can be enormous. They come from the fact that it provides a single easy-to-use interface, which allows quick access to company information and resources such as:

- Customer/supplier information
- Company handbooks and policies
- Employee information
- Workflow systems

There are many ways of categorizing what an Intranet does. They are

- **Information publishing:** using the Intranet to deliver news and other information in the form of directories and web documents.
- **Email:** implementing an email system that integrates seamlessly with the Intranet, allowing information to be both "pushed" and "pulled."
- **Document management:** using the Intranet to allow users to view, print, and work collaboratively on office documents (word-processed documents, spreadsheets, presentations, etc.).
- **Training:** using the Intranet to deliver training at the desktop.
- **Workflow:** using the Intranet to automate administrative processes.
- **Databases and other bespoke systems:** using the Intranet as a front-end to organization-specific system, such as corporate databases.
- **Discussion:** using the Intranet as a means for users to discuss and debate issues.

Resource sharing is not restricted to text documents. Because many companies have high-speed LANs, or Local Area Networks, video clips, animation, and audio can reside on an Intranet server and be delivered to employees' desktops. This is a good way to see highlights of a corporate meeting. What is done with that information in this new Web-enabled environment has a huge impact. Intranets allow an organization to spend less time on things that bring no value such as chasing down the right information to solve a problem.

Productivity increases as corporate knowledge is more accessible and the data is more accurate. Flexibility in time of delivery of knowledge is gained as information is always a click away. Intranets allow for a place where boundaries are lowered and information exchange is encouraged. This leads to more informed employees with the ability to make better, faster decisions. This in turn leads to better productivity and more time for revenue generation.

Traditionally, corporations relied on proprietary hardware and software systems to network its computers, a costly and time-consuming process made more difficult when offices are scattered around the world. Even under the best of conditions, sharing information among different hardware platforms, file formats, and software is not an easy task. By using off-the-shelf Internet technology, Intranets solve this problem, making internal communication and collaboration much easier.

Intranets use HTML to create documents and TCP/IP to transmit information across the network. Information is stored on one or more company servers and accessed by using a web browser, such as Navigator or Internet Explorer. This self-contained, miniature Internet can have all the same features—individual home pages, newsgroups, email—but access is restricted to company employees and contractors.

For employees who are already familiar with surfing the Web, learning how to navigate the company Intranet requires little training. Intranet web pages have the same point-and-click interface. While it is useful for an Intranet to connect to the Internet, it is certainly not essential. Even if they do connect externally, companies restrict access to their Intranet from the Internet by building a firewall. With so much corporate information available on internal servers, security is essential.

3.2 EXTRANET OVERVIEW

Extranets are becoming a very popular means for business partners to exchange information. An Extranet is a private network that uses Internet protocols and the public telecommunication system to securely share part of a business's information or operations with suppliers, vendors, partners, customers, or other businesses. An Extranet can be viewed as part of a company's Intranet that is extended to users outside the company. It creates a business-to-business platform. The same benefits that Web technologies have brought to corporate Intranets are now starting to accelerate business between businesses.

Extranet brings benefits to not only a single organization, but also to its partners, clients, and suppliers. It allows the organization to improve the communication channels, share up-to-date business information with partners, clients, or suppliers, and integrate the business processes.

In order to give an intuitive explanation of Extranet, let us consider the example of a cell phone manufacturer. It is a large enterprise that has locations around the world, sells the products to countries around the world, and purchases parts such as chips also from vendors all around the world. As a modern enterprise, this company has its own Intranet that can share information to its employees. However, some of the information is not strictly constrained inside the enterprise. Interaction with customers and vendors is inevitable. For example, the design department wants to listen to the feedback to their design from customers, so that they can improve their design and make new products. Customers always have opinions on the cell phones they bought from this enterprise, maybe good, maybe bad. The enterprise keeps an eye on its warehouse storage of the core chips that it needs to create cell phones. Whenever the storage drops below a threshold, it needs to set up a new purchase deal with the vendors. But how does this enterprise know if the vendors happen to have this kind of chip available?

Extranet provides the solution to bridge the information between inside and outside an enterprise. As shown in Figure 3.1, there are two sets of routers and firewalls between the Intranet and Internet. The outside firewall and outside router provide a first level of protection from Internet attacks. The inside firewall and inside router provide a second level of protection. Between these two levels of protection is what is normally called the demilitarized zone (DMZ). No pass-through traffic is allowed in this zone. Some servers are set up here that allow access from outside or from inside, but access cannot pass through these servers. Several subnets are possible in this zone, which allows different outside partners to access designated servers. This zone is the Extranet. A web server in the Extranet provides an interface to interact with customers. The main purpose is to collect feedback from customers. The feedback is instantly available to the design department because the web server could be accessed from both inside and outside the enterprise—of course, with different access permission. A database server could also be set up on the Extranet to reflect the dynamic warehouse storage of core chips. This database is valid for access to the

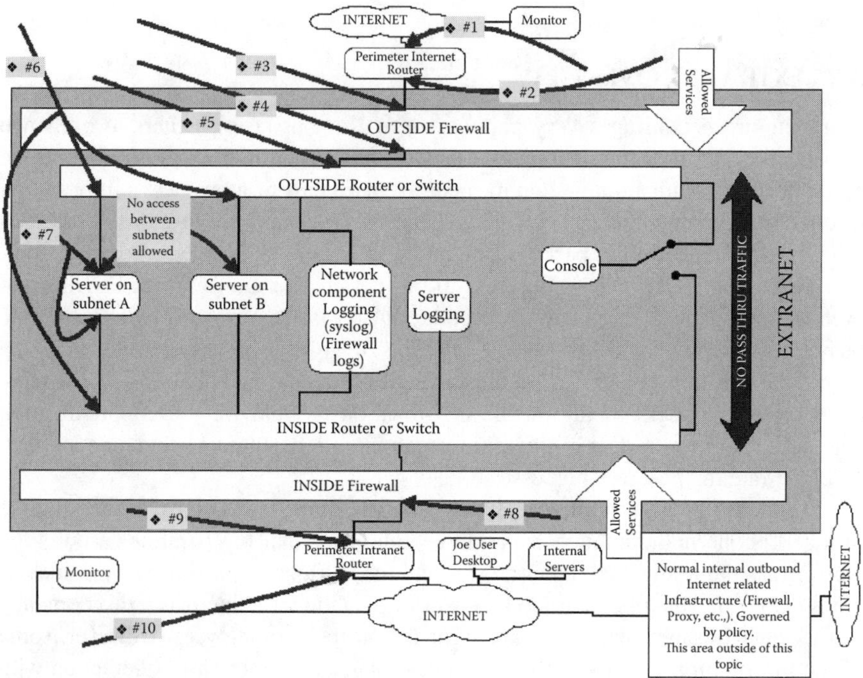

FIGURE 3.1 The configuration of Extranet.

vendors who provide the chips. In this scenario, the information sharing between enterprise and vendors guarantees that the vendors could ship the chips automatically whenever the storage drops under a threshold.

Whereas an Intranet resides behind a firewall and is accessible only to people who are members of the same company or organization, an Extranet provides various levels of accessibility to outsiders. You can access an Extranet only if you have a valid username and password, and your identity determines which parts of the Extranet you can view.

3.3 INTRANET TECHNOLOGIES

The essence of Intranet is the ubiquitously available IP, which is a side effect of widespread Internet access. All the servers and desktops that have the ability to interact with Internet must have an IP protocol stack installed on them.

3.3.1 Web Server

The major appearance of an Intranet is the web-base platform that connects employees and business partners. The web server is obviously the fundamental technology to run the Intranet.

3.3.2 HTML Development

Because of the importance of HTML-based content, the selection of the HTML development tools affects the efficiency of publishing information in the Intranet. There are two types of HTML development tools.

- **WYSIWYG (what you see is what you get):** These tools allow the non-technical user to create sites without the knowledge of HTML code (for example, Microsoft Front Page).
- **HTML editors:** for the more advanced user who knows HTML and wants to have more control of the page structure.

The company may decide to purchase licensing for both types of tools and roll them out according to skill level. It should be noted that these types of tools are slowly converging into one.

3.3.3 Browser

Netscape/Mozilla versus Microsoft; it was the battle of the century. This type of competition should not appear within the organization that is planning the Intranet. The cost to maintain multiple browsers, and make information publication compatible to different browsers is expensive. In an organization Intranet, a single desktop browser is preferred.

However, if the organization wants to open its Intranet to the partners and vendors (Extranet), support for both browsers is required, because the administrator has no control over the partners' network.

3.3.4 Security

More and more corporations are placing sensitive information on an Intranet. Security mechanisms must be carefully deployed to protect sensitive data. The following list shows some common threats for organizations:

- **Snooping or eavesdropping:** The risk of having someone "overhear" data being sent over the Intranet.
- **User impersonation:** The risk of having users gain access by pretending to be someone else.
- **Unauthorized access:** The risk of having users obtain access to confidential data.

Several security techniques could be applied to the Intranet to conquer the above threats as shown below:

- **User authentication:** An LDAP or ODBC compliant central database could be used to store the use authentication information. Passwords must be maintained and updated periodically. Digital signature is another choice for both authentication and data encryption.
- **Access control:** The permission to access all the resources on the Intranet must be carefully planned. Once logged in, users' activities must be tracked throughout the session.
- **Data encryption:** Even if someone from outside the Internet has compromised the security of the Intranet, the sensitive information must still be under the protection of strong data encryption. Using SSL and secure email services may also protect real-time internal communication.

3.3.5 VPN

The simplest Intranet consists of only one LAN. All the users and devices are constrained in a single network. This is suitable for small organizations that have only one location. For those large organizations that have geographically distributed branches, the Intranet must span over different LANs.

A range of options is available to interconnect multiple sites. The most intuitive and most costly solution is leasing dedicated private lines to interconnect all the sites. As the number of sites increases, the number of leased lines increases rapidly in order to build a full mesh network.

The better solution is the emerging secure Virtual Private Network (VPN) technique that uses public Internet as the transfer infrastructure. Tunnels are created in the Internet between Intranet sites such that the Intranet traffic is encapsulated inside the regular IP packets. The source and destination of encapsulating traffic are the gateways that sit at the access point of Intranet sites. Many techniques provide the tunneling function, such as PPTP, L2TP, L2F, and IPSec tunnel mode. To protect the private information transmitted in the tunnel, security mechanisms must be used. A good combination of both tunneling and security is L2TP/IPSec.

3.3.6 Content Management System

Content is king, not only on the Internet but an Intranet as well. Once Web sites are up, systems should be in place to assist publishers in keeping their content up to date. We recommend having this system in place before beginning Intranet development.

Content management systems can be bought off the shelf or custom built. No matter how you to decide to acquire a system for your publishers, follow this wish list for what you should ask for:

- Document check-in/check-out
- Versioning
- Content approval workflow
- Open-standards database and template creation
- Database management and file system management
- Dynamic page generation
- Link management
- Document conversion
- User-friendly content authoring
- Personalization
- Access control or built-in security
- Usage analysis

3.4 EXTRANET TECHNOLOGIES

Four technologies govern the Extranet. These four technologies are the Web, the Internet, groupware applications, and firewalls. The Web's major contribution to Extranets is freedom from specific platforms. A significant limitation of the traditional client/server development approach is the requirement that each workstation execute the same software. It means that when there is a release of a new version of software, it has to be distributed to all the computers that will use it. The cost of the software distribution, manually or automatically, is expensive. Developing applications with a Web-based interface solves the software development and support problem. The upgrade of a web-based application on a Web server is available to all the browsers instantly.

An Extranet requires much stricter security and privacy than Intranet. Extranets require firewall server management, the issuance and use of digital certificates or similar means of user authentication, encryption of messages, and the use of virtual private networks (VPN) that tunnel through the public network. For most companies that are installing Extranets, protection against unwanted intrusion is made up of products centered on a firewall, which is a device that examines incoming or outgoing data to stop any unauthorized transmission of or access to information. In an Extranet environment, more than one party is involved. The overall level of security is as strong as that of the weakest partner. In general, a security-conscious company is likely to appreciate and honor the mutually adopted policies created by the Extranet alliance.

For Intranet, where only one enterprise is considered, most likely the enterprise uses private IP addresses inside the network. For example, the class A private addresses 10.*.*.* are good for a large company. However, if all enterprises use private addresses to build their Intranet, the connection between them when they try to create an Extranet needs to be carefully considered. The machines/networks that need to communicate with partners have to be assigned publicly registered class A or B or C addresses. The address assignment need only be logical and a firewall or router can be used to perform address translation. DNS support has to be provided to the name/address resolution requests from partners.

3.5 THE POWER-LAW RULE FOR INTRANETS AND EXTRANETS

For the Internet, there are power laws that characterize the topologies. Similarly, these power laws extend to intranets as well as extranets. A relationship exists between the degrees of the nodes and the ranks of the nodes when arranged by degrees. This type of relationship is called a power law. It means that there exists a relationship between the degree d_n of a node n in the Intranet or the Extranet and the rank of a node r_n. R is the power of the node rank r_n

$$d_n \propto r_n{}^R \tag{3.1}$$

A power law appears between the degree of a node and the number of nodes that has the degree, frequency. That means that there exits a relationship between the degree of a node, d, and the frequency f_d such as:

$$f_d \propto d_0 \tag{3.2}$$

The relationship between a calculated eigen value λ_i by the connectivity and ranks i that are arranged in ascending order by the eigen value using a close matrix, that if all nodes are connected, it is represented by "1," and if not connected, is expressed by "0" and is known also as a power-law. The eigen value, λ_i and rank, i, have the relationship

$$\lambda_i \propto i_\ell \tag{3.3}$$

Example 3.1

Determine the degree of node 5 in the intranet if the rank of the node is 4 and the power is 3.

Solution 3.1

In this intranet network

$$n = 5; \mathrm{r} = 4; R = 3$$

Therefore, from equation (3.1), $d_n \propto r_n{}^R \rightarrow d_5 = 4^3 = 64$

3.6 THE TOPOLOGY MODELS

There are different types of topology models. Some of these models are as follows: the Tiers, Transit-stub, Barabasi-Albert, and the Waxman models.

3.6.1 The Tiers Model

In this model, the three levels of WAN, MAN, and LAN are treated as a basic policy. It has a hierarchical characteristic. At the MAN and LAN levels, all edges are connected through minimum spanning tree calculation. In the case of the WAN level, the network is made randomly.

3.6.2 The Transit-stub Model

This type of model has the characteristic of hierarchical architecture that is composed of transit and stub connected to each other. It has a graph generation order that can be stated as follows:

- Through a random method, each node of the Intranet/Extranet makes whole transit area graph.
- In the graph, each node is substituted with another connected random graph, which stands for a backbone topology in a transit area.
- In the transit area, each node makes several random graphs that stand for the attached stub area.
- In the pair of nodes, edges are added. There are two kinds of graph involved here. One of those graphs is connected between transit area and stub area. The second graph is connected between one stub area and another.

3.6.3 The Barabasi-Albert Model

In this model, the cause of a characteristic of the Internet appearing as a power law is explained by providing incremental growth and preferential connectivity in a network topology. When that node, i, is added, the connectivity rate of this node that connects with the existing node, j, is determined by the degree of the existing nodes as shown in Equation 3.4, since the connectivity is the first consideration.

$$P(j) = \frac{d_j}{\sum_{k \varepsilon V} d_k} \tag{3.4}$$

where d_j is the degree of the object node and V is the collection of all nodes considered to a network. $\sum_{k\varepsilon V} d_k$ is the sum of all the nodes' degrees, which are already connected to a network.

3.6.4 The Waxman Model

This model is used to add the links in a network considering that the distance between the nodes are arranged in random positions on a 2D grid. If we consider the two nodes u and v, the rates of the existing edge between the two nodes are different depending on the distance of the node from the other node. In addition, it is determined using the connection rate function shown in Equation 3.5. A random network is usually created using the Waxman model.

$$P(u,v) = ae^{-d/\beta L} \tag{3.5}$$

where d is the Euclidean distance from node u to v, L is a longest distance diameter between two nodes in a graph. Waxman's control variables are represented as α and β.

3.7 INTRANET AND EXTRANET APPLICATIONS

There are many Intranet and Extranet applications. Some of the applications are listed and described in this section.

3.7.1 Human Resources Intranet

Human resources departments have been some of the most enthusiastic developers of Intranet applications mainly because of the large amounts of paper-based processes that can be transitioned to the Web.

Examples of possible content and applications:

- Employee handbook
- Telephone/email directory
- Interactive benefits information
- 401K tracking
- Employee surveys
- Recruiting/job listings
- Candidate screening applications
- Organizational charts
- Newsletters
- New employee training
- Employee personalized home pages

3.7.2 Sales and Marketing Intranet/Extranet

In today's very competitive environment, having fast access to accurate information can be crucial for the sales and marketing staff. It can be the difference between making a sale or giving someone time to look elsewhere.

An Intranet/Extranet combination addresses that issue, providing an environment where product descriptions, sales scripts, marketing analysis, and research are all a click away.

Examples of possible content and applications:

- Product demos and scripts
- Pricing charts
- Sales forecasts and reports

- Sales contact management
- Sales lead management
- Market research/search engines
- Sales feedback
- Prospecting
- Press releases
- Sales team collaboration
- Calendars
- Sales multimedia training
- Competitor research

3.7.3 Information Systems Intranet

Many applications are being used to support information system processes. Some applications are used to support the needs of the employee base and some are being extended to partners via Extranets turning IS into a profit center.

Examples of possible content and applications are

- Software and applications development and delivery
- User documentation
- Technical support and help desk
- Network management
- Information and knowledge repositories
- Internet resources
- Resource scheduling
- Technical/security polices and procedures
- Multimedia-based training
- Intranet FAQs, publishing guides
- Web paging or communications systems

3.7.4 Customer Service Intranet/Extranet

This is one area where an Intranet/Extranet allows customer service to peer over the company firewall to directly interact with customers. A good example of how Intranets are being leveraged for customer support is the extension of internal package-tracking databases of UPS and FedEx over the Internet.

An Intranet or Extranet can allow a customer to help themselves or allow support personnel to find an answer faster. Gone are the days of putting a customer on hold while you research the file cabinet for that part number or other solution.

Examples of other possible content and applications are

- Customer information entry and update
- Order entry and tracking
- Online information (databases on customers, inventory, supplies)
- Problem entry and tracking
- Customer FAQs

3.7.5 Finance Intranet

Accounting and finance departments deliver some of the most crucial data across an organization. An accounting Intranet creates a centralized, open-standards platform for publishing that information and new interactive methods for processing transactions with either internal departments, employees, or external partners.

Examples of possible content and applications are

- Accounts payable/receivable support
- Payroll
- Intranet commerce, requisitioning system
- Financial reports
- Policies and procedures
- Budgeting
- Asset management
- Expense reports
- Unit reporting and forecasting

3.8 EXTRANET AND INTRANET DESIGN ISSUES

In order to have an effective Intranet and Extranet structure in any organization, proper design process must be followed.

3.8.1 Intranet and Extranet Design Process

The process must include the following steps:

- You have to design appropriate retrieval aids for your materials.
- Efforts must be made to analyze information and user needs.
- Information must be re-engineered for effective use online.

3.8.2 Intranet and Extranet Design Analysis

In the analysis process, you have to determine how you can design the Intranet/Extranet. The following must be analyzed:

- What are your goals why you need to create the Intranet and /or Extranet?
- Is it possible for this Intranet/Extranet to make the jobs in your organization much easier?
- What kind of problems are your employees experiencing in their jobs?
- Do the users in your organization need the current documentation in the same way?
- Does the information help users meet their true goals?

3.8.3 The Design Standards

It is not that easy to manage Intranets and Extranets because there are multiple authors and multiple groups responsible for their creation and maintenance. You have to create design standards and guidelines for use throughout the organization to ensure that the users consistently design information for easy access. Additionally, one should expect that different types of information may require different standards. However, there should be consistent core standards that are common in all cases. These are as follows:

- Templates should be created for use in cases where there may be similar information.
- The company should get together as one and then create a set of design standards.
- An editor whose job should be only to edit and to create should be involved in the process at an early stage.
- A document process and monitoring system should be established. In addition, it is necessary that the company understands the document purpose and team and individual roles.
- Once the collaborative assignment is completed, the project should be evaluated and a lessons learned list should be developed for future applications.
- The entire team should be allowed to get involved in the design process so that everyone understands why certain decisions have been made.

SUMMARY

1. Companies create Intranet systems for inside the business to improve employee productivity, share data, or update human resources information, and much more.
2. An Extranet is built for use outside the business—either for customers or to let the company communicate better with their vendors.
3. The technologies of both Intranet and Extranet are based on the Internet.
4. Intranet and Extranet are both based on TCP/IP, and the whole family of protocols, such as HTTP, SMTP, FTP, etc.
5. Security is a big issue in Intranet and Extranet, especially for Extranet, where resources are exposed to the outside world.
6. The Tiers, Transit-stub, Barabasi-Albert and the Waxman models are some of the different types of topology models.
7. An Intranet/Extranet combination provides an environment where product descriptions, sales scripts, marketing analysis, and research are all a click away.
8. Design standards and guidelines are to be created for use throughout the organization to ensure that the users consistently design information for easy access.

REVIEW QUESTIONS

3.1 IP technology was adopted in enterprises to set up their private or semi-private networks similar to the Internet.

a. True
b. False

3.2 The term "Intranet" was coined around:
   a. 1965
   b. 1975
   c. 1985
   d. 1995
3.3 One of the best reasons why Intranet rose so quickly is because it leverages the existing technologies based on TCP/IP.
   a. True
   b. False
3.4 HTTP is not the most important application protocol above TCP/IP and is the foundation of Intranet.
   a. True
   b. False
3.5 Most information is formatted in HTTP hypertext and is displayed to users in browsers.
   a. True
   b. False
3.6 With Intranet, enterprises connect their internal resources on a common platform, which does not reduce the operational cost.
   a. True
   b. False
3.7 The simplest Intranet consists of only one what?
   a. LAN
   b. WAN
   c. LAN and WAN
   d. None of the above
3.8 An Extranet does not require much stricter security and privacy than Intranet.
   a. True
   b. False
3.9 An Intranet or Extranet can allow a customer to help themselves or allow support personnel to find an answer faster.
   a. True
   b. False
3.10 A good example of how Intranets are being leveraged for customer support is the extension of internal package-tracking databases of UPS and FedEx over the Internet.
   a. True
   b. False

Answer: 3.1a, 3.2d, 3.3a, 3.4b, 3.5a, 3.6a, 3.7a, 3.8b, 3.9a, 3.10a.

PROBLEMS

3.1 What are the three discernable stages in the development of an Intranet?
3.2 An Extranet requires much stricter security and privacy than Intranet. Why?

3.3 Name some of the benefits and implications of an Intranet.
3.4 List and describe some of the Intranet technologies.
3.5 List and describe some of the Extranet technologies.
3.6 Information systems Intranet has many content applications. Name some of them.
3.7 Finance Intranet has many content applications. Name some of them.
3.8 There are many ways of categorizing what an Intranet does. What are they?
3.9 Customer service Intranet/Extranet has many content applications. Name some of them.
3.10 Sales and marketing Intranet/Extranet has many content applications. Name some of them.
3.11 Human resources has many content applications. Name some of them.
3.12 Determine the degree of node 10 in the Intranet if the rank of the node is 8 and the power is 5.
3.13 If in node 5 of an Extranet, the degree is 64 and power is 3, what is the rank of this node within this Extranet network?
3.14 State the three known power laws that characterize Internet and Intranet/Extranet topologies. Describe these power laws.
3.15 Why are these power laws very significant in the existence and characterization of Internet and Intranet/Extranet topologies?
3.16 State and describe at least four different types of topology models.
3.17 Why are these topology models important in the existence and characterization of Internet, Intranet, and Extranet networks?
3.18 In the transit-stub model, what steps are available in the graph generation order?
3.19 Compare and contrast the different types of models in Internet, Intranet, and Extranet networks.
3.20 Which of the topologies models is better for any Internet, Intranet, and Extranet networks and why?
3.21 What processes, analyses, and design standards are necessary for an effective Intranet and Extranet solutions in an organization?

ACRONYMS

DMZ Demilitarized zone
DNS Domain name resolution protocol
FTP File transfer protocol
HTTP Hypertext transfer protocol
IP Internet protocol
IPSec Internet protocol security
L2F Layer 2 (two) forwarding protocol
L2TP Layer 2 tunneling protocol
PPTP Point-to-point tunneling protocol
SMTP Simple mail transfer protocol
TCP Transfer control protocol
VPN Virtual private network

Part 2

Networking Technologies

4 X.25 and Frame Relay

Poverty has many roots, but the tap root is ignorance.

—L. B. Johnson

X.25 is often regarded as the forerunner of frame relay. It is a communications packet switching protocol designed for the exchange of data over a wide area network (WAN). X.25 was designed to provide error-free delivery using high-error-rate links. It was developed when analog transmission and copper media were the predominant technologies. Frame relay takes advantage of the new, lower-error-rate links made of optical fiber, enabling it to eliminate many of the services provided by X.25. The elimination of functions and fields, combined with digital links, enables frame relay to operate at speeds 20 times greater than X.25. In frame relay, frames are switched to their destination with only a few byte times' delay, as opposed to several hundred milliseconds' delay on X.25.

This chapter provides an introduction to X.25 and frame relay. We will discuss frame relay in more detail since it is the more commonly used.

4.1 X.25 TECHNOLOGY

For roughly 20 years, X.25 was the dominant player in WAN packet switching technology until frame relay, SMDS, and ATM appeared. X.25 has been around since the mid 1970s and so is pretty well debugged and stable. It was originally approved in 1976 and subsequently revised in 1977, 1980, 1984, 1988, 1992, and 1996. It is currently one of the most widely used interfaces for data communication networks. There are literally no data errors on modern X.25 networks.

X.25 is regarded as a standard, a network, or an interface protocol. It is a popular standard for packet-switching networks approved in 1976 by the International Telecommunication Union-Telecommunication Standardization Sector (ITU-T) for WAN communications. It defines how connections between user devices and network devices are established and maintained. Through statistical multiplexing, X.25 enables multiple users to share bandwidth as it becomes available, therefore ensuring flexible use of network resources among all users. An X.25 network transmits data with a packet-switching protocol, bypassing noisy telephone lines. X.25 is also an interface protocol in that it spells the required interface protocols that enable data terminal equipment (DTE) to communicate with data circuit-terminating equipment (DCE), which provides access to the network. The DTE-DCE link provides full-duplex multiplexing allowing a virtual circuit to transmit in either direction.

X.25 is a packet switching protocol designed for the exchange of data between a user device (host), called data terminal equipment (DTE) and a network node, called data circuit-terminating equipment (DCE).

This protocol relies on an elaborate worldwide network of packet-forwarding nodes that can participate in delivering an X.25 packet to its designated address. Networks and dial-up connections support X.25 by using packet assemblers/disassemblers (PADs) and X.25 cards. X.25 utilizes a connection-oriented service that ensures that packets are transmitted in order.

4.1.1 X.25 Devices

X.25 network devices fall into three general categories: data terminal equipment (DTE), data circuit-terminating equipment (DCE), and packet switching exchange (PSE). DTE devices are user end systems that communicate across the X.25 network. They are usually terminals, personal computers, or network hosts, and are located on the premises of individual subscribers. DCE devices are carrier's equipment, such as modems and packet switches, that provide the interface between DTE devices and a PSE and are generally located in the carrier's facilities. PSEs are switches that compose the bulk of the carrier's network. They transfer data from one DTE device to another. Figure 4.1 illustrates the relationships between the three types of X.25 network devices.

X.25 sessions are established when one DTE device contacts another to request a communication session. The DTE device that receives the request can either accept or refuse the connection. If the request is accepted, the two systems begin full-duplex information transfer. Either DTE device can terminate the connection.

The packet assembler/disassembler (PAD) is a device commonly found in X.25 networks. PADs are used when a DTE device is too simple to implement the full X.25 functionality. The PAD is located between a DTE device and a DCE device, and it performs three primary functions: buffering, packet assembly, and packet disassembly. The PAD buffers data sent to or from the DTE device. It also assembles outgoing data into packets and forwards them to the DCE device; this includes adding an X.25

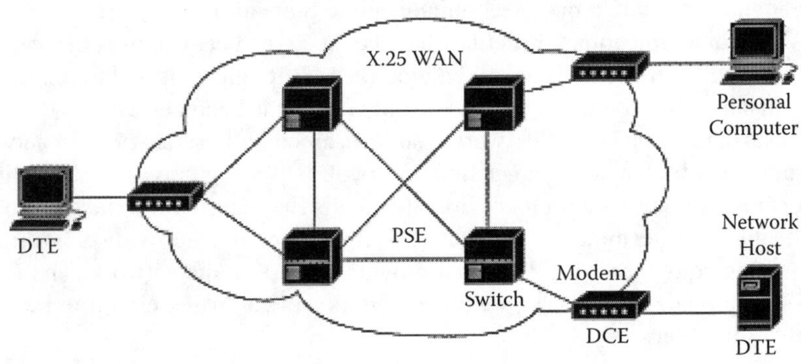

FIGURE 4.1 DTEs, DCEs, and PSEs make up an X.25 network.

header. Finally, the PAD disassembles incoming packets before forwarding the data to the DTE; this includes removing the X.25 header.

4.1.2 X.25 Virtual Circuits

A **virtual circuit** is a logical connection created to ensure reliable communication between two network devices.

A virtual circuit denotes the existence of a logical, bidirectional path from one DTE device to another across an X.25 network. Physically, the connection can pass through any number of intermediate nodes, such as DCE devices and PSEs. Two types of X.25 virtual circuits exist:

- Switched virtual circuits (SVCs): SVCs are very much like telephone lines; a connection is established, data are transferred, and then the connection is released. They are temporary connections used for sporadic data transfers. They require that two DTE devices establish, maintain, and terminate a session each time the devices need to communicate. Each DTE on the network is given a unique address that can be used much like a telephone number.
- Permanent virtual circuits (PVCs): A PVC is similar to a leased line in that the connection is always present. Permanent virtual circuits (PVCs) are permanently established connections used for frequent and consistent data transfers. Therefore, data may always be sent without any call setup.

The basic operation of an X.25 virtual circuit begins when the source DTE device specifies the virtual circuit to be used (in the packet headers) and then sends the packets to a locally connected DCE device. At this point, the local DCE device examines the packet headers to determine which virtual circuit to use and then sends the packets to the closest PSE in the path of that virtual circuit. PSEs (switches) pass the traffic to the next intermediate node in the path, which may be another switch or the remote DCE device. Multiple virtual circuits (logical connections) can be multiplexed onto a single physical circuit (a physical connection). Virtual circuits are demultiplexed at the remote end, and data are sent to the appropriate destinations.

When the traffic arrives at the remote DCE device, the packet headers are examined and the destination address is determined. The packets are then sent to the destination DTE device. If communication occurs over an SVC and neither device has additional data to transfer, the virtual circuit is terminated.

Maximum packet sizes vary from 64 bytes to 4096 bytes, with 128 bytes being a default on most networks. The maximum packet size may be negotiated between DTEs on call setup. X.25 enables up to 4096 virtual circuits to be concurrently active on any one link.

4.1.3 X.25 Protocol Suite

As shown in Figure 4.2, the X.25 protocol encompasses the lowest three layers of the OSI reference model:

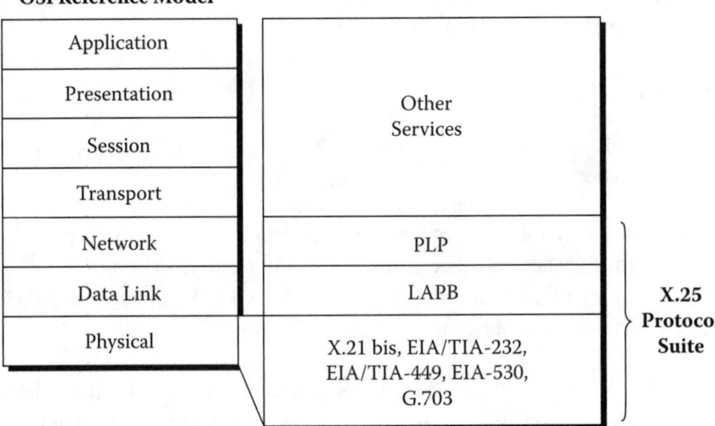

FIGURE 4.2 Key X.25 protocols map to the three lower layers of the OSI reference model.

- Layer 1: The physical layer describes the interface with the physical environment. It is similar to the physical layer in the OSI model. It deals with the electrical, mechanical, procedural, and functional interface between the DTE and the DCE. It includes several standards such as X.21, V.24, and RS234. For example, X.21 is an ITU recommendation for operation of digital circuits. It handles the activation and deactivation of the physical medium connecting DTE and DCE devices. It supports point-to-point connections and full-duplex transmission over four-wire media. The physical connector has 15 pins, but not all of them are used.
- Layer 2: The data link layer provides an error-free link between two connected devices. It is similar to the data link layer in the OSI model. The functions performed by the link level include
  - reliable transfer of data between the DTE and the DCE in an efficient and timely fashion
  - synchronization of the link to ensure that the receiver is in step with the transmitter
  - detection of transmission errors and recovery from such errors
  - identification and reporting of procedural errors to higher levels for recovery
  - The link level for X.25 is an implementation of the ISO standard called link access procedure balanced (LAPB). LAPB is a data link-layer protocol that manages communication and packet framing between DTE and DCE devices. LAPB is a bit-oriented protocol that ensures that frames are correctly ordered and error free. LAPB frames include a header, encapsulated data, and a trailer. Figure 4.3 illustrates the format of the LAPB frame and its relationship to the packet layer protocol (PLP) packet and the X.21bis frame. The following descriptions summarize the fields:

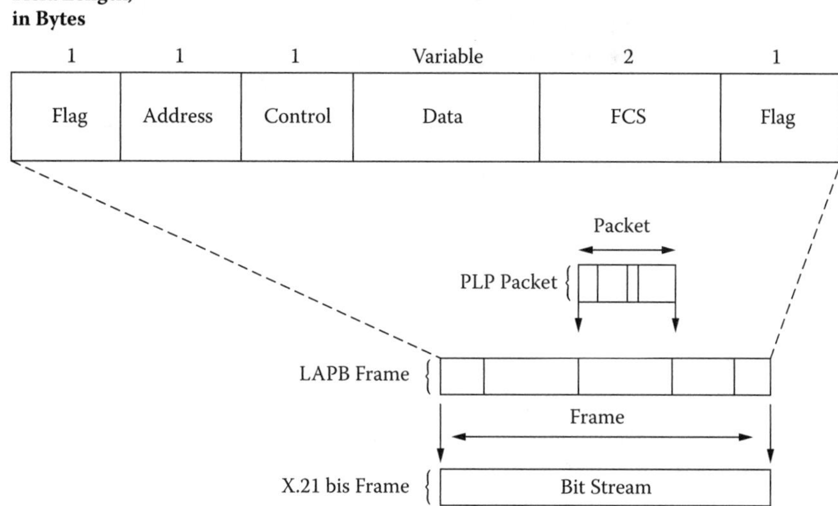

FIGURE 4.3 An LAPB frame includes a header, a trailer, and encapsulated data.

- *Flag*: The flag field (01111110) indicates the beginning and end of the frame. Bit stuffing is used to ensure that the flag pattern does not occur within the body of the frame.
- *Address*: The address fields contain the address of the DTE/DCE and indicate whether the frame carries a command or a response.
- *Control*: The control field contains sequence numbers, commands, and responses for controlling the data flow between the DTE and the DCE and indicates whether the frame is an I-frame (for actual information), an S-frame (for supervision), or a U-frame (unnumbered, used only for control purposes), corresponding to the three kinds of frames. Control frames vary in length depending on the frame type.
- *Data*: The data field contains upper-layer data in the form of an encapsulated PLP packet.
- FCS: *The frame check sequence field* indicates whether or not errors occur in the transmission. It is a variation of the cyclic redundancy code (CRC).

- Layer 3: The network layer (also called the packet level) provides communications between devices connected to a common network. It is similar to the network layer in the OSI model. It creates network data units called packets that contain control information and user data. In the case of X.25, this layer is referred to as the packet layer protocol (PLP) and is primarily concerned with network routing functions and the multiplexing of simultaneous logical connections over a single physical connection. PLP manages packet exchanges between DTE devices across virtual circuits. The standard requires that carriers support a maximum packet length of 128 data bytes. However, it also allows carriers to provide optional maximum lengths from 16 up to 4096 bytes (in powers of 2).

- The packet level provides procedures for handling the following services: virtual circuit (VC), permanent virtual circuit (PVC), and datagrams. The packet level also provides services including the call setup and call clearing procedures required for the VC service. The packet level deals with flow control to ensure that a user (DTE) does not overwhelm the other user with packets, and to maintain timely and efficient delivery of packets.

4.1.4 Benefits and Drawbacks of X.25

X.25 users are typically large organizations with widely dispersed and communications-intensive operations in sectors such as finance, insurance, transportation, utilities, and retail. For example, X.25 is often chosen for zero-error tolerance applications by banks involved in large-scale transfers of funds, or by government utilities that manage electrical power networks. X.25 is the appropriate choice for such organization because they need to guarantee the timely and reliable transfer of information to their customers.

X.25 has its advantages.

- The adoption of a common standard provides an easy way to interface products from different vendors.
- The X.25 standard has gone through numerous revisions and is relatively mature.
- A widely used standard such as X.25 can decrease network costs, since off-the-shelf software and hardware are readily available.
- X.25 provides a higher level of support by defining many operations that enhance the reliability of data transfer between each sending DTE and its DCE (the entrance packet node to the network) and each receiving DTE and its DCE (the exit packet node from the network). In other words, it gives considerably more end-to-end support than a link protocol like HDLC.
- X.25 is reliable in data transfer over a wide range of quality of transmission facilities. Unlike frame relay and ATM technologies, which depend on the use of high-quality digital transmission facilities, X.25 can operate over either analog or digital facilities.
- X.25 is economical. It gives you a high-quality digital network at low cost. In most parts of the world, X.25 is paid for by a monthly connect fee plus packet charges. There is usually no holding charge, making X.25 ideal for organizations that need to be online all the time.
- A useful feature of X.25 is speed matching. Because of the store-and-forward nature of packet switching, plus excellent flow control, DTEs do not have to use the same line speed. So you can have, for instance, a host connected at 56 kbps communicating with numerous remote sites connected with cheaper 19.2 kbps lines.

X.25 does have some limitations. There is an inherent delay caused by the store-and-forward mechanism. On most single networks the turn-around delay is about

0.6 seconds. This has no effect on large block transfers, but in flip-flop types of transmissions the delay can be very noticeable. Another problem for the network is a large requirement for buffering to support the store-and-forward data transfer.

4.2 FRAME RELAY

Frame relay is basically an interface used for wide-area networking. It appeared as an independent protocol in 1989. Before that, frame relay was part of the protocol for narrowband integrated services digital network (ISDN). It is used to reduce the cost of connecting remote sites in any application that would typically use expensive leased circuits. The more locations you have, the greater your savings. With the right carrier service, frame relay network links may be quickly adjusted to meet changes to applications and network topology.

> **Frame relay** is a simplified form of packet switching (similar in principle to X.25) in which synchronous frames of data are routed to different destinations depending on header information.

The main difference between frame relay and X.25 is data integrity (error detection) and network error flow control (error correction). X.25 guarantees data integrity and network managed flow control at the cost of some network delays. Frame relay switches packets end to end much faster, but there is no guarantee of data integrity at all.

Traditional high-speed WANs are built on high-capacity leased lines, which often take months to order and install. Because network managers must react quickly when a new application, organizational unit, business partner, or location is needed, leased-line networks often present a roadblock. Using frame relay, backbone changes can be quickly programmed by the carrier, saving the long installation delays and high costs associated with running physical circuits.

No matter which WAN service you choose (X.25, frame relay, SMDS, ISDN, ATM, leased lines, or the Internet), an end-to-end link is created via three essential components, shown in Figure 4.4. From each site, an access line must be run to the local exchange carrier (LEC) or competitive access provider (CAP) serving the area.

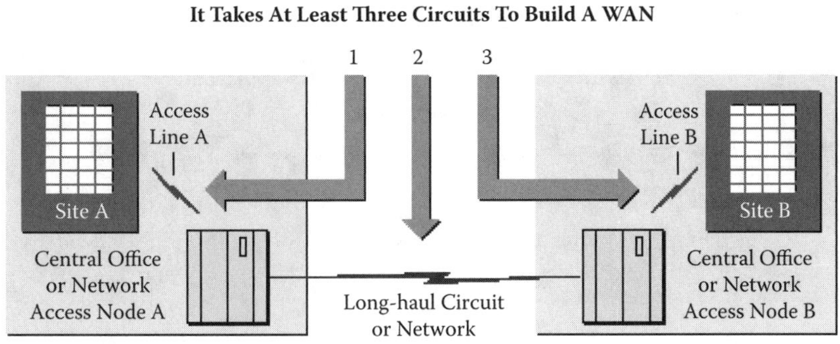

FIGURE 4.4 Basic components of a WAN link.

The third component sits in the middle, joining the two access points. This is the portion supplied by the frame relay backbone.

4.2.1 Frame Relay Technology

Frame relay is an interface, a method of multiplexing traffic to be submitted to a WAN. Carriers build frame relay networks using switches from vendors such as Cascade Communications, Cisco, Alcatel, Nortel, or Bay Networks. As a customer, your devices see only the switch interface, and are blind to the inner workings of the carrier network, which may be built on very high-speed technologies such as T1, T3, SONET, and/or ATM.

The physical layout of a sample frame relay network is depicted in Figure 4.5. The CSU/DSU is the channel service unit/data service unit. This unit provides a "translation" between the telephone company's equipment and your router. The router actually delivers information to the CSU/DSU over a serial connection, much like your computer uses with a modem, only at much higher speed.

All major carrier networks implement permanent virtual circuits (PVCs). These circuits are established via a contract with the carrier and typically are built on a flat-rate basis. Although switched virtual circuits (SVCs) have standards support and are provided by the major frame relay backbone switch vendors, they have not been widely implemented in customer equipment or carrier networks.

Two major frame relay devices are the FRAD and routers:

- Stand-alone frame relay access devices (FRADs) typically connect small remote sites to a limited number of locations. FRAD is also known as frame relay assembler/disassembler since FRAD is a piece of networking equipment that assembles and disassembles data frames. It may perform limited routing functions, but does not have the range of routing protocols that a router does.
- Frame relay routers offer more sophisticated protocol handling than most FRADs. They may be packaged specifically for frame relay use, or they may be general-purpose routers with frame relay software. This software may sometimes be an additional cost over the basic router code.

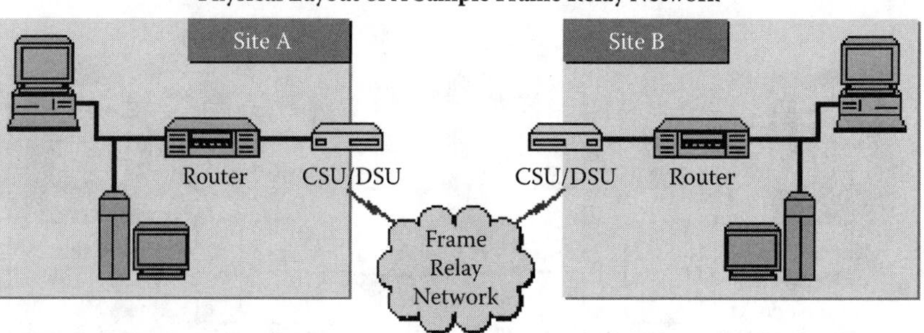

FIGURE 4.5 Physical layout of a typical frame relay network.

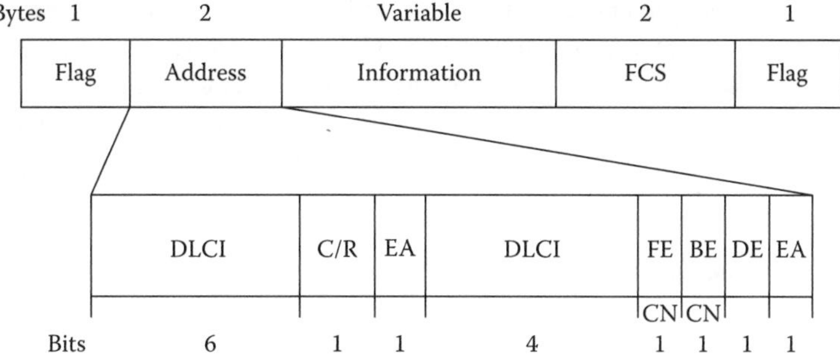

FIGURE 4. 6 Frame relay packet format.

To connect to a frame relay network, you need a device that understands the interface protocol. There are two general approaches: use a stand-alone FRAD to connect equipment with leased-line interfaces, or obtain software for an existing bridge, router, front-end processor, or other WAN device.

4.2.2 Frame Relay Protocol Data Unit

Frame relay uses the synchronous high-level data link control (HDLC) frame format up to 4 kbytes in length. As shown in Figure 4.6, each frame structure consists of the following fields:

- Flag field: The flag is used to perform high-level data link synchronization that indicates the beginning and end of the frame with the unique pattern 01111110. To ensure that the 01111110 pattern does not appear somewhere inside the frame, bit stuffing and destuffing procedures are used.
- Address field: Each address field may occupy either octet 2 to 3, octet 2 to 4, or octet 2 to 5 (1 octet = 1 byte = 8 bits), depending on the range of the address in use. A two-octet address field comprises the EA = ADDRESS FIELD EXTENSION BITS and the C/R = COMMAND/RESPONSE BIT.
- Data link connection identifier (DLCI) bits: The DLCI is used to identify the virtual connection so that the receiving end knows which information connection this frame belongs to. It should be pointed out that this DLCI has only local significance. Several virtual connections can be multiplexed over the same physical channel. DLCI numbers are assigned as follows:
  - DLCI 0 (zero) and 1023 are reserved for management.
  - DLCI 1 to 15 and 1008 to 1022 have been reserved for future use.
  - DLCI 992 to 1007 are reserved for layer 2 management of frame relay bearer service.
  - DLCI numbers 16 to 991 are available for subscribers for each user frame relay network.
- FECN, BECN, DE bits: These are congestion reporting capability bits:
  - FECN = Forward explicit congestion notification bit
  - BECN = Backward explicit congestion notification bit
  - DE = Discard eligibility bit

Frame relay equipment notices congestion when it sees frames marked with the forward explicit correction notification (FECN) and backward explicit correction notification (BECN) bits. These are used to tell the application to slow down, hopefully before packets start to be discarded. They merely indicate an overload within the carrier network. Discard eligibility (DE) provides the network with a signal to determine which frames to discard. The network will discard frames with a DE value of 1 before discarding other frames.

The DE bit may be set by the user on some of its lower-priority frames.

- Information field: This contains the user data. The maximum number of data bytes that may be put in a frame is a system parameter. The actual maximum frame length may be negotiated at call setup time. The information field size varies from 1 octet to a maximum value of 1600 octets, to avoid the need for segmentation and reassembling by end users.
- Frame check sequence (FCS) field: Since the bit error rate of the medium is not completely negligible, it is necessary to implement error detection at each switching node to avoid wasting bandwidth due to the transmission of erred frames. The error detection mechanism used in frame relay is based on the cyclic redundancy check (CRC).

Once the call is established, data is transferred through the network on a hop-by-hop basis. Packets are routed through one or more virtual circuits by using the DLCI in the frame header and routing information at each node as determined during call setup. Each DLCI has a permanently configured switching path to a certain destination. Thus, by having a system with several DLCIs configured, you can communicate simultaneously with several different sites. Currently, only permanent virtual circuit connections are supported. This means that all DLCI connections are set up by the network provider at subscription time.

Example 4.1

This example illustrates the use of the data link connection identifier (DLCI). This number identifies a virtual circuit. For an application, a headend multiplexer might have four DLCIs associated with it, while each of four remotes would have just a single DLCI. The following illustrates how DLCIs might be assigned:

Host Location	Host DLCI #	Remote DLCI #	Remote Location
New York	17	17	Miami
New York	18	17	Atlanta
New York	19	17	Houston
New York	20	17	St. Paul

In this example, the New York host location might be a single 56- or 64-kbps line into a frame relay host multiplexer. At each remote location there might be a multiplexer. The DLCI numbers are not necessarily assigned in any order by the carrier, although it is commonly done.

4.2.3 Virtual Circuit Size Parameters

Carriers usually program their frame relay switches to allow traffic to pass through the network. The following are the essential parameters we must understand:

- Access rate is the rate at which your access circuits join the frame relay network. These are typically at 56 kbps, T1 (1.536 Mbps), or fractional T1 (a multiple of 56 kbps or 64 kbps). It is not possible to send data at higher than access rate.
- Port speed is the rate at which the port on the frame relay switch is clocked. It is not possible to send data at higher than port speed.
- Committed information rate (CIR) is the amount of data that the network will receive from the access circuit. When you subscribe, you will specify the line speed (e.g., 56 kbps or T1) and a committed information rate (CIR) for each DLCI. This value specifies the maximum average data rate that the network undertakes to deliver under "normal conditions." If you send faster than the CIR on a given DLCI, the network will flag some frames with a discard eligibility (DE) bit. The network will do its best to deliver all packets but will discard any DE packets first if there is congestion. Many inexpensive frame relay services are based on a CIR of zero. This means that every frame is a DE frame, and the network will throw any frame away when it needs to.
- Committed burst size (Bc) is the maximum amount of data that the network will transfer in a burst, defined over a (short) interval. All frames received at this level will be passed.
- Admission control. In addition to CIR and Bc, admission control is the congestion control element used in frame relay networks. This is the principal mechanism used in frame relay to ensure that resource requirement, once accepted, can be guaranteed. The network decides whether to accept a new connection request based on the relation of the requested traffic descriptor and the network's residual capacity. The traffic descriptor is defined as a set of parameters communicated to the switching nodes at call setup time or service subscription time.
- Excess burst size is the amount of data above Bc that the network will try to deliver. Frames submitted at this level may be marked as "Discard Eligible," indicating that they may be dropped.
- A data-link connection identifier (DLCI) identifies a virtual circuit to your equipment, as mentioned earlier. This is an end point for a PVC.
- Oversubscription: An instance where the sum of CIRs from multiple PVCs to a given location are higher than the port or access channel rate.

Port speed is typically the most costly parameter to increase, though access rates can jump dramatically if new local loops are involved, such as a move from 56 kbps to T1. Individual PVC fees are next most costly. Once a PVC is established, the additional cost to increase CIR is typically small and can be done in small (4 kbps) increments.

4.2.4 Major Advantages and Disadvantages

The success of the frame relay protocol is based on the following two underlying factors. First, because virtual circuits consume bandwidth only when they transport data, many virtual circuits can exist simultaneously across a given transmission line. Second, the improved reliability of communication lines and increased error-handling sophistication at end stations allows the frame relay protocol to discard erroneous frames and thus eliminate time-consuming error-handling processing. These two factors make frame relay a desirable choice for data transmission.

Frame relay is the fastest growing WAN technology in the United States. In North America it is fast taking on the role that X.25 has had in Europe. It is used by large corporations, government agencies, small businesses, and even Internet service providers (ISPs). The major advantages of selecting frame relay include

- Costs are not sensitive to location, as are leased lines.
- It supports higher bandwidth with less overhead and latency than X.25.
- It is more widely available than switched multimegabit data service (SMDS), and does not require costly, specialized data service units (DSUs) for access links. Payback is faster than SMDS for small-to-moderate-sized networks that do not require frequent connections in a fully meshed configuration.
- Changes may be made quickly. Most carriers can process requests to increase port speed, committed information rate (CIR), and other key parameters in a matter of hours.
- It supports a wide range of protocols via standard encapsulation techniques.

The demand for frame relay services is exploding, and for two very good reasons—speed and economics. Frame relay is consistently less expensive than equivalent leased services and provides the bandwidth needed for other services like LAN routing, voice, and fax.

However, frame relay has the following drawbacks:

- There is no flow control on frame relay. The network simply discards frames it cannot deliver.
- Frame relay performs error detection only, not error correction. The network delivers frames, whether the CRC check matches or not. It does not even necessarily deliver all frames, discarding frames whenever there is network congestion. In practice, however, the network delivers data quite reliably and very few frames are discarded by the network. Unlike the analog communication lines that were originally used for X.25, modern digital lines have very low error rates.

Vendors, carriers, users, and consultants committed to the implementation of frame relay in accordance with national and international standards formed the Frame

Relay Forum in 1991. Those interested in Forum membership, meeting schedules, or in ordering marketing or technical materials should contact the Forum at:

Frame Relay Forum
39355 California St., Suite 307
Fremont, CA 94538
Email: info@frforum.com
Website: www.frforum.com

Example 4.2

In this example, we consider situations in which you may (or may not) choose to use frame relay. Each is listed with benefits and drawbacks to help focus on the real need.

Situation One
You have been accessing online services by using a modem, but would like to start using more advanced services that are not available from the provider you have now. You may be thinking about World Wide Web. You may just want faster connections to the services you use now.

Response
For service at this level, frame relay probably is not appropriate. You probably do not have a LAN in your organization, so taking advantage of frame relay would be a substantial investment both in new equipment and upgrades to what you currently have.

Situation Two
You are the CEO of your company and you have some type of LAN within your organization. You are getting really serious about providing Internet access for your staff. If you have a lot of phone lines that are being used to access online services that would be available through a WAN or the Internet, you may see cost savings in frame relay.

Response
If the services you need to access are on the Internet, you can use frame relay to establish a permanent connection with an Internet service provider (ISP). Frame relay works very well for this type of connection because it can be purchased at any level you need and increased with minimal additional expense. Frame relay becomes cost-effective fairly rapidly when compared to a large number of standard phone lines (more than five or six), as long as your ISP does not charge too much.

Situation Three
You want to provide a service to others outside your organization. This may be a central online catalog that is shared, an electronic index that is maintained by your library, or even a WWW server. As soon as you have a service to offer, you need permanent connectivity so that others may use your services anytime they wish. This may be in the form of a WAN, or simpler point-to-point connections.

Although Internet connectivity may be an eventual goal for a WAN, it does not have to start out connected to the Internet. A WAN may even be more cost-effective than multiplexers for terminal access in branch offices. As long as you plan for Internet connectivity, adding it later is fairly painless.

Response

If the computer that runs the service you want to offer is already on a LAN, extending your reach with a WAN may be the best solution. With relatively low incremental costs to add additional nodes in a frame relay network, the total cost is usually less than if you were to use multiplexers and terminals across point-to-point connections.

Example 4.3

In this example, we follow a step-by-step guide for designing a frame relay network:

1. Laying out an initial topology

As a way of getting started, lay out each of your sites on a map. Because frame relay tariffs are not typically distance sensitive, you can disregard the geographic location in laying out a basic topology. This initial topology will probably be based on your current locations and expected host computer placement.

2. Analyzing your bandwidth needs

You will need a proper analysis of your bandwidth needs in order to size the network correctly and provide the appropriate expected level of service to your clients. Application modeling is itself a mix of science and art. After collecting data on network events generated by applications, design a topology and validate the results of the analysis using a small test network.

3. Taking inventory of your applications

It is theoretically possible to run real-time voice or video across a frame relay link, but you have to make sure that your end-to-end latency is not too high to degrade quality. For voice, many people can notice delays of under 100 ms, and everybody will notice when it exceeds 300 ms.

The key item to watch is the amount of traffic in kilobytes offered to the network over the sample period. If it is far beyond a level that the WAN will support (e.g., hundreds of kilobytes per second), you may have to reconsider the application's architecture before you continue.

4. Building a Test Network

A sample network for testing is the fastest way to determine what application performance will really be like. It is the place to try out new applications before they get deployed.

Your test network will only display the effects of increased latency due to a WAN bandwidth constraint. It would not show propagation and internodal delays that may be rampant in a WAN. It will catch bad applications before they get out, but it would not predict response times.

5. Making Decisions

Now that you understand your applications, you can place them on the overall topology map. Based on what you have learned, you may want to move application locations if you can. Before you start drawing the network, consider each of the applications and the ways that it might be optimized.

Draw PVCs between sites as dictated by the application spreadsheet. Do not assign bandwidth numbers yet. If high-priority traffic exists, consider using dedicated PVCs.

Finally, determine access link rates (56 kbps or T1) for each of your locations. If you have existing digital facilities in place, you may wish to consolidate them into a single T1 link to reduce cost and provide additional channels for future growth. If you conclude that a location on the frame relay network is already overutilizing 56 kbps access circuit, consider installing T1 service to avoid installation delays and duplicate cost down the road.

4.3 FRAME RELAY STANDARDS

Several organizations are involved in the development of frame relay standards. Two of such organizations are the International Telecommunication Union and the Frame Relay Forum.

The International Telecommunication Union–Telecommunications Standardization Sector (ITU-T) (www.itu.int/ITU-T/) is an agency of the United Nations and is based in Geneva, Switzerland. Due to the number of groups involved, the process of setting standards is slow but democratic. The standards developed are known as recommendations. The following ITU-T standards define frame relay:

- Recommendation I.233 describes a recommended set of frame mode bearer services.
- Recommendation I.122 defines those terms and vocabulary essential to the understanding and application of the principles of ISDN.
- Recommendation I.370 defines congestion management for the ISDN frame relaying bearer service.
- Recommendation Q.922 specifies the frame structure, elements of procedure, format of fields, and procedures of the data link layer to support frame mode bearer services.
- Recommendation Q.933 specifies signaling protocol in frame relay to support switched virtual connections.

The Frame Relay Forum (www.frforum.com) was organized in 1991 and is a global consortium of carriers, vendors, and users. The objective of the consortium is to facilitate the development and implementation of frame relay services and configurations that operate in agreement with national and international frame relay standards. In addition, the forum develops implementation agreements (IAs), which support reliable transmission of voice, video, and data via a frame relay infrastructure. IAs also specify approaches for internetworking frame relay with broadband technologies such as ATM. Here are some of the IAs:

- FRF.2.2—Frame Relay Network-to-Network Interface (NNI) Implementation Agreement. This IA identifies the standards and recommendations to be followed and describes the application of bidirectional procedures and the application of event-driven procedures at the NNI.

- FRF.6—Frame Relay Service Customer Network Management Implementation Agreement. This IA provides information that allows end customers to obtain performance monitoring, fault detection, and configuration information about their frame relay service.
- FRF.10.1—Frame Relay Network-to-Network SVC Implementation Agreement. This IA provides a frame relay-switched virtual connection (SVC) network-to-network interface (NNI).
- FRF.11.1—Voice over Frame Relay Implementation Agreement. This IA describes the internetworking necessary to transport digital voice over frame relay.

4.4 FRAME RELAY VERSUS X.25

While X.25 is defined for layers 1, 2, and 3 of the ISO model, frame relay is defined for layers 1 and 2 only. This means that frame relay has significantly less processing to do at each node, which improves throughput by orders of magnitude. Some of the functions associated with packet transport, such as error correction and flow control, are still formed, but on an end-to-end basis by the end-user devices, instead of by the network.

X.25 prepares and sends packets, while frame relay prepares and sends frames. X.25 packets contain several fields used for error and flow control, none of which is needed by frame relay. The frames in frame relay contain an expanded address field that enables frame relay nodes to direct frames to their destinations with minimal processing.

Unlike X.25, frame relay supports scalable transmission rates at speeds ranging from 56 kbps, T1 (1.544 Mbps), and E1 (4.048 Mbps) to T3 (44.736 Mbps) and E3 (34.368 Mbps) for enabling data, voice, and video transmission.

X.25 packets contain several fields used for error and flow control, none of which is needed by frame relay. Frame relay limits the functionality inside the network based on the premise that the probability of frame error is very small. Other functions, such as error recovery, flow control, etc. can be performed at higher layers by users.

Frame relay is considered as a cost-effective replacement for leased line services in addition to being technically superior to the X.25 services. Frame relay is cost effective partly due to the fact that the network buffering requirements are carefully optimized. Compared to X.25, with its store and forward mechanism and full error correction, network buffering is minimal. Table 4.1 makes further comparison between X.25 and frame relay.

4.5 VOICE OVER FRAME RELAY

Recently, data networks have been growing at a much faster rate than voice networks, mainly due to the growth of the Internet. Soon the amount of data traffic will exceed that of voice traffic. As a result of this trend, more and more voice is being sent over data networks (voice over frame relay, voice over IP, and voice over ATM) than data is being sent over voice networks. Of the three popular packet/cell technologies (frame relay, IP, and ATM), frame relay is the most widely deployed. Frame relay is

TABLE 4.1
Comparing Frame Relay with X.25

Network Characteristic	Frame Relay	X.25
Propagation Delay	Low	High
Error Correction	None, done by the terminal equipment at each end of the link	Node to Node
Protocol family	HDLC	HDLC
Good for interactive use?	Yes	Barely acceptable. Rather slow with one second or more round trip delay.
Good for LAN file transfer?	Yes	Slow
Good for voice?	Good	No
Ease of implementation	Easy	Difficult

Source: http://www.dcbnet.com/notes/framerly.html.

commonly used in corporate data networks due to its flexible bandwidth, widespread accessibility, support of a diverse traffic mix, and technological maturity.

Voice over frame relay (VoFR) is an approach to passing voice across a frame relay network.

Several network managers are interested in the economic advantage of sending voice traffic over frame relay networks. Network consolidation results when we combine voice and data and that results in potential savings. Although a manager can use a frame relay network exclusively for voice, it is not cost effective.

The device for handling voice over a frame relay network is a voice over frame relay access device (VFRAD), which multiplexes voice, data, and fax traffic onto frame relay virtual circuits. A VFRAD may be integrated into other CPE or may be stand-alone. To be able to carry voice on frame relay, network managers install a gateway between their telephone networks and their data networks. The gateway converts digital voice signals coming from the private telephone switch into frames and drops the frames onto the frame relay data network. The voice frames can then travel anywhere the data network goes. When they arrive at the desired site, the voice frames pass through another gateway that turns them back into regular digitized voice traffic. That traffic is sent to the private voice switch at the second site. The switch directs the call to the proper telephone.

Since frame relay was originally designed for data traffic, adding voice capability poses some challenges. Unlike most data applications, voice is very sensitive to delay. Good voice quality provides a faithful recreation of the conversation, with the same tone, inflection, pauses, and intonation used by the speakers. Long and variable delays between packets result in unnatural speech and interfere with the conversation. These problems do not necessarily negate the value and promise of VoFR. Significant advances in digital signal processors and compression algorithms often provide voice at a level approaching toll quality, for a fraction of the cost of public service. Also, a growing market awareness of its advantages and increasing work by standards groups is ameliorating the problems.

In 1998, the Frame Relay Forum endorsed an implementation agreement (IA) for enabling frame relay voice transmissions. The agreement defines procedures for transmission of compressed voice within a frame relay frame payload. The Frame Relay Forum also determined methods of prioritizing the transmission of voice and data frames entering the network.

Frame relay service providers have also begun to offer different quality of service (QoS). The highest quality of service, real-time variable frame rate, is for voice. The second quality of service, non-real time variable frame, is typically for LAN-to-LAN, Internet, and intranet traffic. The lowest quality of service, available/unspecified frame rate, is used for email, file transfer, and residential Internet traffic. In addition, the VFRAD can be configured to assign less sensitive traffic with a Discard Eligibility (DE) bit. These frames will be dropped first in case of network congestion.

Voice over frame relay (VoFR) is considered here because voice is "king." We may also consider fax over frame relay and video over frame relay. Many of the issues are similar to those of transmitting voice.

4.6 CONGESTION CONTROL

The explosive growth in traffic experienced by computer networks has led to severe congestion problems. Therefore, a common problem confronted by data networks, such as frame relay, packet switching, and ATM networks is that of congestion. For example, it is common to see internet gateways drop 10% of the incoming packets due to local buffer overflow.

What causes congestion? Congestion occurs when the demand is greater than the available resources. Congestion may be due to the following reasons:

1. Shortage of buffer space: The congestion problem will be solved when the cost of memory becomes cheap enough to afford a large memory. Increasing the size of the buffers is not a solution, because excessive buffer size can lead to excessive delay, which some applications cannot tolerate.
2. Congestion is caused by slow links: The congestion problem will be solved when high-speed links become available. However, higher speed links may make the network more unbalanced and can make the congestion condition in the switch even worse.
3. Congestion is caused by slow processors: The congestion problem will be solved when processor speed is improved. Faster processors will transmit more data per unit time. If several nodes begin to transmit to one destination simultaneously at their peak rate, the target will soon be overwhelmed. Congestion is a dynamic problem, and any static solutions are therefore not sufficient to solve the problem.

Contrary to these beliefs (known as the "myths" about congestion control), without proper protocol redesign, the above developments may lead to more congestion and thus reduce performance. Congestion may be caused by the fact that frame relay implements no flow control between switches. Congestion typically occurs where multiple links feed into a single link, such as where internal LANs are connected to WAN links. Congestion also occurs at routers in core networks where nodes are subjected to more traffic than they are designed to handle. Congestion control is therefore necessary to ensure that users get the negotiated quality of service (QoS).

The objectives for frame relay congestion control, according to the ITU-T Recommendation I.370, should:

- mimimize frame discard
- maintain an agreed quality of service (QoS)
- minimize the possibility that one end user will monopolize the network resources
- be simple to implement
- create minimal additional network traffic
- distribute network resources fairly among end users
- limit spread of congestion to other networks
- minimize the variance in QoS delivered

Frame relay has a way of handling congestion. It informs the end-user protocol of a congestion problem using the FECN (forward explicit congestion notification) and BECN (backward explicit congestion notification) bits in a frame. FECN and BECN are types of messages sent on a frame relay network. In a frame relay network, FECN is a header bit transmitted by the source (sending) terminal requesting that the destination (receiving) terminal slow down its requests for data. BECN is a header bit transmitted by the destination terminal requesting that the source terminal send data more slowly. FECN and BECN are intended to minimize the possibility that packets will be discarded (and thus have to be resent) when more packets arrive than can be handled.

4.7 FRAME RELAY APPLICATIONS

For companies with geographically dispersed locations, frame relay provides a cost-effective way of providing a secure private network. Some companies use virtual private networks (VPNs) over the Internet for intercompany communications, but that option does expose the companies to some serious security issues. In contrast, frame relay privacy is guaranteed by the nature of the network, backed up by legislation.

Some of the applications of frame relay include the following:

- Voice applications: Voice over frame relay (VoFR) enables you to consolidate voice traffic on your existing frame relay network and realize significant cost savings. VoFR offers telecommunication and network managers the opportunity to consolidate voice and voice-band data (e.g., fax and analog modems) with data services over the frame relay network. This has been recently enjoying the general acceptance of any efficient and cost-effective technology.
- LAN internetworking: Frame relay enables peer-to-peer, meshed LAN internetworking without the expense of fully meshed leased line networking.

Initially, frame relay gained acceptance as a means to provide end users with a solution for LAN-to-LAN connections and to meet other data connectivity requirements. Frame relay's compelling benefit is that it lowers the cost of ownership compared to competing technologies. Frame relay is a protocol standard for LAN internetworking that provides a fast and efficient method of transmitting information from a user device to LAN bridges and routers. This service is a cost-effective alternative to private lines that allows your business to link LANs and enables application sharing and high-speed data transmission.

- TCP/IP implementation: The TCP/IP suite comprises two components, the IP operating at layer 3 and the transmission control protocol (TCP) operating at layer 4. IP is a best-effort delivery protocol, relying on the transmission control mechanisms supported by TCP. IP traffic is typically bursty in nature, making it an ideal network-layer protocol for frame relay WANs. A TCP/IP implementation over frame relay would require the mapping of layer 3 IP addresses to a DLCI.
- CAD/CAM applications: Computer-aided design/manufacturing (CAD/CAM) applications are used by companies involved with engineering, architecture, and manufacturing. Frame relay is seen as a cost-effective alternative to private lines, which enables your business to share these mission-critical applications and efficiently transfer the memory-intensive files you create.
- Imaging applications: Several companies are transmitting their images over frame relay. It has the bandwidth and transmission speed to make it an effective and economic medium for image processing. Any company with dispersed locations that needs to transfer photographs, X-rays, technical drawings, forms, signatures, or other visuals may benefit from frame relay.
- Financial services: The bank employs frame relay to provide automated transaction processes coordinated between all sites for customers, tellers, customer service agents, and traders. Frame relay is also used to improve internal communications through the use of email and enhance security and speed up check processing by using imaging technologies to view signatures. Monthly telecommunications charges are also reduced and the network is simplified.

Other applications include ATM-internetworking, manufacturing, government, education, health care, retail, wholesale, insurance, and real estate.

Example 4.4

An insurance company has independent agents throughout Texas. The office in Houston processes claims and consults with claims adjusters based in San Antonio. Each of the groups needs access to the latest reports, account numbers, and policy information to provide the best possible services.

The insurance company found frame relay had a positive impact on their business by enabling them to

- reduce the amount of paperwork by submitting claims electronically
- improve customer service by using high-speed transmission capabilities to transfer medical images or photographs of accident damage between healthcare providers, claims assessors, and agents
- improve internal communications and coordination between claims processors, adjusters, agents, and customer service representatives
- share applications and data between remote offices so every user has access to the latest information

SUMMARY

1. X.25 is a packet switched data network protocol that defines an international recommendation for the exchange of data between a user device (DTE) and a network node (DCE).
2. Virtual circuit is the end-to-end connection (through a network) between two user devices (DTEs).
3. Frame relay is a packet switching technology that relies on low-error-rate digital transmission links and high-performance processors.
4. X.25 was designed to provide error-free delivery using high-error-rate links. Frame relay takes advantage of the new, lower-error-rate links, enabling it to eliminate many of the services provided by X.25.
5. Congestion control is about controlling traffic entry into telecommunication networks.
6. Voice over frame relay (VoFR) technology consolidates voice and voice-band data (e.g., fax and analog modems) with data services over the frame relay network.
7. The most popular frame relay application provides companies with LAN-LAN internetworking. In addition, frame relay finds applications in voice over frame relay (VoFR), business, education, medical care, and industry.

REVIEW QUESTIONS

4.1 What kind of technology is X.25?
  a. Circuit switched
  b. Packet switched

4.2 Permanent virtual circuits do not require that sessions be established and terminated. Therefore, DTEs can begin transferring data whenever necessary, because the session is always active.
  a. True
  b. False

4.3 A packet of 5000 bytes can be carried successfully on X.25 without the need of breaking it up.
   a. True
   b. False
4.4 What does the flag field 01111110 indicate?
   a. Data
   b. Control information
   c. Beginning/end of frame
   d. Frame check sequence
   e. Invalid packet
4.5 Which of these packet technologies is the most widely deployed?
   a. X.25
   b. Frame relay
   c. IP
   d. ATM
4.6 The frame-relay network uses the two lowest layers of the OSI model.
   a. True
   b. False
4.7 Whenever there is congestion on the line, which part of the frame is used to decide which frames to discard?
   a. Flag
   b. DLCI
   c. FE
   d. BE
   e. DE
4.8 Implementation Agreements are developed by:
   a. ITU-T
   b. American National Institute
   c. Frame Relay Forum
   d. Internet Engineering Task Force
4.9 Which of the following recommendations for frame relay was developed by ITU-T?
   a. FRF.2
   b. FRF.6
   c. FRF.10.1
   d. Q.933
4.10 Frame relay is different from X.25 in this respect:
   a. X.25 was designed to provide error-free delivery using high-error-rate links.
   b. X.25 is defined for layers 1, 2, and 3 of the ISO model, while frame relay is defined for layers 1 and 2 only
   c. X.25 is packet switched, while frame relay is circuit switched.
   d. X.25 prepares and sends packets, while frame relay prepares and sends frames.

Answer: 4.1b, 4.2a, 4.3b, 4.4c, 4.5b, 4.6a, 4.7e, 4.8c, 4.9d, 4.10c

PROBLEMS

4.1 What are the minimum and maximum lengths an X.25 packet can have?
4.2 What is a PAD? What role does it play in X.25 protocol?
4.3 Describe the function of each of the three X.25 layers.
4.4 Explain the terms DTE, DCE, SVC, and PVC.
4.5 What is the maximum speed for X.25?
4.6 Using Web resources, describe X.75. How is it related to X.25?
4.7 Using Web resources, find out the difference between ITU recommendations X.3 and X.28.
4.8 Do some research on X.21 and X.21bis and mention the difference between them.
4.9 What are the merits and demerits of X.25 networks?
4.10 Explain the following frame relay terms:
  a. Access speed
  b. Committed information rate (CIR)
  c. Data-link connection identifier (DLCI)
  d. Oversubscription
4.11 What are forward explicit congestion notification (FECN) and backward explicit congestion notification (BECN) packets? How do they affect performance?
4.12 Frame relay has a variable size frame. What are the minimum and maximum lengths a frame can have?
4.13 How is flow control performed in a frame relay network?
4.14 Under what circumstances would you recommend frame relay to an organization?
4.15 Distinguish the data formats of the following protocols:
  a. X.25
  b. Frame relay
4.16 What is FRAD? How is it different from a PAD?
4.17 Explain the function of each field in the data format of the frame relay.
4.18 Which two bits in the frame relay header are used to signal the user device that congestion is occurring on the line?
4.19 How does frame relay provide quality of service (QoS)?
4.20 What are the differences between frame relay and X.25?
4.21 What are the "myths" about congestion control?
4.22 How is congestion controlled in a frame relay?

5 Fiber Channel

When something can be read without effort, great effort has gone into its writing.

—Enrique Jardiel Poncela

Fiber channel is part of the recent revolution in computer communications caused by incorporating serial transmission line design techniques and technologies into application in computer architectures that have conventionally used bus-based types of data transport. Current data transfer protocols (e.g., HIPPI and SCSI) have problems that limit their usefulness in mass storage. The limitations include limited speed, limited distance between devices, and limited number of devices supported. These limitations caused the industry to start thinking about alternatives. The alternative is fiber channel.

Fiber channel is a highly reliable, gigabit interconnect technology that serves as a means of quickly transferring data between supercomputers, mainframes, workstations, desktop computers, storage devices, and other peripherals. It allows concurrent communications among those computer devices using SCSI (small computer systems interface) and IP protocols. The standard allows payload bit rates of 200, 400, or 800 Mbps.

In this chapter, we first discuss the basic features of fiber channel. Later, we discuss its technology, topology, classes of service, benefits, and applications. Finally, we will wrap up with an application for fiber channel technology—storage area network (SAN). We will try as much as possible not to bore you with details.

5.1 BASIC FEATURES

There are basically two types of data communication between processors and peripherals: networks and channels. A network consists of distributed nodes and supports interaction among these nodes. A channel provides a direct or switched connection between devices over very short distances. While a channel is hardware intensive and transports data at high speed with low overhead, a network is software intensive, with high overhead and slower speed than a channel. A fiber channel combines the best features of the two methods of communication into an I/O interface that meets the needs of both the network users and channel users. It has the flexibility and interconnectivity characterizing protocol-based networks while maintaining the simplicity and speed associated with channel connections. Although it is called fiber channel, both optical and electrical media are supported. (Thus "fiber" will be used as a generic term to mean an optical or copper cable.) Also, its architecture represents neither a channel nor a network.

Fiber channel is a network technology for linking servers to storage systems that move data more efficiently than existing network pipes.

Fiber channel was developed by the American National Standards Institute (ANSI). It is being promoted by the Fiber Channel Industry Association (FCIA), an industry consortium. It is well-suited for companies that desire to build their own SANs (storage area networks).

With development started in 1984 and ANSI standards approved in 1994, fiber channel was designed to remove the barriers of performance existing in legacy LANs and channels. The performance-enhancing features of fiber channel for networking include:

- Unification of technologies: Fiber channel is geared toward unifying LAN and channel communications. It achieves this by defining an architecture with enough flexibility and performance to fulfill both sets of requirements.
- Bandwidth: Fiber channel provides more than 100 Mbps for I/O and communications on current architectures.
- Inexpensive implementation: By limiting low-frequency components, fiber channel allows design of receivers using inexpensive CMOS VLSI technology.
- Flexible topology: Physical topologies are defined for point-to-point links, packet-switched network topologies, and shared-media loop topologies.
- Flexible service: Fiber channel offers classes of services including dedicated bandwidth between port pairs, multiplexed transmission, and multiplexed datagram transmission.
- Standard protocol mappings: Fiber channel can operate as a data transport mechanism for multiple protocols with defined mappings for IP, HIPPI (high-performance parallel interface), intelligent peripheral interface (IPI), small computer systems interface (SCSI), ESCON (enterprise serial connection, developed by IBM), and ATM AAL5.
- Support for distances up to 10 km.
- Complete support for traditional network self-discovery.
- Availability of both real and virtual circuits.
- Use of 8B/10B encoding, which allows design of gigabit receivers using inexpensive CMOS VLSI technology.
- Low overhead providing efficient usage of the bandwidth.
- Full support for time-synchronous applications like video using fractional bandwidth virtual circuits.
- Efficient, high-bandwidth, low-latency transfers using variable length frames (0–2 kB).
- Two data paths to provide redundant connections; full duplex links with each having two fibers.
- A serial optical and coaxial system operating from 100 Mbps to 800 Mbps.

Fiber channel is a network that performs channel operations or a channel with network extensions. It uses an asynchronous mode of transmission permitting 100 percent utilization of the available bandwidth. It is scalable over a broad range of transfer rates, media types, and protocols.

5.2 FIBER CHANNEL TECHNOLOGY

The fiber channel standard (FCS) is made simple in order to minimize implementation cost and enhance throughput. It consists of five levels as shown in Figure 5.1. The idea of levels is similar to the layers in networks. The conventional physical and media access control (MAC) layers are divided into five hierarchical protocol levels. Each level defines a function or a set of related functions. Levels allow elements in each level to change (with technology or other needs) without affecting adjacent levels. The five levels are summarized in Table 5.1.

FC-0 level defines the physical portion of the fiber channel including the transmission media, connectors, transmitters, and receivers and their interfaces, and optical and electrical parameters for a variety of data rates. The media can be optical cable with laser or LED transmitters for long-distance transmission, copper coaxial cable for high speeds over short distances, or shielded twisted pair for low speed over short distances. The data transfer rates are defined as 100, 200, 400, and 800 Mbps. FC-0 operates with a BER of less than 10–12. The FC-0 specification provides for a large variety of distances, cable plants, and technologies. Each variant is described with one code from each of the four characteristics as shown in Figure 5.2. For example, code 100-SM-LL-L indicates speed is 100 Mbps, medium is single-mode fiber, transmitter is long-wave laser, and distance is long.

FC-1 level describes the transmission protocol that includes byte serial encoding/decoding, error control, and synchronization. This layer is responsible for establishing rules for transmitting information on the network. It handles the 8B/10B encoding scheme shown in Figure 5.3. An 8B/10B scheme is used just as the 4B/5B technique used in FDDI, but it is more powerful than 4B/5B. The information transmitted over a fiber is encoded 8 bits at a time into a 10-bit transmission character. The information received is recovered 10 bits at a time and those transmitted

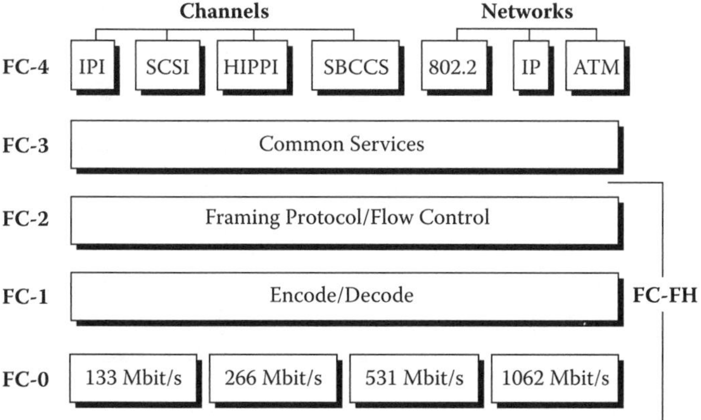

FIGURE 5.1 Structure of fiber channel standards.

TABLE 5.1
Fiber Channel Levels

Level No.	Function	Content
FC-0	Physical interface	Optical/electrical, cable plant
FC-1	Byte encoding/decoding	8B/10B encoding, link controls
FC-2	Framing and signaling	Framing, flow control, service class
FC-3	Common services	Under construction
FC-4	Upper-layer protocol interface	SCSI-3, IP, etc.

characters are decoded into one of the 256 8-bit combinations. The rationale for the transmission code is to facilitate error correction and bit synchronization. In other words, the scheme allows the protocol to synchronize data transmission and provides a retransmission mechanism when errors are detected. In addition, the 8B/10B encoding/decoding scheme is simple to implement and provides DC balance for the transmitted bit stream. The 8B/10B scheme was developed and patented by IBM for use in their ESCON interconnect system.

FC-2 level is the framing and signaling protocol layer. (The frames are of variable length, with fixed overhead.) This level is the transport mechanism specifying the rules required to transfer data. It provides for point-to-point, arbitrated loop, and switched environments. It also handles flow control, error control, congestion management, and class-of-service designations. This is the most complex portion of fiber channel. It provides three classes of service between ports (dedicated connection,

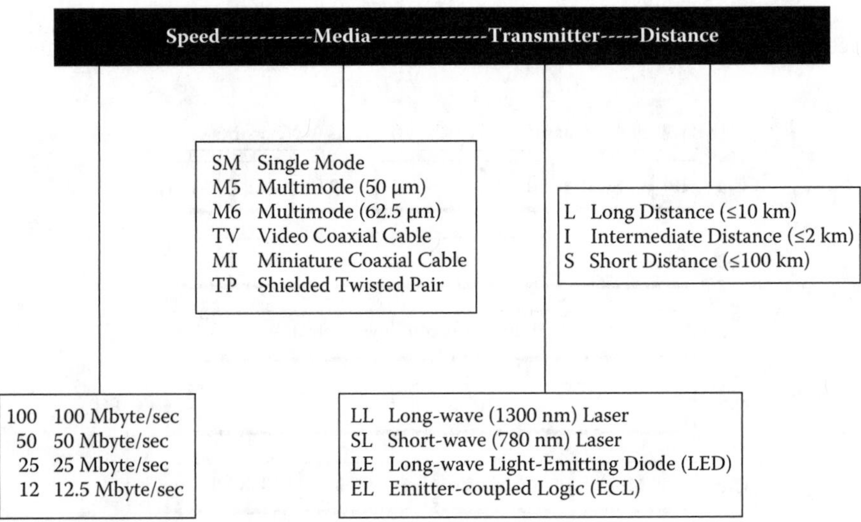

FIGURE 5.2 Fiber channel FC-0 nomenclature.

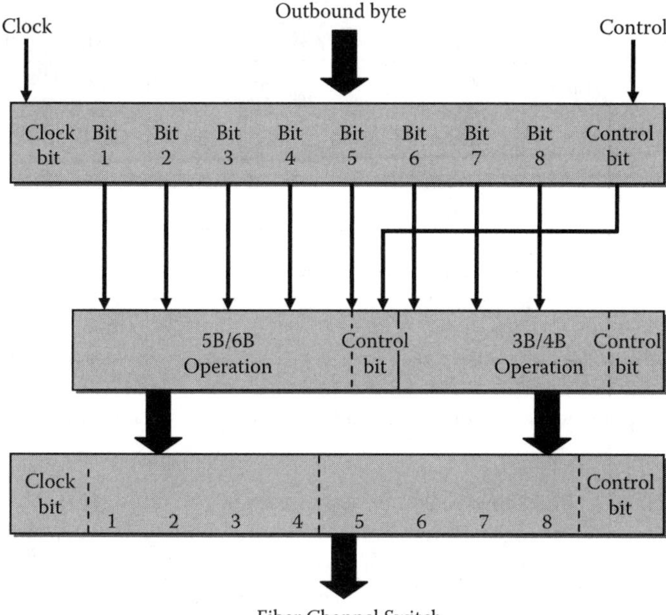

FIGURE 5.3 Fiber channel 8B/10B encoding scheme.

multiplex connection, and connectionless service), error detection, packetization and sequencing, segmentation and reassembly, and flow control between ports. The following building blocks are defined in FCS to aid in the transport of data across the link:

- Ordered sets: These are four-byte transmission words containing data and special characters that have special meaning. The frame delimiters, the start-of-frame (SOF) and end-of-frame (EOF) ordered sets, immediately precede or follow the contents of a frame. An Idle primitive indicates a port is ready for frame transmission or reception. The primitive Receiver Ready indicates that the buffer is available.
- Frame: A frame contains the information (payload), the SOF and EOF delimiters, the address of the source and destination ports, and link control information. The frame structure is shown in Figure 5.4. There are two types of frames: data frames and link control frames. It is the responsibility of the FC-2 level to break the data into frame size and reassemble the frames.
- Sequence: A sequence is formed by a series of one or more frames transmitted unidirectionally from one port (the sequence initiator) to another (the sequence recipient). Each sequence is uniquely specified by the sequence identifier (Seq_ID), as in Figure 5.4. The Seq_cnt in Figure 5.4 is a counter field that uniquely identifies frames within a sequence; the first frame starts with 0 and Seq_cnt is incremented for subsequent frames.

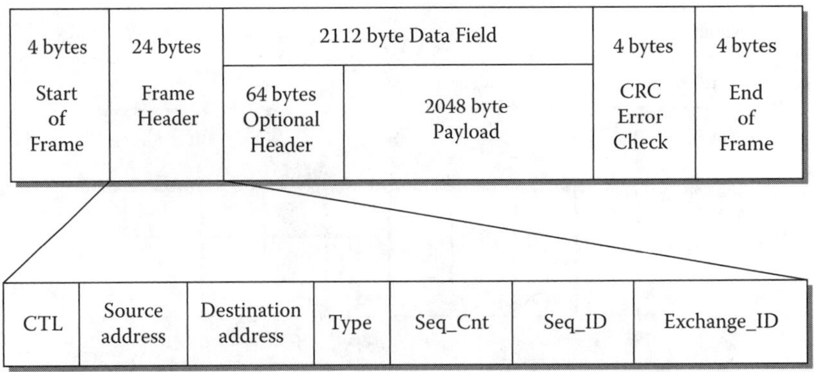

FIGURE 5.4 Frame structure for fiber channel.

- Exchange: An exchange comprises a series of one or more nonconcurrent sequences flowing unidirectionally or bidirectionally between two ports.
- Protocol: A protocol is a set of frames that may be sent in one or more exchanges to achieve a specific purpose. Possible protocols include primitive sequence protocol, port login protocol, port logout protocol, fabric login protocol, and data transfer protocol.

FC-3 level provides a set of services common across the N ports of a fiber channel node. Three such services are defined as

- multicasting, which delivers a single transmission to multiple destination ports
- stripping, which uses multiple ports in parallel to transmit a single piece of information across multiple links simultaneously
- hunt group, which is the ability for more than one port to respond to the same alias address. Hunt groups are like a telephone hunt group where an incoming call is answered by any one of the phones; frames addressed to a hunt group are delivered to any available port within the hunt group

FC-4 level defines the mapping of various channels and application interfaces that can execute over fiber channel. Fiber channel supports the following channel and network protocols:

- intelligent peripheral interface (IPI)
- small computer system interface (SCSI)
- high-performance parallel interface (HIPPI)
- single-byte common code set mapping (SBCCS)
- IEEE 802.2 local area network (LAN)
- Internet protocol (IP)
- ATM (asynchronous transfer mode) adaptation layer (AAL5)
- link encapsulation

The lower three levels (FC-0, FC-1, and FC-2) together form the fiber channel physical layer FC-PH, while the upper two layers are for common services and protocol interface. Several standards have been produced for level FC-4 to map a variety of channel and network protocols to lower levels. Fiber channel can support data, video, and voice at the same time, although each type of information has different characteristics and makes demands on the sender and receiver of information.

5.3 FIBER CHANNEL TOPOLOGIES

Fiber channel devices are known as nodes. Each node has at least one port to access other ports in the other nodes. A node may be a computer, workstation, a disk drive, TV camera, or a display. A port is the hardware entity within a node that permits data communication over a link. Each port uses a pair of fibers; one carries information into the port while the other carries information out. This pair of fibers is known as a link. An N_Port is a port at the end of a link and "N" stands for node. An F_Port is an access point of the fabric and "F" stands for fabric.

In fiber channel, the switch connecting device is called a "fabric." In other words, a fabric is an entity that allows interconnection of any N_ports to it; that is, it provides switching and routing. The fabric is self-sustaining; nodes do not need station management. This greatly simplifies implementation. The link is the two unidirectional fibers transmitting to opposite directions with their associated transmitter and receiver.

A **topology** is a set of hardware components such as media, connector, and transceivers connecting two or more N_ports.

As shown in Figure 5.5, there are three topologies defined for fiber channel:

- Point-to-point topology: This is the simplest topology. It is basically a simple, direct connection between two N_ports. Before data transmission can take place, the two N_ports must perform a login so that N_port addresses can be assigned. After login, a connection is maintained. This topology guarantees maximum bandwidth availability and freedom from collision or congestion, but it is usually a relatively inefficient usage of hardware.
- Switch topology or fabric: This is a network of N_ports each connected to an F-port into a switching network or fabric. A fabric is a mesh of connections. When connected to a fabric, an N_port can access all the other N_ports connected to the fabric, making it possible to establish many connections. Fabrics provide full bandwidth per port. The fabric topology provides the connectivity and link bandwidth utilization flexibility, but can be a quite expensive topology due to the complex switches involved.
- Ring topology: This is known as "fiber channel arbitrated loop" (FC-AL), which is a simple closed serial loop that allows multiple N_port interconnections without switching elements. It is a shared, gigabit transport. That is, instead of each port having its own link, all ports share a single media. As the fiber channel standard progressed, it became clear that dedicating

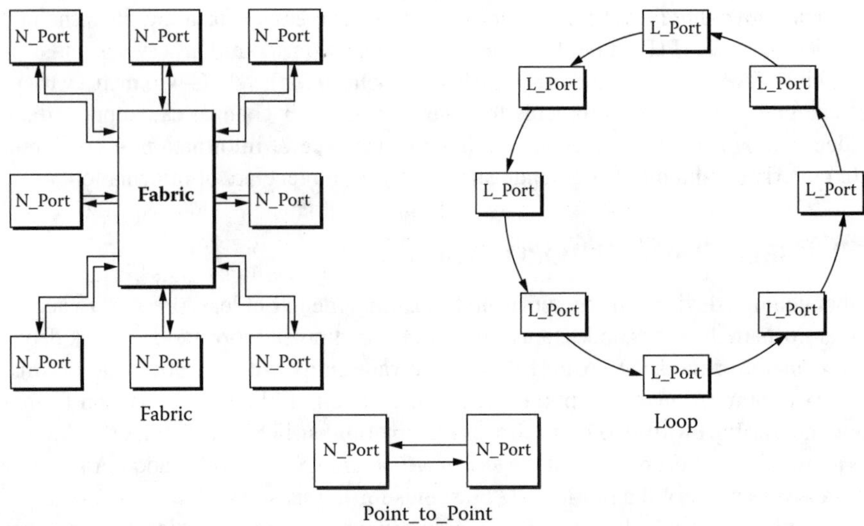

FIGURE 5.5 Fiber channel topologies.

> a switch port to every device might be neither economically feasible nor necessary in many applications. For example, a single host might need to connect with a number of disk drives, and that scenario cannot justify a switch. The fiber channel arbitrated loop (FC-AL) was developed for just such applications. The loop can have up to 126 ports connected in series in a loop form. The ports in the loop must combine the functions the N_ports and F_ports making them NL_ports. However, only two connected ports can communicate at any time. Loop initialization allows each device to select a unique physical address, thereby avoiding addressing conflict on the loop. With loop arbitrated topology, the signal passes through each port on the loop, with each port acting as a repeater.

The fabric topology is the most general topology. The arbitrated loop topology is a simple, low-cost scheme. Both fabric and arbitrated loops provide more flexibility and support for more devices. Advances in fabrics and arbitrated loop technology have enabled more flexibility and functionality than point-to-point link provides. Each topology has its own standard dealing with issues unique to the topology.

5.4 CLASSES OF SERVICES

To ensure efficient transmission of different types of traffic, FC defines several classes of service. Users select service classes based on the characteristics of their applications (like packet length and transmission duration) and allocate the services by the fabric login protocol. These classes of service are defined in the FC-2 level. The classes range from 1 to 6, except that class 5 is not defined.

- Class 1 service: This provides dedicated connections. This is also called circuit-switched service and is similar to ATM constant bit rate (CBR). Once established, a class 1 connection is retained and guaranteed by the fabric. This service guarantees the maximum bandwidth between two N_ports, so this is best for sustained, high-throughput transactions. In class 1, frames are delivered to the destination port in the same order as they are transmitted. Class 1 service is useful for applications in which large blocks of data must be transmitted.
- Class 2 service: This provides a multiplexed service. This is also called connectionless service similar to ATM's variable bit rate (VBR) service. Class 2 is a frame-switched, connectionless service that allows bandwidth to be shared by multiplexing frames from multiple sources onto the same channel or channels. The fabric may not guarantee the order of the delivery and frames may be delivered out of order. This service class can be used when the connection setup time is greater than the latency of a short message. Both class 1 and class 2 send acknowledgment frames confirming frame delivery. If delivery cannot be made due to congestion, a busy frame is returned and the sender tries again. Class 2 is well suited for networks that are not sending time-sensitive information, such as storage area networks (SANs).
- Class 3 service: This is known as datagram type of service. This is also known as unacknowledged connectionless service. It is a connectionless packet-switched service that does not guarantee frame delivery. Class 3 service is identical to class 2, except that the frame delivery is not confirmed. This type of transfer provides the quickest transmission by not sending confirmation. This service is useful for real-time broadcasts, where timeliness is key and information not received in time is valueless.
- Class 4 service: This service provides dedicated connection that guarantees quality of service parameters. It is an acknowledged connection-oriented service. It is a newer connection-oriented service, which works to provide better quality of service quantities for bandwidth and latency than class 2 or 3. Class 4 service is a "fractional service" because it uses a fraction of the bandwidth of the link between two ports for communication between ports. Class 4 service is suitable for time-sensitive and real-time applications such as videoconferencing.
- Class 6 service: This service provides a unidirectional dedicated connection between an originator and a multicast group. This is unidirectional dedicated connection service. It is also a reliable, multicast service. It is similar to class 1 service. When a station desires to transmit frames to more than one N_port at a time, it sets up a class 1 dedicated connection with the multicast server within the fabric. Class 6 service is commonly used for video broadcast and for real-time transfer of large amounts of data.

Although classes 4 and 6 are defined, they are not widely implemented. The five classes and their services are summarized in Table 5.2. The type of flow control offered by FC depends on the class of service (CoS) that is designated for particular network traffic that is sent.

TABLE 5.2
Classes of Service

Class	Service
Class 1	Dedicated connection service Connection-oriented Acknowledged delivery
Class 2	Multiplexed service Connectionless Acknowledged delivery
Class 3	Datagram service Connectionless Unacknowledged delivery
Class 4	Fractional service Connection-oriented
Class 6	Unidirectional dedicated connection service Connection-oriented

5.5 BENEFITS AND APPLICATIONS

Fiber channel's distinguishing marks are speed and distance. Currently, fiber channel moves data at "gigabit speed," i.e., 2 or 4 Gbps. Fiber channel can connect devices over relatively long distances. There may be a distance of up to 500 m between a device and a hub and up to 10 km between hubs. Besides speed and distance, fiber channel enjoys the following advantages:

- multiple topologies meet application requirements
- multiple protocols can all take advantage of the high-performance, reliable technology
- high efficiency due to very little overhead
- it is scalable—from point-to-point links to integrated enterprises with hundreds of servers

Fiber channel has found application in storage area networks (SANs), which will be fully discussed in the next section. SAN provides faster, more reliable access to storage devices compared with accessing them as a node on a conventional network. Besides SAN, fiber channel is used in the following areas:

- high-performance CAD/CAE network
- quick-response network for imaging applications
- nonstop corporate backbone
- movie animation
- storage backup systems and recovery
- server clusters
- campus backbones

5.6 STORAGE AREA NETWORKS

An interesting application for fiber channel technology is the storage area networks (SANs). The high-speed and flexible networking characteristics of fiber channel have made it a suitable networking technology for storage area networks (SANs). A SAN is based on the principle of networking devices. As shown in Figure 5.6, a SAN consists of a group of mass storage devices connected indirectly to host computer systems and each other by means of interconnecting devices such as hubs, bridges, and switches. Storage devices include disk drives (nicknamed JBOD for "just a bunch of disks"), disk arrays, tape drives, and tape libraries.

> **Storage area network** (SAN) is a high-speed network consisting of servers and storage devices designed to handle large amounts of data traffic.

A SAN is a means to allow multiple servers to have direct access to common storage devices or a storage pool. A SAN can also be viewed as moving all the problems of handling data from hardware to software. A SAN is a software-managed solution that is optimized for moving large amounts of data. It consists of a separate, dedicated high-performance network that moves data between heterogeneous servers and storage resources.

Prior to SANs, the management of a variety of storage and backup media had been complicated. Paper tape was once a fashionable place to store data. As technology developed, external storage became available in the form of magnetic tapes and disks. Local area networks (LANs) were created to enable the connection of

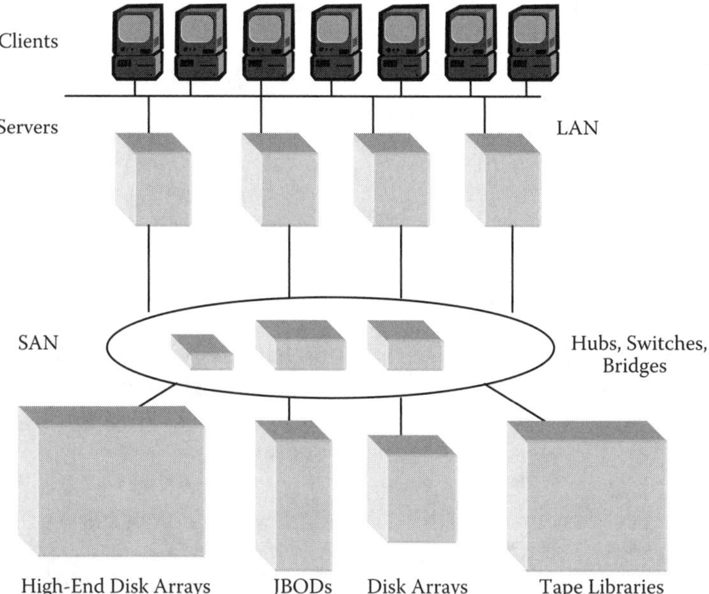

FIGURE 5.6 A typical storage area network.

one computer to another to allow users to access data stored in different databases. For over a decade, the small computer system interface (SCSI) has provided storage interconnectivity for (LANs). Effective in its time, SCSI is struggling to cope in the face of the data explosion. It is limited in speed, distance between devices, and number of devices supported. SAN combines the latest concepts in the storage, backup, and management of data across a variety of media.

SANs generally comprise servers (hosts), storage devices (tapes and disk arrays), and bridges and multiplexers, all connected to fiber channel switches. As with LANs or WANs, the switches furnish the backbone for all connected devices, with one or more switches acting as a fiber channel switching fabric. SAN switch fabrics allow attachments of thousands of nodes. SANs can also be formed using fiber channel arbitrated loop (FC-AL), a type of shared-media network. The FC-AL architecture allows for up to 126 devices per loop, either attached directly to fiber channel switches or to hubs that in turn connect to switches.

Storage area networks are designed to incorporate the best of both storage and networking interfaces: fast, efficient communication, optimized for efficient transfer of large amounts of data. Although a SAN is based on the principle of networking devices together, it is different from LANs and WANs. The main difference between SANs and other networks (such as LANs, MANs, and WANs) is that in a SAN, communication within the network is well managed, well controlled, and predictable. For example, the differences between LAN and SAN are twofold: due to network design and due to protocol design. Due to network design, switched-based transmission versus shared medium, the quality of service can be higher. Due to protocol design, performance relates to the higher levels of protocol design. LANs in general access through a TCP/IP protocol stack, whereas SANs are accessed through a simpler SCSI protocol stack.

Storage area network (SAN) offers a number of benefits, which include

- improved bandwidth: operation at high speeds, many times faster than the fastest SCSI server can operate
- scalability: unlike its SCSI counterpart, the capacity of a SAN can expand quickly to meet your storage requirements; thousand of devices may be added to a SAN
- availability and fault tolerance: ensuring mission-critical data is required as needed
- integration: management of a SAN is simple, while retaining powerful management options
- lowered costs: the SAN offers many cost benefits over the traditional SCSI interface, including the ability to add capacity cost effectively
- flexibility: SAN has great flexibility in managing data resources—it is very easy to reallocate disk storage, for example

The storage area network (SAN) is driving down the cost of storing data, eliminating the need for costly SCSI systems. Even with all these benefits, SANs are not the cure-all for all storage management problems. Like any new technology, solving one set of problems tends to highlight additional issues that need to be understood and resolved.

The major challenge facing the widespread implementation of fiber channel storage networks is the high cost and complexity of deployment. For example, gigabit fiber channel switches cost twice as much as gigabit Ethernet switches, not to mention training and support costs.

SUMMARY

1. Fiber channel is a networking technology (developed by ANSI) whose transmission media can be copper wires or optical fibers.
2. The fiber channel standard defines five protocol levels with each level defining a function or a set of related functions.
3. Fiber channel supports three topologies: point-to-point link, fabric, and ring.
4. The Fiber Channel Industry Association (FCIA) is an industry consortium that promotes fiber channel.
5. Storage area networks (SANs) are gigabit-rate networks that rely on fiber channel for high throughput, great distances, and connectivity options between servers and storage devices.

REVIEW QUESTIONS

5.1 Fiber channel only supports optical fiber.
   a. True
   b. False

5.2 Fiber channel was developed by:
   a. IEEE
   b. ITU-T
   c. FCIA
   d. ANSI

5.3 Which of these is not a feature of fiber channel for networking?
   a. Unification of technologies
   b. Flexible service
   c. Flexible topology
   d. High quality of service
   e. Standard protocol mappings

5.4 Fiber channel does not support which topology?
   a. Fabric
   b. Ring
   c. Star
   d. Point-to-point link

5.5 The conventional physical and medium access control MAC layers are divided into five protocol levels in fiber channel.
   a. True
   b. False

5.6 Which of these is not an advantage of fiber channel?
   a. Speed
   b. Distance

c. Scalability
d. High efficiency
e. High cost

5.7 Which fiber channel level is responsible for framing, flow control, and service class?
a. FC-0
b. FC-1
c. FC-2
d. FC-3
e. FC-4

5.8 Which service class is not defined for fiber channel?
a. Class 1
b. Class 3
c. Class 4
d. Class 5
e. Class 6

5.9 Which class of service is a unidirectional connection?
a. Class 1
b. Class 3
c. Class 4
d. Class 5
e. Class 6

5.10 Which of these is not a benefit of SAN?
a. Cost
b. Improved bandwidth
c. Scalability
d. Integration

Answer: 5.1b, 5.2d, 5.3d, 5.4c, 5.5a, 5.6e, 5.7c, 5.8d, 5.9e, 5.10a

PROBLEMS

5.1 What are the basic features of fiber channel?
5.2 What is a fabric?
5.3 What functions are defined at the FC-2 layer of the fiber channel?
5.4 Why is fiber channel so popular in the enterprise?
5.5 Explain why the term "fiber channel" is a misnomer.
5.6 What are the networking classes of service for the fiber channel?
5.7 Describe the different kinds of media supported by fiber channel.
5.8 What are the basic data rates on fiber channel?
5.9 Give a description of the frame format of fiber channel.
5.10 Mention some of the benefits of implementing fiber channel.
5.11 Mention some of the applications of fiber channel.
5.12 Compare fiber channel with Ethernet LAN.
5.13 How is fiber channel different from IEEE LANs?
5.14 Describe a storage area network.

5.15 Mention some of the benefits of SANs.
5.16 How are SANs different from LANs?
5.17 What is the future for fiber channel SANs?
5.18 Using Web resources, find out the difference between fiber channel and ISCSI (Internet small computer system interface).
5.19 Using Web resources, write briefly about the fiber channel standard protocols.
5.20 What is SATA tunneling over fiber channel? Use Web resources to find out.

6 Synchronous Optical Network (SONET)

Education, then, beyond all other devices of human origin, is the great equalizer of the conditions of men, the balance-wheel of the social machinery.

—Horace Mann

Exchange Carriers Standards Association (ECSA) developed synchronous optical network (SONET) for ANSI. Synchronous digital hierarchy (SDH) is the international equivalent of SONET. These standards define the optical telecommunication transport for the U.S. and European Telecommunications. The SONET and SDH are technologies for network transport. They use synchronous operations in facilitating real-time data, voice, and video transmission through fiber optic cabling. The rates range from 51.84 Mbps (OC-1 or optical carrier level 1) to 13.271 Gbps (OC-255). High-speed communications from different vendors are interconnected by applying the SONET standard. The ability of SONET to transport many different digital signals using a standard synchronous transport signal (STS) format is its major attribute. We will discuss the SONET and SDH in this chapter. The SONET signal, frame, components, topologies, and standards will also be discussed.

6.1 SONET SIGNAL

A method is defined by SONET that can be used to carry many signals from various sources and at various capacities via a synchronous, optical hierarchy. The ability of SONET to transport many different asynchronous and synchronous digital signals using a standard synchronous transport signal (STS) format is its major attribute. SONET's base-level signal is called synchronous transport level 1 (STS-1). It operates at 51.84 Mbps. With reference to the OSI model, SONET establishes interface specifications at the physical layer or layer 1.

Information transport is carried by SONET by first defining the optical carrier (OC) levels and the electronically equivalent signal levels that are called synchronous transport signal (STS) levels. Once an STS-1 is created within SONET architecture, it is possible then to realize the full family of transmission speeds by having multiples of the STS-n where n assumes values from 1 to n. Example of the SONET signal hierarchy is as shown in Table 6.1. It should be noted that more rates exist. We have only tried to give examples of the most implemented ones.

TABLE 6.1
The SONET Signal Hierarchy

Electrical Signal	Optical Number	Speed	Media	Capacity
STS-1	OC-1	51.84 Mbps	Optical Fiber	28 DS-1 or 1 DS-3
STS-3	OC-3	155.520 Mbps	Optical Fiber	84 DS-1 or 3 DS-3
STS-12	OC-12	622.08 Mbps	Optical Fiber	336 DS-1 or 12 DS-3
STS-24	OC-24	1.244 Gbps	Optical Fiber	672 DS-1 or 24 DS-3
STS-48	OC-48	2.488 Gbps	Optical Fiber	1344 DS-1 or 48 DS-3
STS-192	OC-192	9.95 Gbps	Optical Fiber	5376 DS-1 or 192 DS-3
STS-256	OC-256	13.271 Gbps	Optical Fiber	7168 DS-1 or 256 DS-3
STS-768	OC-768	40 Gbps	Optical Fiber	21504 DS-1 or 768 DS-3

6.1.1 SONET Signal Rates

The digital transmission hierarchy that currently exists consists of DS-0 (64 Kbps) via DS-4NA (139.264 Mbps) signals. These signals, called payloads, are transported by SONET in a synchronous way over fiber transmission systems. A structure called virtual tributaries (VTs) is used to map these signals. With this type of structure arrangement, it is then possible for the DS-3 signal to be directly mapped into the STS-1. A byte interleaved multiplexing scheme is used to form the SONET signal architecture. The term interleaving means the transmission of pulses from two or more digital sources in a time-division on sequence over a single path. The add/drop multiplexer (ADM) network element can add or drop the payloads from the carrier.

6.2 SONET FRAME

The frame format that is used to produce the STS-1 basic rate (51.84 Mbps) is defined by SONET. A frame in the basic format is 90 bytes in columns wide and 9 bytes high in rows. This results in the application of an 810 byte frame for the basic STS-1 signal. The occurrence of the frame is about 8000 times per second. Mathematically, this can be expressed as follows:

$$810 \text{ bytes} \times 8 \text{ bits/byte} \times 8{,}000 \text{ frames/sec} = 51.84 \text{ Mbps}$$

Figure 6.1 shows the SONET STS-1 frame structure. The transport overhead is the first part of the structure. It uses the first three columns and all the nine rows resulting in 27 bytes of transport overhead. This transport overhead is divided into two pieces using the first three columns and the first three rows (9 bytes) for the section overhead and the remaining six rows in the first three columns (18 bytes) for the line overhead. The synchronous payload envelope (SPE) consists of 783 (87 columns × 9 rows) bytes, which includes the path overhead. The addition of the transport overhead and the payload gives the sum of 810 bytes. This is contained in the 125 ms frame.

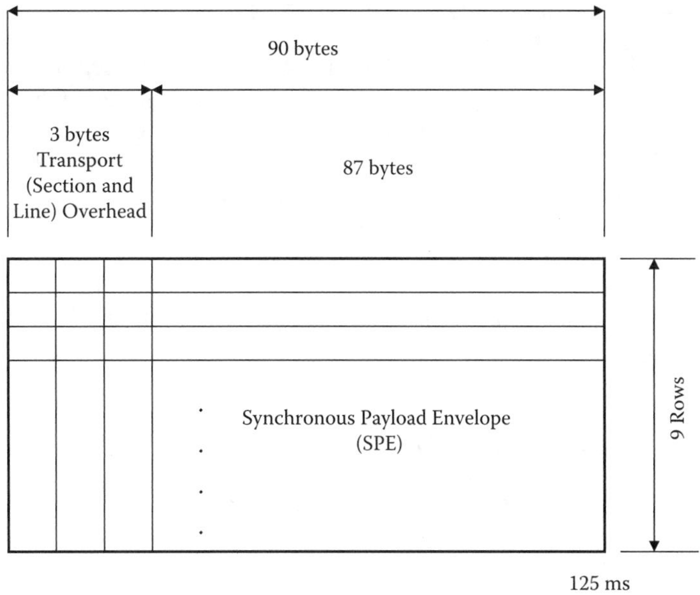

FIGURE 6.1 The SONET STS-1 frame structure.

6.3 SONET COMPONENTS

The four SONET components are listed as follows:

- line overhead (LOH)
- section overhead (SOH)
- transport overhead (TOH)
- path overhead (POH)

6.3.1 Line Overhead (LOH)

The LOH serves as a pointer for frequency justification and data communications channel. LOH is processed at all nodes. In the first three columns in the SONET frame, the LOH occupies the bottom six octets. The entire data communications channel octets are used for line communication. They are part of the 576-kbps message used for alarms, monitoring, maintenance control, etc. Higher-level components are used to communicate with LOH. Examples of these higher-level components are

- switches
- terminals
- digital cross-connects
- multiplexers

6.3.2 Section Overhead (SOH)

In the case of the section overhead (SOH), it is processed at each node and each regenerator. Performance monitoring and framing are part of the functions of the SOH. The framing bits are 16 bits, which are the first two octets. They are provided with all the STS-1 via STS-N signals. It has a bit pattern that is consistently 1111011000101000 in binary or F628 in hexadecimal. In summary, the functions of the section overhead are

- a local orderwire
- framing information
- data communications channels for operations, administration, maintenance, and provisioning (OAM&P) messages
- performing monitoring

6.3.3 Transport Overhead (TOH)

Bit error monitoring, alarm, and data communications overhead are some of the functions of the TOH portion of the STS-1 frame. These help in the improvement and enhancement of the security and reliability of the SPE between nodes in the synchronous network.

6.3.4 Path Overhead (POH)

Inside the SPE is the POH. At all the points of the SONET system, the POH bytes (9 bytes) are processed by the path-terminating equipment. The four classes of functions at the POH are

- class A: independent functions of payload
- class B: dependent functions mapping
- class C: overhead functions that are user specific
- class D: future-use functions and growth

Functions supported by the POH are as follows:

- path status
- signal label the individually mapped payloads
- path trace
- performance monitoring of the STS SPE

6.4 SONET TOPOLOGIES

The SONET network has various topologies implemented using different multiplexers. These topologies include

- point-to-point
- point-to-multipoint
- hub and spoke
- ring

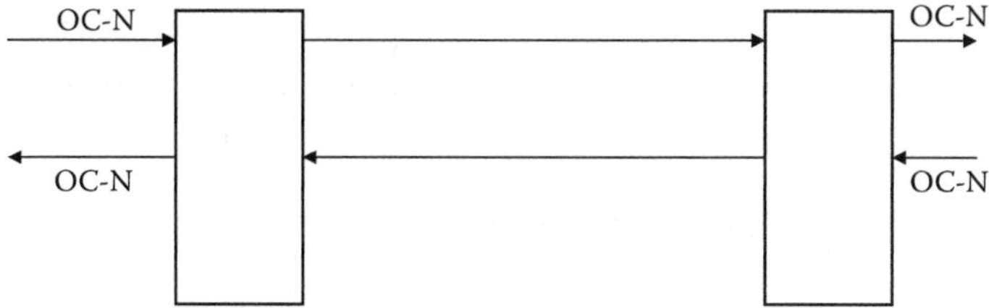

FIGURE 6.2 SONET point-to-point topology.

These topologies are making it possible for the telecommunication companies to be able to access both the public networks and the private networks. They do it by installing highly interoperable and reliable equipment at the public and private networks.

6.4.1 SONET Point-to-Point Topology

An example of SONET point-to-point topology is shown in Figure 6.2. As a simple point-to-point circuit, with an optical fiber, two devices are connected with all the necessary repeaters strategically placed at different points. Since this may be an entry-level point into a SONET, it acts as a concentrator—SONET multiplexer device for multiple communications channels like DS-1 and DS-3.

6.4.2 SONET Point-to-Multipoint Topology

The SONET point-to-multipoint topology is used mainly in situations where there is a task. Without demultiplexing the whole broadband (high-speed) signal, it drops circuits out. This kind of technique is often regarded as "a form of add-drop multiplexing that is used to drop circuits off." The ability to perform add-drop multiplexing is a major benefit of SONET.

This is because there are various vendors that supply the different network elements even though they may be compatible at the OC-N level. Using the add-drop multiplexer (ADM), it is then possible to decide which one to use (add or drop) at the appropriate time during the process of signal transmission in a network such as a WAN. Figure 6.3 shows the ADM installed to enable the drop off or addition of signals when needed.

6.4.3 SONET Hub and Spoke Topology

In the unforeseen growth in the architecture of the SONET, there are some times when there may be some uncertainties. The SONET hub and spoke topology helps to alleviate such uncertainties by providing some flexibility in such cases. It is also known as the star network. In a SONET multiplexer network, they are all concentrated into a digital cross-connect, hence the name hub, before they are then forwarded on to the next node as shown in Figure 6.4.

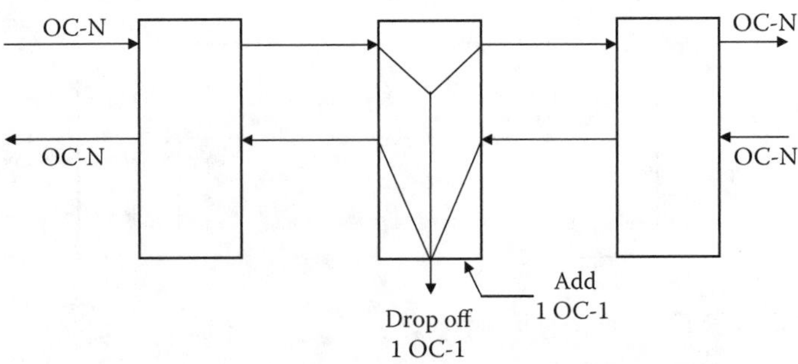

FIGURE 6.3 SONET ADMs installed in a point-to-multipoint topology.

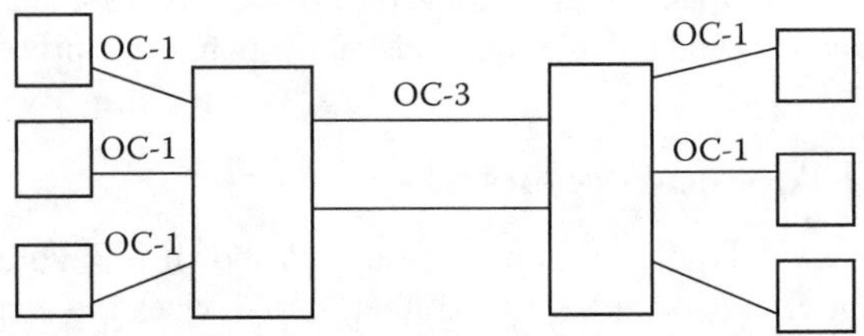

FIGURE 6.4 SONET hub and spoke topology.

6.4.4 SONET Ring Topology

In the SONET ring topology, between adjoining ADMs, a series of point-to-point links are installed. It has both bidirectional and unidirectional capabilities installed that provide the network with much robustness. The SONET ring has as its main advantage the fact that it can survive in case there is a failure in the network or a cable cut. There is the possibility of the system using alternative paths where more than one fiber link is installed. However, the recovery time must be within milliseconds. Figure 6.5 shows the diagram of the SONET ring topology. Figure 6.6 shows the overall SONET network overview showing the ADMs and the other features of

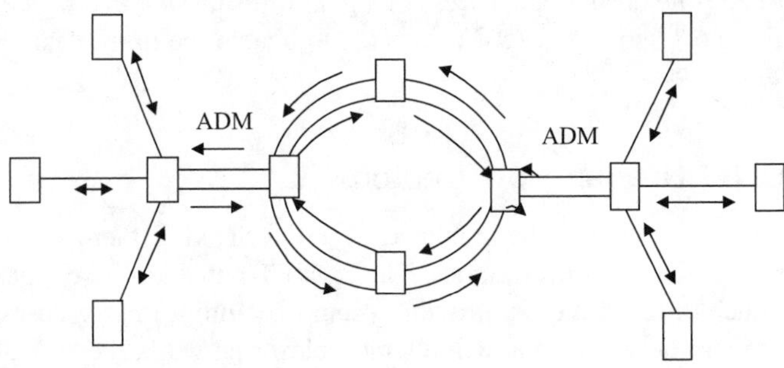

FIGURE 6.5 Diagram of the SONET ring topology.

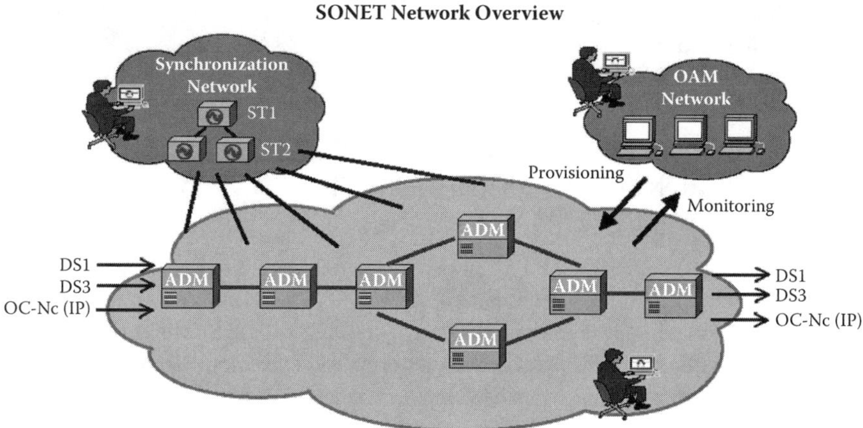

FIGURE 6.6 Overall SONET network overview showing the ADMs.

the network such as the quadrature amplitude modulation (QAM) and synchronization networks.

6.5 ADVANTAGES AND DISADVANTAGES OF SONET

SONET has numerous advantages and some disadvantages.

Advantages

- Using SONET helps in achieving interoperability of different vendor equipment over fiber optic systems.
- Using one facility, SONET has the capability to combine, segregate, and consolidate traffic from various locations.
- The SONET network standard is well integrated to where different types of traffic can be transported.
- North America and Europe are merged at higher rates because of SONET.
- In an era of homeland security, SONET can offer end-to-end network performance monitoring and network management.
- Because of SONET, new switching standards like B-ISDN and ATM are provided.
- SONET has helped to facilitate significant cost efficiencies and performance improvements.

Disadvantages

- SONET is mainly used as high-speed backbones and as carrier networks.
- SONET may not be one of those networks that can be used by everyone because it is well suited for high-speed backbones and as carrier networks.

- SONET is very expensive and therefore cannot be afforded by many companies, more so startups.
- Because SONET cannot be afforded by all, it is not available easily and everywhere.

6.6 SYNCHRONOUS DIGITAL HIERARCHY

In synchronous digital hierarchy (SDH), the 155,520 kbits/s synchronous transport module 1 (STM-1) is the basic building block. The STM-1 signal consists of payload and overhead bytes that are organized in a 125-ms frame structure. Serial transmission is the way in which the information that goes through any of the selected media such as optical fiber is conditioned. The rate is always synchronized to the network. Information is processed on a byte-wide basis at 19.44 MHz at each of the nodes where the signal is demultiplexed.

SDH was mainly developed to mitigate some of the major deficiencies associated with plesiochronous (almost synchronous) digital hierarchy (PDH). Because PDH was not synchronizing properly, it resulted in jitter problems in the network. Hence, the SDH standards that emerged in 1992.

> **Definition:** SDH like SONET is a network transport technology that uses synchronous operations for facilitating real-time data transmission, voice, and video through fiber optic cabling at rates ranging from 51.84 Mbps (optical carrier-level 1 (OC-1)) to 13.21 Gpbs (OC-255).

Based on the synchronous transfer mode-N (STM-N) frame format, SDH forms a multiplexing rate, where N is the number rates. The basic STM-N frame is similar to SONET OC-1 and consists of

$$270 \text{ columns} \times 9 \text{ rows} = 2{,}430 \text{ octets}$$

$$9 \text{ columns} \times 9 \text{ rows} = 81 \text{ octets section overhead}$$

There are about 2349 octets remaining that are used to create the payload. As the value of N increases, you obtain the higher rate frames. The synchronous transport signal level N (STS-N) and the SDH level N (SDH-N) are defined in SDH standards. The important definitions are the optical rate of the multiplexing and the electrical rate of the multiplexing. They work at the rate of speed required to map and multiplex the higher rates of speed. The standards for SDH and SONET are basically the same. The only major difference is that SDH reflects the European version while SONET reflects the North American version of the standard.

6.6.1 SDH Virtual Container

The SDH virtual container (VC) is the information structure used in STM-1. This VC consists of information payload and path overhead information fields organized in a block frame structure that repeats every 125 or 500 ms. In SDH, pointers provide the alignment information needed in the identification of the VC frame start.

6.6.2 SONET/SDH Pointers

Definition: A pointer is a value that defines the frame offset of a VC with respect to the STM-1 frame reference.

The VCs can transport signals from the existing PDH. These VCs are already defined by ITU/CCITT standard organization. A VC-4 and its associated pointer are called an administrative unit 4 (AU-4). A VC-12 and its associated pointer are called a tributary unit 12 (TU-12). A VC-12 has the capability of transporting about 2048 kbits/s.

In SONET/SDH networks, pointer processors allow payloads to have an arbitrary phase within the STM-1 transport frame. Between the start of the STM-1 frame and the start of each payload frame, the pointers in the SONET/SDH STM-1 frame overhead encode the offset in bytes. If the input and output timing signals in any given network element has an offset in phase, then the phase relation between the output STM-1 frame and an individual payload frame can change over time. The way to do the adjustment mechanism for this type of situation is to output more or fewer payload bytes and change the appropriate pointer value in the STM-1 frame. This results in a requantization process of the payload phase. Jitter and wander can be produced as a result of this requantization process.

6.7 NEXT-GENERATION SONET (NGS)

SONET can be classified now as either traditional or next-generation SONET (NGS). The traditional SONET first emerged in the mid-1980s. The SONET standard enabled multiple transmission rates, transport of lower-level asynchronous payloads, higher-order signals to carry lower-order payloads, interoperability among multiple equipment manufacturers, and centralized fault discovery and management. All these were done using a byte-interleaved multiplexing scheme. Virtual tributary 1.5 (VT-1.5) and synchronous transport signal 1 (STS-1) were and continue to be the basic building blocks of the SONET hierarchy.

However, while it is true that traditional SONET has been very effective in transporting voice and T1 private line circuits, Ethernet and other data traffic types became very dominant challenges. Therefore, at the dawn of the 1990s, the various standard organizations began to address these concerns. This gave rise to the new equipment that addressed these challenges and the birth of the next-generation SONET (NGS). This next-generation SONET equipment features typically Ethernet interfaces and a switch core capable of grooming both VT-1.5 and STS-1 payloads. Lower capital costs and much higher densities are some of the characteristics of the next-generation equipment.

The key traditional global service technologies are

- Plesiochronous digital hierarchy (PDH) is used to carry telephone traffic. In North America, Europe, and Japan, PDH is based on the T-n, E-n, and J-n standards, respectively.
- Asynchronous transfer mode (ATM) is a connection-oriented fixed-cell switching technology. It has a large install based around the world due to its traffic flexibility. ATM has the ability to provide quality of service (QoS).

- Packet over SONET/SDH (PoS) refers to the encapsulation of IP packets into PPP frames and then transporting them over a SONET/SDH network. Both point-to-point (PPP) and SONET/SDH are provisioned point-to-point, making them ideal complements of each other. PoS as an alternative to IP over ATM over SONET/SDH, eliminates not only the need to maintain both an IP and an ATM network, but also the inefficiencies of the large ATM header (the cell overhead tax). PoS is also used as a transport mechanism in most of the emerging technologies such as many service/edge and core routers utilizing MPLS over SONET.
- Time division multiplexing (TDM) is a method of putting multiple data streams in a single signal by separating the signal into many segments, each having a very short duration. Each individual data stream is reassembled at the receiving end based on the timing. The circuit that combines signals at the source (transmitting) end of a communications link is known as a multiplexer. It accepts the input from each individual end user, breaks each signal into segments, and assigns the segments to the composite signal in a rotating, repeating sequence. The composite signal thus contains data from multiple senders. At the other end of the long-distance cable, the individual signals are separated out by means of a circuit called a demultiplexer, and routed to the proper end users. A two-way communications circuit requires a multiplexer/demultiplexer at each end of the long-distance, high-bandwidth cable. An asset of TDM is its flexibility. The scheme allows for variation in the number of signals being sent along the line, and constantly adjusts the time intervals to make optimum use of the available bandwidth. The Internet is a classic example of a communications network in which the volume of traffic can change drastically from hour to hour.

The NGS global service technologies are

- Multiple protocol label switching (MPLS) adds a small, fixed-length label to packets that enter the network, thereby allowing packets with the same class of labels to be forwarded in the same manner by the MPLS nodes in the core of the network. MPLS is independent of layer 2 and layer 3 in the open systems interconnection (OSI) model, and can support ATM, frame relay, PPP, and Ethernet as a data link layer. MPLS provides IP virtual private networks (VPNs) service.
- Storage area network (SAN) is a high-speed network that enables direct connections between servers and storage devices located in LANs. SANs seamlessly connect storage devices located in different geographic areas using the existing public and private SONET/SDH transport infrastructure.
- Ethernet over SONET/SDH (EoS) is a technology that provides a cost-effective way for service providers to provide Ethernet-based LAN service using existing SONET/SDH infrastructures. EoS allows service providers to preserve the proven quality and performance of SONET/SDH while encapsulating Ethernet frames at the network ingress and egress points.

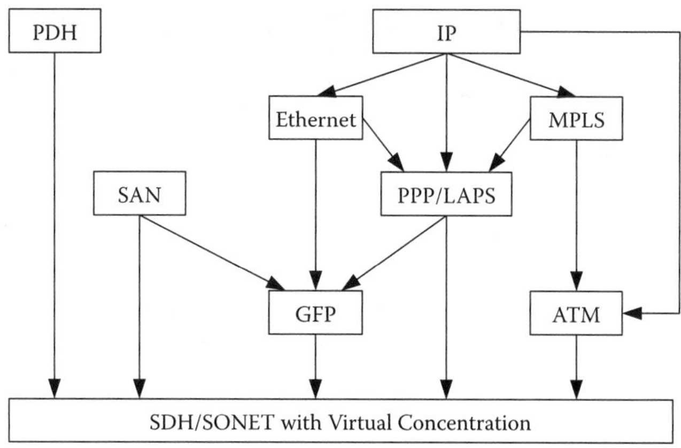

FIGURE 6.7 Topology (protocol) mapping of traditional and next-generation SONET.

It should be noted that the most common encapsulation techniques for EoS are generic framing procedure (GFP), link access procedure-SDH (LAPS), PPP, with high data link control (HDLC) -like framing.

- Virtual concatenation is a seamless bridge between the present and the future. It provides both flexibility and bandwidth efficiency using the service providers' current SONET/SDH equipment.

Figure 6.7 shows the mapping of the various NGS topology (protocol). The seamless integration with infrastructure, bandwidth efficiency, significant cost savings, flexible and future-ready networking solutions, and new revenue-generating services without the need to re-engineer existing networks are some of the major attributes of the NGS. Table 6.2 shows the comparison of the traditional and next-generation SONETs.

TABLE 6.2
Comparison of the Traditional and Next-Generation SONET

Capability	Traditional SONET	Next-Generation SONET (NGS)
DCC Functionality Space Requirements	OSI multiple shelves/racks	IP and OSI/IP single-shelf or chassis architecture
Key Global Service Technologies	TDM, PDH, ATM and PoS	MPLS, SAN, EoS and virtual concatenation
Data Capabilities	None or very limited	Ethernet, virtual concatenation, link-capacity adjustment scheme
Switch Core	STS-1 multiplexing	STS-1 and VT-1.5 grooming

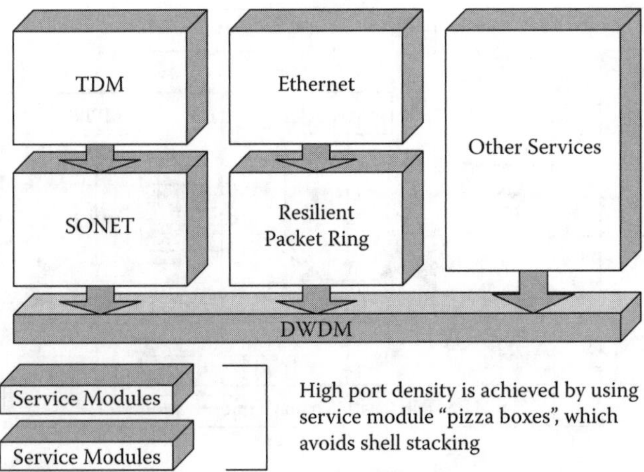

FIGURE 6.8 The architecture of the next-generation SONET platform.

6.7.1 Next-Generation SONET Platform

Figure 6.8 shows the architecture of the NGS platform. Higher transmission bit rates of about 2.5 Gbps, 10 Gbps, etc. and dense wavelength division multiplexing (DWDM) capabilities are implemented on a single NGS platform. This supports multiple client interface types covering a large range of bit rates. Additionally, new methods of carrying data services are created, such as resilient packet ring (RPR), which ensures an efficient use of the bandwidth through the implementation of a layer 2 self-learning Ethernet switch in the metropolitan optical network. Therefore, the NGS platform enables the evolution and enhancement of existing networks such as metro networks and sets a new economic benchmark for network efficiency while increasing broadband service potential.

6.7.2 Next-Generation SONET Requirements

The NGS has key requirements. These NGS requirements are that it must have

- flexible and scalable aggregation
- unrestricted bandwidth management
- seamless integration with existing infrastructure, i.e., it must be in compliance with existing standards
- ability to support the traditional SONET functionalities and services
- ability to support both voice and data traffic on the same platform
- low power consumption
- ability to provide continued efficient use of bandwidth for data and voice
- with a small footprint, ability to reach high-density service termination

6.7.3 Advantages of Next-Generation SONET

Next-generation SONET has many advantages. The advantages are

- The platforms of the NGS help eliminate the need for a head-end node that terminates each specific collector ring.
- NGS architectures and Ethernet-over-fiber afford the scalability, operational simplicity, and data transport efficiency to offer service providers a superior return on investment (ROI).
- It can sustain a high number of collector rings with various bit rates.
- Service providers are enabled by the NGS portfolio to manage customer services from source to destination and through multiple rings at the finest granularity while maintaining a low operating cost and better bandwidth management capabilities.

6.8 SONET AND SDH STANDARDS INCLUDING NGS

The two international transmission standards whose signal hierarchies allow compatibility between the European signal hierarchy and the North American signal hierarchy are the SONET and synchronous digital hierarchy (SDH) standards. SDH is based on 2048 kbits/s E1 signals—the European signal hierarchy—while SONET is based on 1544 kbits/s T1 signals—the North American signal hierarchy. In addition to other things, SONET and SDH standards define synchronous frame structures for the multiplexed digital traffic. The American National Standards Institute (ANSI) and the Exchange Carriers Standards Association (ECSA) are the two standard organizations that have developed SONET standards.

There are four protocol layers considered in the SONET standards. They are as follows:

- Layer 1: The photonic layer, which is responsible for the physical connectivity like transport of optical signals and conversion.
- Layer 2: The section layer, which is responsible for encapsulating data and making sure that data are sent in the right order.
- Layer 3: The line layer, which is responsible for monitoring for transmission problems, error recovery, and signal switching.
- Layer 4: The path layer, which is responsible for mapping signals into the right channel and for making sure there is channel reliability.

Within ANSI, other subcommittees such as ANSI T1X1 play a crucial role in the SONET standard development. In the case of the ANSI T1X1 subcommittee, it develops optical interface specifications. The ANSI SONET specification standard, which was formally called ANSI T1.106, established the guidelines for SONET operations that support short-haul, mid-range, and long-haul applications over fiber optic links. In the development of all these standards, Bellcore played a significant role in them. SDH was adopted by ITU-T in 1988. It became a group of three recommendations,

TABLE 6.3
Next-Generation SONET Standards

Region	Standard	Description
ANSI	T1.105	VCAT – Identical text to G.707/G.783 but with SONET terminology. GFP, LCAS – References G.707/G.7042
ETSI	EN300 417-9-1	Defines VCAT as per G.707/G.783.
GLOBAL	IEEE 802.17	Defines resilient packet ring MAC.
	G.709/G.798	Definition + equipment specification for VCAT.
	G.7042	Defines LCAS. Refers to G.707 and G.783 for VCAT function.
	G.783	Detailed equipment specification (that include VCAT support requirements).
	G.707(Sect11)	Defines VCAT. Also defines contiguous concatenation.
	G.7041	Defines GFP.

namely G.707, G.708, and G. 709. However, it was not until 1992 that the actual SDH standards emerged.

The standards for SDH and SONET are basically the same. The only major difference is that SDH reflects the European version while SONET reflects the North American version of the standard. There are also standards for the NGS such as the ones shown in Table 6.3.

Note:

- Recommendation G.707: Synchronous Hierarchy Bit Rates
- Recommendation G.708: Network Node Interface for Synchronous Digital Hierarchy
- Recommendation G.709: Synchronous Multiplexing Structure

Table 6.3 shows a brief list of some of the NGS standards. In addition, it shows the region where the specific standard is implemented.

SUMMARY

1. In this chapter, we have discussed the synchronous optical network (SONET).
2. SONET has a many-signal hierarchy. These signals are characterized according to their various speeds and capacities.
3. The SONET frame has a basic format of 90 bytes in columns and 9 bytes high in rows.
4. The LOH, SOH, TOH, and POH make up the various components of SONET.
5. The four protocols of SONET are point-to-point, point-to-multipoint, hub and spoke, and ring.
6. SONET has many advantages as well as some disadvantages.
7. In SONET standards, four protocol layers are considered.
8. Synchronous digital hierarchy (SDH) was mainly developed to mitigate some of the major deficiencies associated with plesiochronous (almost synchronous) digital hierarchy (PDH).

9. SONET can be classified now as either traditional or next-generation SONET (NGS).
10. The key traditional global service technologies are TDM, PDH, ATM, and PoS.
11. The key NGS global service technologies are MPLS, SAN, EoS, and virtual concatenation.
12. NGS platform enables the evolution and enhancement of existing networks such as metro networks and sets a new economic benchmark for network efficiency while increasing broadband service potential.

REVIEW QUESTIONS

6.1 SONET is a synchronous optical network.
   a. True
   b. False

6.2 SONET rates range from 51.84 Mbps OC-1. to 13.271 Gbps OC-255.
   a. True
   b. False

6.3 The add/drop multiplexer (ADM) network element cannot add or drop the payloads from the carrier.
   a. True
   b. False

6.4 The digital transmission hierarchy that currently exists consists of DS-0 64 kbps via DS-4NA 139.264 Mbps signals.
   a. True
   b. False

6.5 A frame in the basic format is 90 bytes in columns wide and 9 bytes high in rows.
   a. True
   b. False

6.6 The four protocols of SONET are point-to-point, point-to-multipoint, hub and spoke, and ring.
   a. True
   b. False

6.7 The LOH does not serve as a pointer for frequency justification and data communications channel.
   a. True
   b. False

6.8 Bit error monitoring, alarm, and data communications overhead are some of the functions of TOH portion of the STS-1frame.
   a. True
   b. False

6.9 The SONET hub and spoke topology does not alleviate uncertainties by providing some flexibility in the cases where uncertainties exist.
   a. True
   b. False

6.10 Jitter is not a major problem in the SDH network.
   a. True
   b. False

Answer: 6.1a, 6.2a, 6.3a, 6.4a, 6.5a, 6.6a, 6.7b, 6.8a, 6.9a, 6.10b

PROBLEMS

6.1 Define SONET. What is significant about SONET?
6.2 Describe the major signal hierarchy of SONET in terms of the electrical signal, the optical number, speed, media, and capacity.
6.3 If a SONET frame signal has 910 byte, calculate the speed (basic rate) of transmission for this particular frame.
6.4 Name the different components of SONET. Describe their various functions.
6.5 Describe the different topologies of SONET.
6.6 What are the major differences between each of the SONET topologies?
6.7 What are the advantages of SONET?
6.8 Name some of the disadvantages of SONET.
6.9 How many protocol layers are considered in the SONET standards? Describe them.
6.10 Describe SDH. What is the major advantage of SDH? How does SDH relate to SONET?
6.11 If you observed that SONET uses 9 rows and 90 columns, does each row correspond to a phone column or not?
6.12 If we were to use SONET for data, would it be limited to multiplexing the data streams the same way that the voice streams would be multiplexed?
6.13 Draw the basic SONET STS-1 frame infrastructure. Identify all the major components.
6.14 Describe the SONET hub and spoke topology. What is the major function of this system?
6.15 What kind of users could benefit from the traditional SONET and the NGS transport system? Develop the list of users.
6.16 Describe the LOH in a SONET system. What are the functions?
6.17 What are the functions of SOH in a SONET system? What is SOH?
6.18 What are the specific functions of the TOH that make it important in the security improvement of the STS-1?
6.19 What are the four classes of functions POH in a SONET system? What are the supported functions that can be associated to a POH system?
6.20 Compare and contrast the differences between LOH, SOH, TOH, and POH in a SONET system.
6.21 What are the platforms and the requirements for the NGS systems?
6.22 What are the advantages of the NGS?
6.23 What SONET technologies can be classified as traditional SONET and NGS? Describe each of them.

7 Virtual Private Network (VPN)

> Success doesn't discriminate. It's an equal opportunity employer—available to everyone willing to pay the price.
>
> —Anonymous

The concerns most people, governmental agencies, and corporations have for security issues have rapidly increased the use of virtual private networks (VPNs). For many decades, VPN services have been available, but not widely used until recently. Network issues have become increasingly important as Internet usage grows exponentially. Multiple hosts can communicate securely over a public network using a VPN. There are great inadequacies in today's infrastructures to cover the potential needs that small groups have or the new requirements that come up on a daily basis from the increasingly constant demands for more sophisticated services. In a VPN, users tunnel through the public network in a way that allows the tunnel participants to benefit from a secure connection in its setup just as if they were on a typical private network. Figure 7.1 shows an example of a typical VPN.

Definition 1: A VPN is a communications environment where access is controlled to permit peer connections within a defined community of interest only and it is constructed via some kind of partitioning of a common underlying communications medium, where this underlying communications medium provides services to the network on a non-exclusive basis.

Definition 2: A network that uses a private Internet protocol (IP) address space that operates over another network's infrastructure is called a VPN. The address space may or may not be registered.

In this chapter, we will consider the different types of VPN, the VPN infrastructure, and virtual networks' requirements. The VPN general architecture, current VPN advantages and disadvantages, provisioning models, and the VPN standard will also be considered.

7.1 TYPES OF VPN

There are different types of VPN. Each type depends on the particular function it performs as well as the various methods for constructing it. The main idea is that VPNs provide an abstraction of a network infrastructure that is used by groups or enterprises to cover their needs. The different types of VPNs are place-oriented VPN (PO-VPN),

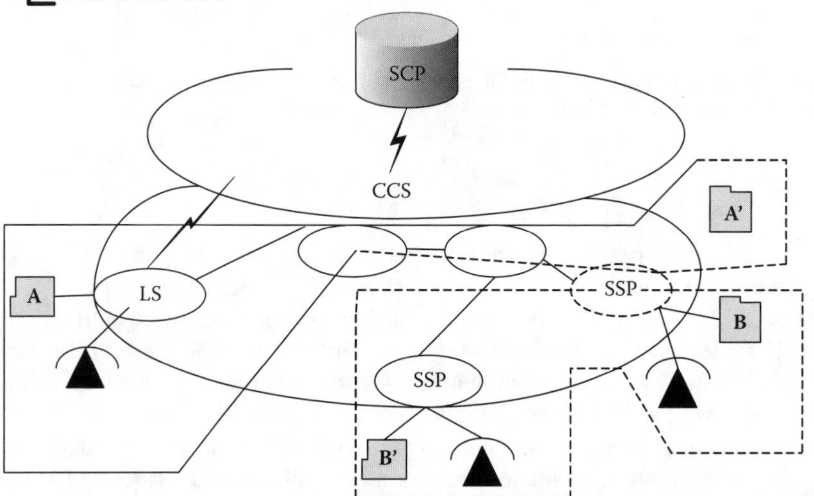

FIGURE 7.1 Virtual private network (VPN).

software VPN (SVPN), dial-in VPN (DVPN), secure and manageable VPNs (SMVPN), fault-tolerant VPN (FVPN), legacy VPN, and secure active VPN (SAVPN).

7.1.1 Place-Oriented VPN (PO-VPN)

A PO-VPN as shown in Figure 7.2 can be regarded as an alternative to VPN. Its users manage its own resources in a policy-related way. PO-VPNs are built on top of existing physical networks and controlled agents. PO-VPNs satisfy the following requirements:

- Interoperability: It adapts to current network infrastructures, which are heterogeneous. Since PO-VPNs are based on agent technology and these agents are themselves transport and computer independent, they therefore promote interoperability among systems and software.
- New service deployment: Agents are used for the implementation of new protocols that are not currently in existence or not currently standardized.
- Resource management: Locations have interfaces to the agency's resource manager or each has its own resource manager.
- Flexibility/adaptivity: Ability to dynamically make changes to the configuration of the PO-VPN and services it offers.
- Security, privacy, and admission control: Security is a must. It has an extensive authentication/authorization scheme that is supported for the agents that want to execute in locations of this PO-VPN.

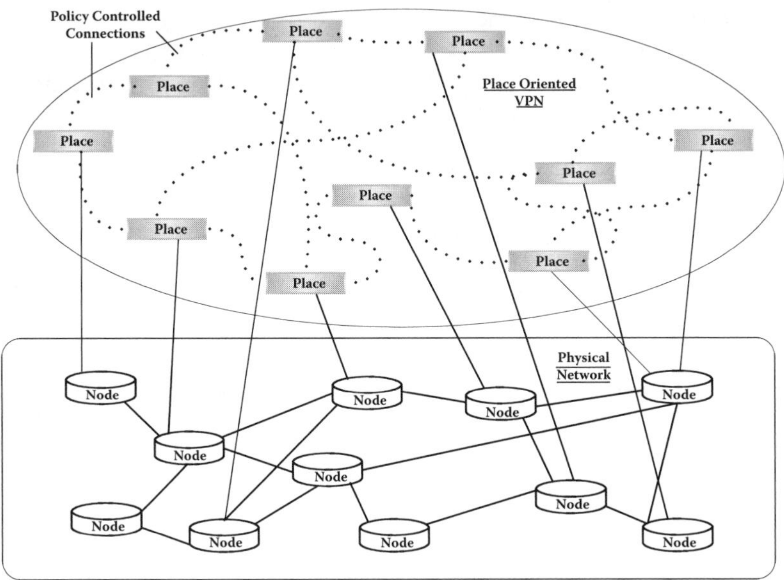

FIGURE 7.2 Place-oriented VPN (PO-VPN).

- Effectiveness and performance: In having application/task-specific optimization and adaptation of the network to these requirements, it is possible to obtain optimal use of the underlying infrastructure and boost performance. Some examples are the ability to compress the traffic flow between nodes whenever there is low capacity of the line, providing alternative routes whenever there is a malfunction, and, according to various security bulletins, have the ability to update the security components of the nodes.
- Cost matters: Since most small and medium enterprises (SMEs) and other smaller groups cannot afford such high costs as bigger enterprises, this has the ability to provide opportunities to minimize cost and it can be automatically implemented by the user in a PO-VPN environment.
- Communication: Protocols can be provided to make it possible for users to communicate. This can be done by agents updating the software of the underlying infrastructure and implementing the new protocols.
- Information path selection: Through the agents, the user has some control over the desired messages/data. The administrator has the capability of selecting the network topology in a strategic manner that would make the desired information path possible.

Those who can benefit from PO-VPNs are as follows:

- Individuals who cannot afford to build their own VPNs or in low populated groups.

- Groups whose requirements over time can change in an unpredictable manner.
- Those companies that have VPNs and a heterogeneous infrastructure and want to cooperate with other companies for a project lifetime.

7.1.2 Software VPN (SVPN)

Software VPN (SVPN) is a VPN implemented in software. It is an alternative to the hardware-implemented VPN solutions and has great accessibility and economic advantage compared to the hardware-implemented VPN. This kind of VPN has great performance impact, has limiting network throughput, and high CPU usage. Depending on the type of SVPN implemented (fast network, 100Mb/s or slow network, 10.3 kb/s), performance can be observed in terms of network throughput and CPU usage.

7.1.3 Dial-In VPN (DVPN)

In the case of dial-in VPN (DVPN), a phone is used as an accessible connection to a computer network service or a VPN service. The service may offer a number of dial-up lines, since they can be busy at any time, and different numbers may be designated to modems offering different speeds. The DVPN service allows users in a corporate network the ability to dial local (or toll-free) numbers, which will be routed to the nearest points of presence (POPs). They can then be connected to their corporate network using some form of tunneling. The encapsulating of a network transmission in an IP packet for secure transmission over a network is called tunneling. The two common forms of tunneling are layer 2 (data link layer), for example, L2TP; and layer 3 (network layer), for example, mobile IP.

The infrastructure components required to implement a DVPN are a remote node, remote access server, tunnel management system, gateway, customer premises equipment, and authentication server. The topology of the DVPN service is shown in Figure 7.3.

- Remote node (RN): The user can access the DVPN service (DVPNS) using the remote node device. In most cases, a personal computer (PC) is used; however, a router with additional networks and nodes behind it can be used.
- Remote access server (RAS): The public switched telephone network's (PSTN) access into the service provider's network is made possible by the remote access server (RAS). In a POP, there may be multiple remote access servers. The mobile IP foreign agent is used in the layer 3 solution implementation.
- Tunnel management system (TMS): The host server that runs the tunnel management system software is the tunnel management system server. It is a database that authenticates when requested to provide information necessary to allow a user to have access to the user's corporate network.
- Gateway: The gateway makes it possible for the corporate network to have access into the service provider's network. The virtual network connections into the corporate network are handled by this gateway. A typical

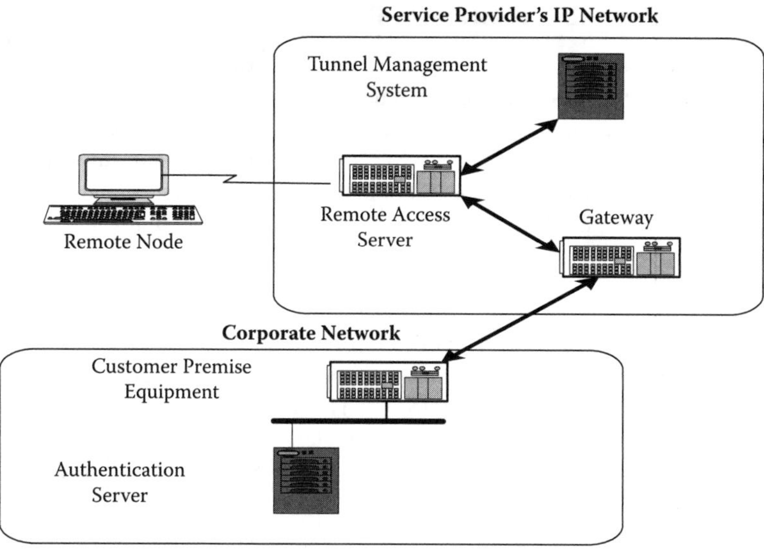

FIGURE 7.3 Dial-in VPN service topology.

example is the frame relay, ATM, and dedicated serial line. The mobile IP home agent is used in the layer 3 solution implementation.

- Customer premises equipment (CPE): The corporate network is accessed by the service provider via the CPE. The dial-in nodes, static route for subnet(s) are used. In the layer 3 implementation, a generic router is used.
- Authentication server: An authentication protocol such as the RADIUS is used. It is a host-based authentication server that communicates to the RAS through the gateway.

7.1.4 Secure and Manageable VPNs (SMVPN)

The secure and manageable VPNs can be used to enable secure remote accesses to the end users. This is done by the end users constructing VPNs from their homes to their corporations or other organizations. With SMVPNs, it is possible to receive mails from mail servers securely inside any corporate network.

7.1.5 Fault-Tolerant VPN (FVPN)

The underlying routing instability in an Internet network is one of the key factors that lead to the Internet's inherent fault-intolerant properties. The FVPN is used to mitigate against these fault-intolerant properties. The idea is to perform a precalculation and install a set of redundant minimally overlapping paths between the endpoints that can be used as backup routes as primary paths fail. The FVPN provides a continuous VPN availability for multiple users and applications.

7.2 VPN GENERAL ARCHITECTURE

The VPN general architecture is shown in Figure 7.4. In this case, it shows the integration of a corporate LAN/intranet, the users, the service provisioning component (SPC), the firewalls, and the encryption devices. The VPN has three main components: a public component, a service provision component, and an information systems operator component. This specific VPN architecture uses special monitor and co-ordination processes. This architecture helps in the production and achievement of efficient delivery of services and activities.

- Public component: This is used for the dissemination of information. Because this component supplies information to external users, it can be considered as a preliminary to a category of VPN services.
- Service provision component (SPC): This component is used in the interfacing of the VPN. It concentrates the interface on use, thus collecting the requests and distributing the results. It is the soul of the system. As shown in Figure 7.5, it has key subcomponents such as interpreter, application software, and the dispatcher and control console.
  - Interpreter: In the VPN, the interpreter subcomponent transforms documents from internal format to the specific format of each workstation.
  - Application software: Inside the VPN, the application software subcomponent is specifically dedicated to the automatic grouping of services that share common resources or exchange information with similar data types.
  - The dispatcher: Performing monitoring and scheduling operations of the VPNs services are the key goals of the dispatcher subcomponent. It consists of three modules: the interface module, the scheduling module and the monitoring module.

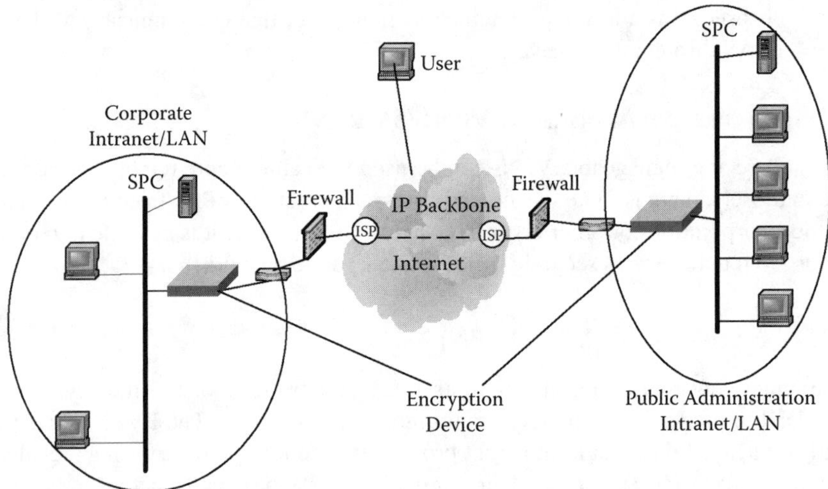

FIGURE 7.4 VPN general architecture.

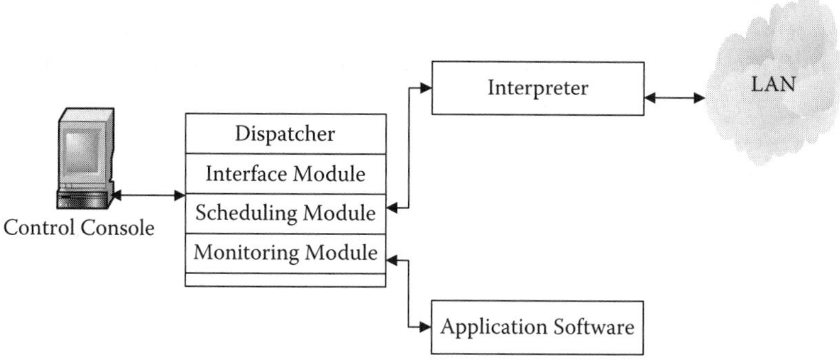

FIGURE 7.5 The service provisioning component (SPC).

- Control console: Through the interface module of the dispatcher, the control console that is based in a workstation with which the body responsible for a service can monitor and configure the workstation.
- VPN component: it is the VPN component that connects different actors who are involved in the management and implementation of services, such as public administration organizations and different enterprises. The component is used to receive service requests from users and distribute service activities among different bodies.

7.3 CURRENT VPN ADVANTAGES AND DISADVANTAGES

As has been shown in our discussions that VPN has many advantages and there are a few disadvantages. In this section, some of the current VPN advantages and disadvantages will be discussed.

7.3.1 Advantages

The advantages are as follows:

- Satisfaction capability of high-peak bandwidth demands: VPN is designed to handle the high-peak bandwidth demands of most communication networks.
- Reduction in access line charges: VPN provides the capability for reduced charges, especially those that come about as a result of long-distance charges. This is made possible by bypassing long carrier originating and terminating end-of-call access charges.
- Extension of private networks: With the aid of VPNs, the reach of private networks is extended to public networks transparently and easily.
- Use of switches in meshed networks: In the fully meshed private point-to-point systems, using VPN only requires one dedicated local loop to the

VPN switch. However, in the case of private line networks, one circuit is required for each switch in the meshed network.

- VPN economics: Economically, VPN has become an alternative to the leased lines for building a private network.

7.3.2 Disadvantages

The disadvantages are as follows:

- Flexibility: Some of the current VPNs do not exhibit flexibility.
- Inadequate dynamics: Most are not dynamic enough as to allow instant changes within a network, for example, changes due to rapid environmental situations and the inability of the user to be able to control the VPN.
- Deployment issues: The current VPNs have lengthy setup and deployment times.
- Software issues: It is difficult and always a very slow process to change software, especially, where there is an evolution of the VPN software. An example is when current VPNs bring forth different networks with different policies. It becomes difficult to manage and therefore slows the process:

7.4 VPN SECURITY ISSUES

In any network, security is always an issue and the VPN is no exception. Security is an issue because you do not want unauthorized users to have access to the VPN nor should abuse of access be tolerated. To mitigate against unauthorized use and access, adequate security measures must be in place. The VPN security requirements include the following:

- Service usage authorization check: The user must be checked to see if adequate authorization is in place before that user can get access.
- Access and authentication control to user directory: It is important to check if the user is a member of the organization being protected before access can be granted. The authentication is done by checking the personnel directory of the organization.
- Access and authentication control for a customer directory: The customers that belong to a certain organization are in a customer directory. To grant access, the customer must be authenticated using the existing customer directory. The VPN features that customers can use are as follows:
    - site and end point management
    - installation and management of closed user groups
    - setting authorization codes for end-users
    - equipment and resources of the communications services management
    - the billing service customization policy for the service
- The security recommendations of the Open Systems Interconnection/International Standards Organization (OSI/ISO) must be adhered to if adequate security services are to be obtained. Some of these services are

- access control
- authentication service
- data confidentiality and integrity
- VPN audit requirements

7.5 VPN STANDARDS

The standards for VPN have not yet matured. There have been several alliances by several interested companies such as Microsoft and 3Com. However, a good number of requests for comments (RFC) that are of importance have been proposed and are currently being considered. Examples of these RFCs for VPN are

- RFC2547: Border gateway protocol/multiprotocol label switching (BGP/MPLS) VPNs.
- RFC2685: VPN identifier.
- RFC2735: NHRP support for VPN.
- RFC2764: A framework for IP-based VPNs.

SUMMARY

1. Multiple hosts can communicate securely over a public network using a VPN.
2. Since there are different types of VPN, each type depends on the particular function it performs as well as the various methods for constructing it.
3. The VPN has three main components: a public component, a service provision component, and an information systems operator component.
4. Access control, encryption, and authentication are some of the key security components of the VPN.
5. VPNs provide an abstraction of a network infrastructure that is used by groups or enterprises to cover their needs.

REVIEW QUESTIONS

7.1 Interoperability is one of the requirements of the place-oriented VPN VO-VPN.
   a. True
   b. False

7.2 One of the infrastructure components required to implement a DVPN are
   a. Remote node
   b. Remote access server
   c. Tunnel management system
   d. Gateway
   e. All of the above

7.3 The secure and manageable VPNs cannot be used to enable secure remote accesses to the end user.
 a. True
 b. False
7.4 The three main components of VPN are public, service provision, and information systems operator.
 a. True
 b. False
7.5 Reduction in access line charges is not one of the advantages of VPN.
 a. True
 b. False
7.6 Some of the current VPNs do not exhibit flexibility qualities.
 a. True
 b. False
7.7 Some of the recommendations of the Open Systems are:
 a. Access control
 b. Authentication service
 c. Data confidentiality and integrity
 d. VPN audit requirements
 e. All of the above
7.8 VPN does not yet have a matured standard.
 a. True
 b. False
7.9 VPNs provide an abstraction of a network infrastructure that is used by groups or enterprises to cover their needs.
 a. True
 b. False
7.10 Inadequate dynamics is one of the disadvantages of the VPN.
 a. True
 b. False

Answer: 7.1a, 7.2e, 7.3b, 7.4a, 7.5b, 7.6a, 7.7e, 7.8a, 7.9a, 7.10a

PROBLEMS

7.1 What is the definition of a VPN?
7.2 State and describe at least five different kinds of VPN.
7.3 What are the requirements that a PO-VPN must satisfy?
7.4 Describe the various infrastructure components required to implement a DVPN.
7.5 Discuss the general architecture of the VPN.
7.6 Describe the advantages of the VPN.
7.7 What are the disadvantages of the VPN?
7.8 Discuss the security issues of a VPN.
7.9 What are the VPN security requirements?
7.10 Describe some of the RFCs now being considered for VPN.

7.11 What vulnerabilities are addressed by considering the security issues of a VPN?
7.12 What vulnerabilities are unique to or heightened by considering the security issues of a VPN?
7.13 VPN raises some performance issues. What are those performance issues?
7.14 Crypto issues are relevant in the VPN context. What are they?
7.15 Is it true that VPNs are used for specific kinds of applications or environments? If so, what are some examples of where and why would VPNs be deployed?
7.16 What are those applications or environments in which VPNs would really be detrimental?
7.17 What kind of policies and procedures need to be developed for VPN?
7.18 Describe the relationship between firewalls and VPN.
7.19 What is the advantage of using VPN technologies instead of WAN technologies?
7.20 What are the concerns when using a VPN with cellular/wireless connections?

8 ISDN and BISDN

He who asks a question is a fool for five minutes; he who does not ask a question remains a fool forever.

— Chinese Proverb

Integrated Services Digital Network (ISDN) is a set of ITU-T standards approved in 1984 for digital transmission over ordinary telephone copper wire as well as over other media. ISDN involves the digitization of the telephone network, which permits voice, data, text, graphics, music, video, and other source material to be transmitted over existing telephone wires.

There are two categories of services offered by ISDN. First, there is narrowband ISDN, which has been around for a while. Second, the real excitement is found in broadband ISDN (BISDN). Due to the growing demand for broadband services, BISDN will become the universal network of the future. BISDN employs asynchronous transfer mode (ATM) as the switching and multiplexing technology. ATM has emerged as the most promising technology for supporting broadband multimedia communication services. Although ATM-based networks are not widely deployed yet, they are already generating much curiosity and excitement.

The aim of this chapter is to provide an introduction to the rapidly evolving ISDN and BISDN and their related emerging services and technologies. Although ATM will be briefly mentioned in this chapter for the sake of completeness, its full discussion will be in the next chapter.

8.1 ISDN

ISDN stands for integrated services digital network. (To you that may mean "incredibly speedy data now" and high speed Internet access.) The phrase "digital network" means that the user is given access to a telecom network ensuring high-quality transmission via digital circuits. Digital signaling produces connections of the highest possible quality and carries digital data at higher speeds than analog conversion. The phrase "integrated services" refers to the simultaneous transmission of voice and data services over the same wires. This way, computers can connect directly to the telephone network without first converting their signals to an analog form using modems. This integration brings with it a host of new capabilities combining voice, data, fax, and sophisticated switching. And because ISDN uses the existing local telephone wiring, it is equally available to home and business customers. ISDN was intended to eventually replace the traditional "plain old telephone service" (POTS) phone lines with a digital network that would carry voice, data, and video.

ISDN is a high-speed communication network, which allows voice, data, text, graphics, music, video, and other source material to be transmitted simultaneously across the world using end-to-end digital connectivity.

8.1.1 ISDN Devices and Interfaces

ISDN devices include terminals, terminal adapters (TAs), network-termination devices, line-termination equipment, and exchange-termination equipment. They are explained as follows.

- **TE1:** Terminal equipment type 1. This is the ISDN telephone, computer, ISDN fax machine, or whatever it is that we connect to the ISDN phone line. TE1s connect to the ISDN network through a four-wire, twisted-pair digital link.
- **TE2:** Terminal equipment type 2. This is the old analog telephone, old-style fax machine, modem—whatever we used to hook up to the analog phone line. It is a non-ISDN terminal (such as DTE) that predates the ISDN standard. TE2s connect to the ISDN network through a TA.
- **TA:** Terminal adaptor. This lets old TE2 stuff talk to the ISDN network. It also adapts other kinds of equipment, like Ethernet interfaces, to ISDN. The ISDN TA can be either a stand-alone device or a board inside the TE2. TAs are designed to facilitate equipment with data rates lower than the 64 kbps per B-channel.
- **NT1:** Network termination type 1. This is the end of the line for the local phone company, and the beginning of your house's phone network. In North America, the NT1 is a customer premises equipment (CPE) device. In most other parts of the world, the NT1 is part of the network provided by the carrier.
- **NT2:** Network termination type 2. In most homes, this may not exist. The NT2 is a more complicated device that typically is found in digital private branch exchanges (PBXs) and that performs Layer 2 and 3 protocol functions and concentration services. An NT1/2 device also exists as a single device that combines the functions of an NT1 and an NT2.
- **LT:** Line termination. This is the physical connection to the phone company.
- **ET:** Exchange termination. This is the local phone company's logical connection from your telephones to "the phone network."

ISDN specifies a number of reference points that define logical interfaces between functional groupings, such as TAs and NT1s. ISDN reference points include the following:

- *R*: The reference point between non-ISDN equipment and a TA.
- *S*: The reference point between user terminals and the NT2.
- *T*: The reference point between NT1 and NT2 devices.
- *U*: The reference point between NT1 devices and line-termination equipment in the carrier network. The U reference point is relevant only in North America, where the NT1 function is not provided by the carrier network.

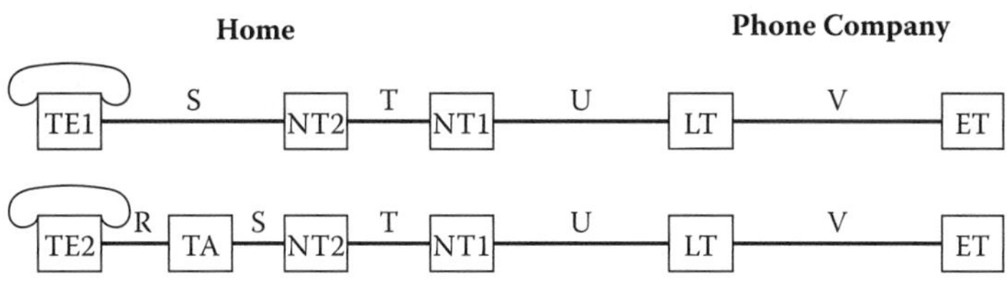

FIGURE 8.1 Two common reference configurations.

(R, S, T, and U are simply consecutive letters of the alphabet, chosen by the ITU to designate ISDN reference points.) Figure 8.1 shows two configurations although the actual networks will look more complicated than these.

In the United States, the telephone company (called telco in the business community) provides its home customers with a *U interface*. The U interface is a two-wire (single pair) interface from the phone switch. It supports full-duplex data transfer over a single pair of wires, therefore only a single device (NT1) can be connected to a U interface. In other parts of the world, the telco is allowed to supply the NT1, and thereby the customer is given an S/T interface.

The NT1 is a relatively simple device that converts the two-wire U interface into the 4-wire *S/T interface*. The S/T interface supports multiple devices (up to seven devices can be placed on the S/T bus). Today, many devices have NT1s built into their design. This has the advantage of making the devices less expensive and easier to install, but often reduces flexibility by preventing additional devices from being connected.

Technically, ISDN devices must go through a *network termination 2* (NT2) device, which converts the T interface into the S interface. (Note: the S and T interfaces are electrically equivalent.) Virtually all ISDN devices include an NT2 in their design.

A typical sample ISDN configuration is illustrated in Figure 8.2, where three devices are attached to an ISDN switch at the central office. Two of these devices are ISDN-compatible, so they can be attached through an S reference point to NT2 devices. The third device (a non-ISDN telephone) attaches through the reference point to a TA. Any of these devices also could attach to an NT1/2 device, which would replace both the NT1 and the NT2.

8.1.2 Services

There are two basic types of ISDN services: basic rate interface (BRI) and primary rate interface (PRI). The basic rate interface is intended for home use, while primary rate interface is designed for businesses with larger data needs, or with the need to set up their own local telephone system.

The *basic rate interface* (BRI) service is intended to meet the needs of most individual users. To better suit voice applications, this 144-kbps channel is partitioned into subchannels: two 64-kbps B (for *bearer*) channels and one 16-kbps D

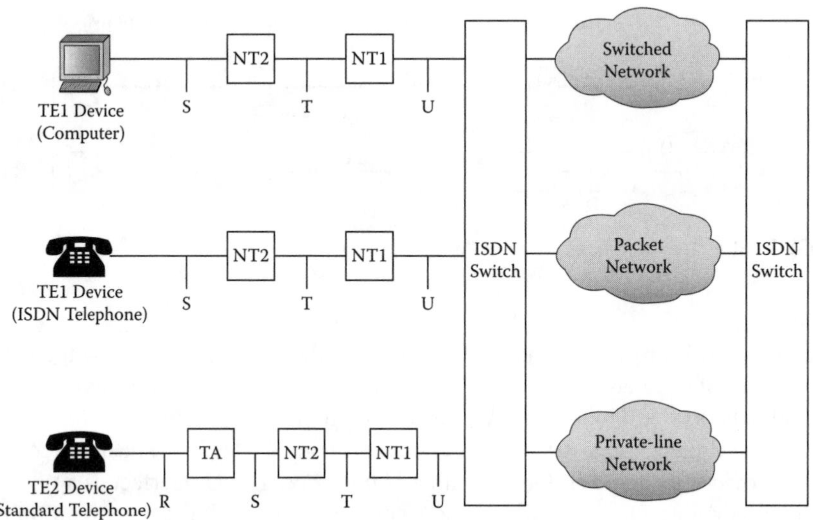

FIGURE 8.2 A typical ISDN configuration showing the relationships between devices and reference points.

(for *data*) channel. You can combine the two B channels together to form a single 128-kbps data channel through a process called *bonding*. Thus, the BRI consists of two *B channels* and one *D channel* (*2B + D*). The B-channel service operates at 64 kbps and is meant to carry user data. (Note that, in ISDN terminology, "k" means 1000 (10^3), not 1024 (2^{10}) as in many computer applications; therefore, a 64-kbps channel carries data at a rate of 64,000 bits per second.) The D-channel service operates at 16 kbps and is meant to carry control and signaling information, although it can support user data transmission under certain circumstances. The D channel is used to control the attachment of the one to eight devices to the two B channels. It is referred to as the packet-switched setup and signaling connection shared by all users of ISDN. BRI also provides for framing control and other overhead, bringing its total bit rate to 192 kbps.

To access BRI service, it is necessary to subscribe to an ISDN phone line and have ISDN devices such as ISDN terminal adapters. Customer must be within 18,000 feet (about 3.4 miles or 5.5 km) of the telephone company central office for BRI service; beyond that, expensive repeater devices are required, or ISDN service may not be available at all.

There are three kinds of connections that can be set up over a B channel:

1. Circuit-switched: In this case, the user makes a call and a circuit-switched connection is established using common-channel signaling. There is an end-to-end circuit between the users.
2. Packet-switched: In this case, the user is placed in a packet-switched mode and data is exchanged with other users via X.25 LAPB.
3. Semipermanent: This is equivalent to a leased line.

The *primary rate interface* (PRI) service is intended for users with greater capacity requirements. It consists of 23 B channels and one D channel (23B + D) in North America and Japan, yielding a total bit rate of 1.544 Mbps (the PRI D channel runs at 64 kbps). The PRI in Europe, Australia, and other parts of the world provides 30 B channels plus one 64-kbps D channel (30B + D) and a total interface rate of 2.048 Mbps.

One can join several B channels into one channel called *H channel.* In other words, H channels provide a way to aggregate B channels. There are different H-channels and are implemented as:

- H0 = 384 kbps (6 B channels)
- H10 = 1472 kbps (23 B channels)
- H11 = 1536 kbps (24 B channels)
- H12 = 1920 kbps (30 B channels) - International (E1) only

Typically, each B channel is set up for only one of these types of traffic: data or voice. Typically you can get both B channels for data, or one for voice and the other for data. In order to facilitate this, North American phone companies use an optional part of the ISDN standard to identify each TE1 or TA you use. The phone company assigns a service profile identifier (SPID) to each of these devices. A SPID looks like a telephone number with extra digits. However, depending on which kind of switch you are served from and how you are going to use the ISDN service, you may not need a SPID or you may need a SPID for each B channel or each device.

8.1.3 Architecture

ISDN has definitions for three basic layers, which are explained as follows.

Layer 1

As shown in Figure 8.3, ISDN physical-layer (layer 1) frame formats differ depending on whether the frame is outbound (from terminal to network) or inbound (from network to terminal). The frames are 48 bits long, of which 36 bits represent data. The bits of an ISDN physical-layer frame are used as follows:

- *F*: Provides synchronization
- *L*: Adjusts the average bit value
- *E*: Ensures contention resolution when several terminals contend for a channel
- *A*: Activates devices
- *S*: Unassigned
- *B1, B2, and D*: Handles user data

Multiple ISDN user devices can be physically attached to one circuit. In this configuration, collisions can result if two terminals transmit simultaneously. ISDN therefore provides features to determine link contention. When an NT receives a D bit from the TE, it echoes back the bit in the next E-bit position. The TE expects the next

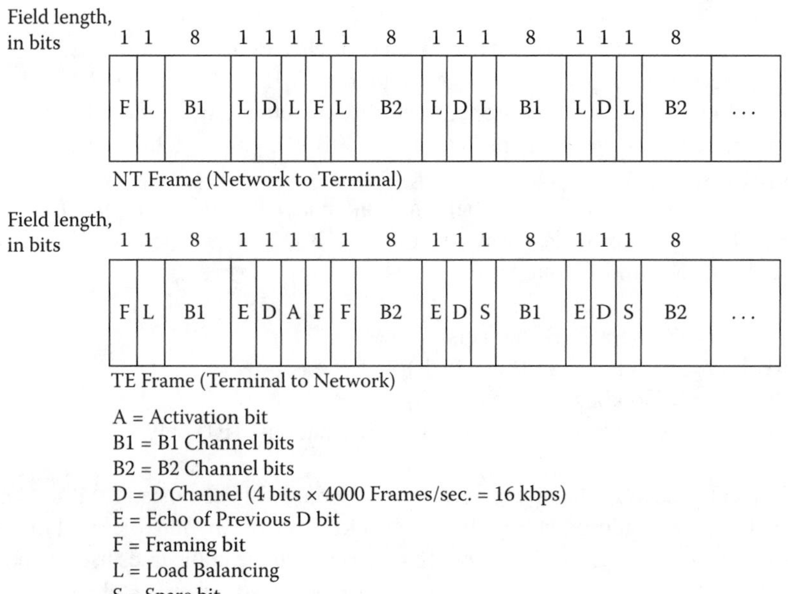

FIGURE 8.3 ISDN physical-layer frame formats differ depending on their direction.

E bit to be the same as its last transmitted D bit. If the TE detects a bit in the echo (E) channel that is different from its D bits, it must stop transmitting immediately. This simple technique ensures that only one terminal can transmit its D message at one time. After successful D message transmission, the terminal has its priority reduced by requiring it to detect more continuous ones before transmitting. Terminals cannot raise their priority until all other devices on the same line have had an opportunity to send a D message. Telephone connections have higher priority than all other services, and signaling information has a higher priority than nonsignaling information.

Layer 2

Layer 2 of the ISDN signaling protocol is *link access procedure, D channel* (*LAPD*). This layer is used across the D channel to ensure that control and signaling information flows and is received properly. As shown in Figure 8.4, the LAPD frame format is almost identical to the X.25 LAPB protocol (see Figure 2.3).

Address: 1 or 2 bytes
SAPI (service access point identifier): 6-bits
C/R (command/response): bit indicates if the frame is a command or a response
A0 (address extension): bit indicates whether this is the final octet of the address or not
TEI (terminal endpoint identifier): 7-bit device identifier
EA1 (address extension): same as EA0

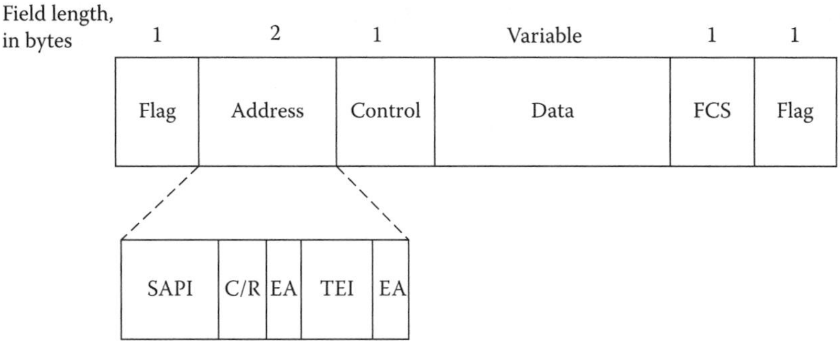

SAPI = Service Access Point Identifier (8 bits)
C/R = Command/Response bit
EA = Extended Addressing bits
TEI = Terminal Endpoint Identifier

FIGURE 8.4 LAPD frame format is similar to LAPB.

Control (2 bytes): The frame level control field indicates the frame type (information, supervisory, or unnumbered) and sequence numbers as required
Information: Layer 3 protocol information and user data
CRC (2 bytes): Cyclic redundancy check is a low-level test for bit errors on the user data
Flag (1 byte): This is always $7E_{16}$ (or $0111\ 1110_2$)

The LAPD *address* field can be either 1 or 2 bytes long. If the extended address bit of the first byte is set, the address is 1 byte; if it is not set, the address is 2 bytes. The first address-field byte contains the service access point identifier (SAPI), a 6-bit field that identifies the point where layer 2 provides a service to layer 3. The *terminal end-point identifiers* (TEIs) are unique IDs given to each device (TE) on an ISDN S/T bus. The field identifies either a single terminal or multiple terminals. A TEI of all ones indicates a broadcast. The value may be assigned statically when the TE is installed, or dynamically when activated.

Layer 3

Layer 3 is used for the establishment, maintenance, and termination of logical network connections between two devices. A variety of call-establishment, call-termination, information, and miscellaneous messages are specified, including SETUP, CONNECT, RELEASE, USER INFORMATION, CANCEL, STATUS, and DISCONNECT. These messages are functionally similar to those provided by the X.25 protocol. Figure 8.5 shows the typical stages of an ISDN circuit-switched call.

ISDN has the following features and benefits:

- Speed: There is an upper limit to the amount of information that an analog telephone line can hold. Currently, it is about 56 kbps. ISDN allows multiple digital channels to be operated simultaneously through the same

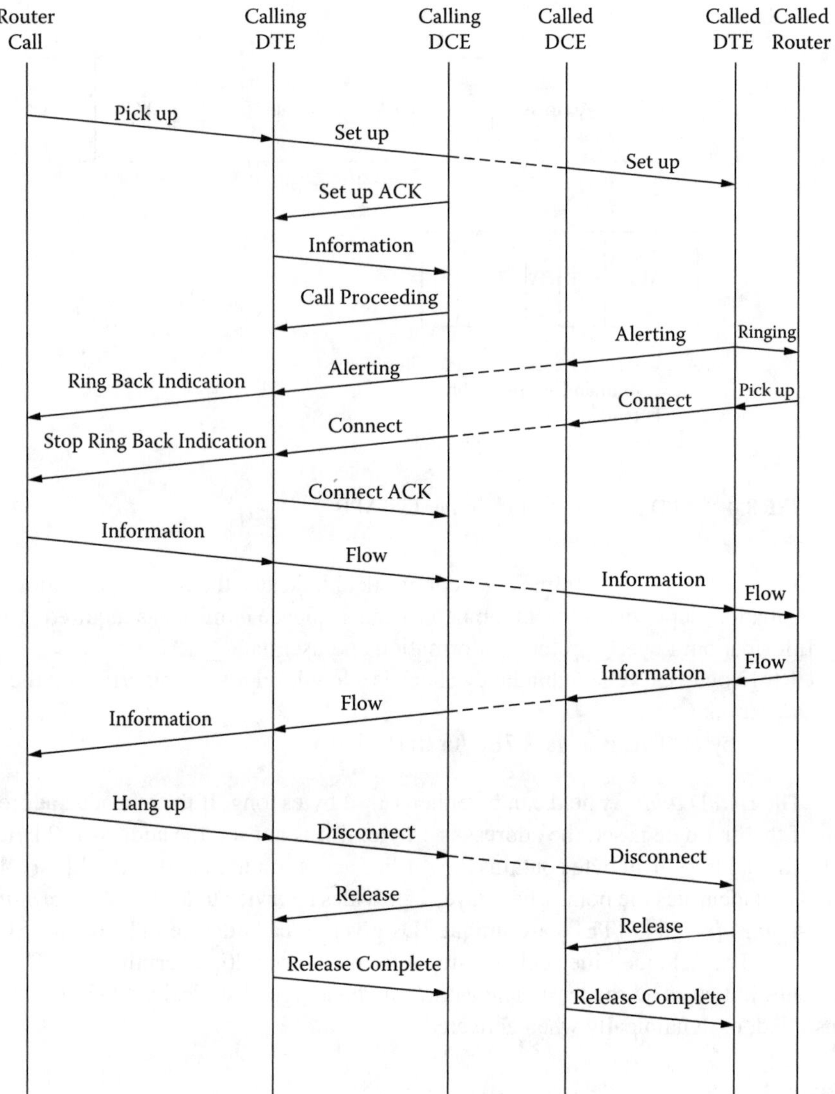

FIGURE 8.5 An ISDN call moves through various stages to its destination.

regular phone wiring used for analog lines but permits a much higher data transfer rate than analog lines. By combining your two B channels you have access to up to 128 kbps—more than four times as fast as a 28.8 kbps modem on a standard phone line. Thus, ISDN allows you to send and receive data four times faster than traditional phone lines and modems.

- Multiple devices: Previously, it was necessary to have a phone line for each device you wished to use simultaneously. For example, one line each was required for a telephone, fax, computer, bridge/router, and live video

conference system. ISDN technically refers to a specific set of digital services provided through a single, standard interface. Because it is a digital network, ISDN is capable of transmitting all forms of data, text, image, video, and digitized voice over a single network transparently.
- High productivity: Due to the fact that ISDN has two communication paths, you can perform two communication tasks simultaneously. For example, you can send a data file and receive a phone call at the same time without the phone call disrupting the data transmission. This capability makes you more productive, which could save you time and money.
- Capacity: ISDN has two channels, or paths, for communication. These channels can be combined to send extremely large or data-intensive files, such as pictures, graphs, audio, or even video. By combining both channels, high-speed Internet access can also be achieved.
- High quality: ISDN is digital, which translates to fewer errors when sending or receiving files. Digital also means that voice and sound are crystal-clear. There is no noise or static on a digital line.
- Signaling: The phone company sends a ring voltage signal to ring the bell in your phone ("in-band signal"). ISDN sends a digital packet on a separate channel ("out-of-band signal"). The out-of-band signal carried by the D channel does not disturb established connections, and call setup time is very fast. The signaling also indicates who is calling, what type of call it is (data/voice), and what number was dialed. The ISDN device is then capable of making intelligent decisions on how to direct the call.

ISDN has its detractors, most of whom rally behind other methods for going digital directly to the subscriber. The alternatives include existing copper-wire digital services, such as T1 (at 1.544 Mbps), frame relay (at 56 Kbps to 1.544 Mbps), asynchronous transfer mode (at 25 Mbps to 100 Mbps), and switched multimegabit data service (at 35 Mbps). Most of these services cost several hundred to several thousand dollars per month. ISDN is not as fast as everyone would like, but it is a lot faster and cheaper than what we have got.

ISDN service is available today in most major metropolitan areas and probably will be completely deployed throughout the United States very soon. Many internet service providers (ISPs) now sell ISDN access. However, the idea of using existing copper wiring to provide this network decreased ISDN capabilities in reality. That is why broadband ISDN (BISDN) was born.

Example 8.1

Consider a minimal ISDN setup connecting two computers, illustrated in Figure 8.5. The incoming twisted pair enters a telco-provided box called the *network terminator* (NT1), which breaks the 144-Kbps channel into the two B and single D subchannels.

As shown in Figure 8.6, a single four-wire cable carries the 2B + D channels to another box called the terminal adapter (TA). Unlike the NT1, which provides only a single function (creating the 2B + D channels), the TA can do many things. Its job is to connect any and all of your terminal equipment (TE)—computers, fax machines, LANs, or telephone sets—to one or both of the B channels. In this example, the TA

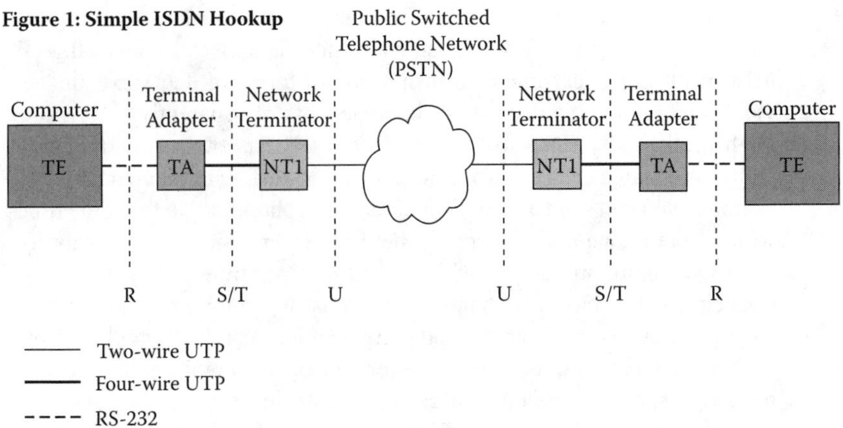

FIGURE 8.6 A simple ISDN configuration; for Example 8.1.

is shown as a separate unit, but it could easily be contained within the computer (as an add-in card or integrated feature).

Figure 8.6 also shows the external ISDN reference points, labeled R, S/T, and U. Each interface point requires an electrically different device connection and cabling. The U reference point is the incoming unshielded twisted pair (UTP); the S/T reference point is a four-wire UTP cable. RS-232 serves as an interface between data terminal equipment (DTE) and data communications equipment (DCE) employing serial binary data interchange.

Example 8.2

Up to eight devices can share access to the channels using a feature of ISDN called passive bus. Passive bus uses a second kind of network terminator (NT2) to let up to eight separate TAs share a single 2B + D circuit.

Figure 8.7 shows a passive bus with a dozen computers and four fax machines sharing an ISDN circuit. You need one TA for every two pieces of terminal equipment. Whenever a computer or fax machine wants to use a B channel, its associated TA checks to see if a channel is available, and if so, dedicates it to the requesting TE. The example shows maximum device sharing, but the cost of additional 2B + D circuits is low enough that you will likely have fewer devices on a single bus.

For incoming ISDN calls, the telco's central office (CO) sends a call setup message to the TA via the D channel, indicating that a call is available to be picked up (if multiple TAs are connected via passive bus, any TA can pick up the call). The TA can answer the call and assign it to an available B channel or if both B channels are in use, it can free a channel by placing an active call on hold and making the new call active. These calls can be either data or voice. Thus, a single TA could have as many as 15 simultaneous calls in progress, with any two of those calls active (i.e., actually communicating).

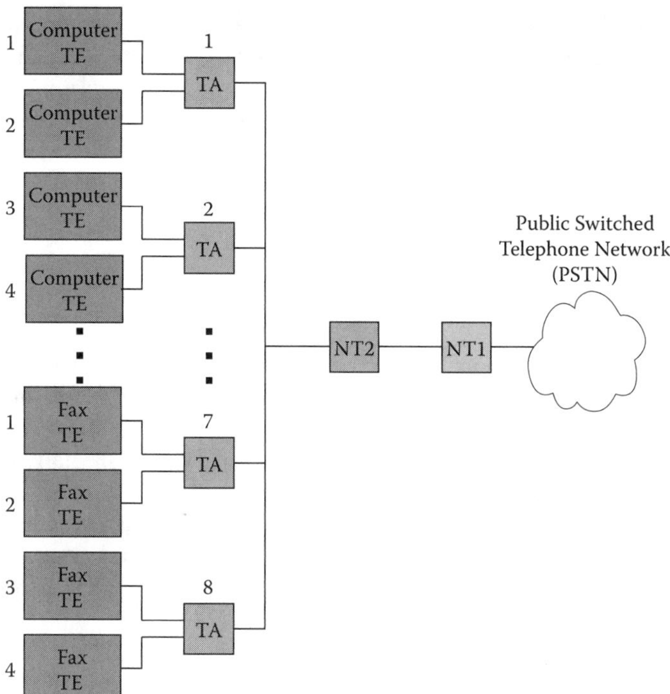

FIGURE 8.7 ISDN passive bus channel sharing; for Example 8.2.

Example 8.3

This example illustrates layer 3 call setup. It demonstrates the steps that occur when an ISDN call is established. In the following example, there are three points where messages are sent and received: (1) the caller, (2) the ISDN switch, and (3) the receiver.

1. Caller sends a SETUP to the switch.
2. If the SETUP is OK, the switch sends a CALL PROCeeding to the caller, and then a SETUP to the receiver.
3. The receiver gets the SETUP. If it is OK, then it rings the phone and sends an ALERTING message to the switch.
4. The switch forwards the ALERTING message to the caller.
5. When the receiver answers the call, it sends a CONNECT message to the switch.
6. The switch forwards the CONNECT message to the caller.
7. The caller sends a CONNECT ACKnowledge message to the switch.
8. The switch forwards the CONNECT ACK message to the receiver.

Complete. The connection is now up.

Example 8.4

To ensure that the pattern 01111110 does not appear inside the frame except in the flags (thus destroying synchronization), a procedure known as bit stuffing is used. Between the transmission of the starting and ending flags, the transmitter inserts a 0 bit after each occurrence of five 1s. An example is presented below.

Original Data:

111111101111111111100111111

After bit stuffing:

11111**0**11011111**0**11111**0**10011111**0**1

8.2 BISDN

When the digital video systems started to develop in the 1980s, it was soon noticed that the maximum bandwidth (2.048 Mbps) of the basic ISDN (narrowband) is not enough. Moreover, the problem with the basic ISDN is that it allows only the use of multiples of the basic bit rate (64 kbps). At the same time, especially in the optical transfer media's sector, the development had been very fast and it was thought that there is the need to develop the so-called broadband integrated services digital network (BISDN). Thus narrowband integrated services digital network (NISDN) provides services that will be carried by channels based on 64 kbps up to 1.5 Mbps (or 2 Mbps for European standard).

Broadband ISDN (BISDN) is a second generation of ISDN. In some respect, BISDN works the same way as the NISDN; BISDN supports multiple B channels that are controlled by a D channel. The difference is the amount of information that can be transferred. Table 8.1 shows the range of data rates for narrowband and broadband ISDN. In other respects, BISDN is not only an improved ISDN, but also a complete redesign of the NISDN. It consists of ITU-T communication standards designed to handle high-bandwidth applications such as video. The key characteristic of BISDN

TABLE 8.1
ITU-T Narrowband and Broadband Channels

	Channel Date Rate	Application
Narrowband		
D	16–64 kbps	Signaling, data
B	64 kbps	Data, voice, fax
H0	384 kbps	Data, voice, fax, compressed video
H11	1.536 Mbps	PBX access, compressed video, high-speed data
H12	1.926 Mbps	PBX access, compressed video, high-speed data
Broadband		
H2	30–45 Mbps	Full-motion video, video telephone
H3	60–70 Mbps	Not defined
H4	120–140 Mbps	Enhanced video, bulk data transfer, fax

Source: M. L. Moore, *ISDN Strategies,* IDG Books, 1995, p. 168.

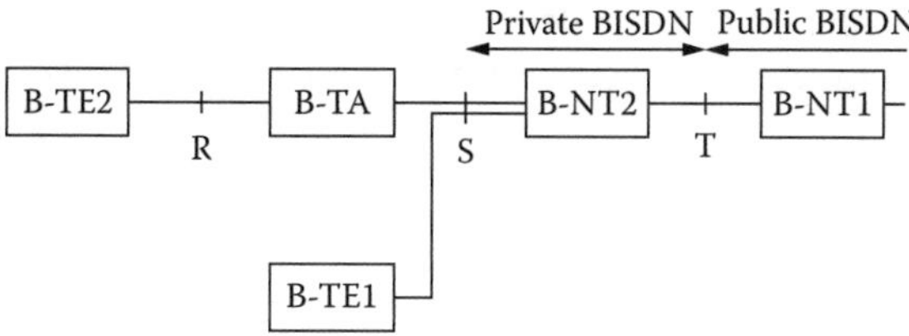

FIGURE 8.8 BISDN interfaces and terminals.

is that it provides transmission channels capable of supporting rates greater than the primary ISDN rate. Broadband services are aimed at both business applications and residential subscribers.

> **BISDN** is a digital network operating at data rates in excess of the 2.048Mbps—the maximum rate of standard (narrowband) ISDN.

BISDN's foundation is cell switching, and the international standard supporting it is asynchronous transfer mode (ATM). While BISDN is a concept of broadband data transfer, ATM is a technology that is used to implement BISDN.

8.2.1 Interfaces and Terminals

The interfaces and terminals are similar for NISDN and BISDN, but functional sets begin with the letter B, such as B-TA, B-NT2, B-TE, etc. As illustrated in Figure 8.8, BISDN uses the following interfaces and network terminals.

B-TE1: Broadband terminal equipment 1.
B-TE2: Broadband terminal equipment 2.
B-TA: Broadband terminal adapter.
B-NT1: Broadband network termination 1. This terminal is used to terminate the network at the physical layer. This element is needed between two networks when they are owned by different operators. This is the situation when, for example, private and public ATM networks are connected.
B-NT2: Broadband network termination 2. This terminal is used to terminate the network at the physical layer and also at the higher (ATM) level. This element requires more intelligence than B-NT1 because it has to also operate at the higher protocol layers.
S: An interface between BISDN terminal and private network.
T: An interface between BISDN private and public networks.
R: An interface between B-TE2 and B-TA.

8.2.2 ATM Technology

Asynchronous transfer mode (ATM) is a fast packet-oriented transfer mode based on asynchronous time division multiplexing. The phrase "transfer mode" indicates that this technology is a specific way of transmitting and switching through the network. The term "asynchronous" refers to the fact that the packets are transmitted using asynchronous techniques (e.g., on demand), and the two end-points need not have synchronized clocks. ATM will support both circuit-switched and packet-switched services.

> **Asynchronous Transfer Mode** (ATM) is the ITU-T standard for broadband ISDN, which supports integrated voice, data, and video communications.

ATM (asynchronous transfer mode) has been accepted all over the world as the transfer mode used for BISDN. ATM can handle any kind of information, i.e., voice, data, image, text, and video in an integrated manner. An ATM network is made up of an ATM switch and ATM endpoints. An ATM switch is responsible for cell transit through an ATM network. An ATM endpoint (or end system) contains an ATM network interface adapter. Examples of ATM endpoints are workstations, routers, digital service units (DSUs), LAN switches, and video coder-decoders (CODECs). An ATM network consists of a set of ATM switches interconnected by point-to-point ATM links or interfaces. ATM switches support two primary types of interfaces: User-network interface (UNI) and network-network interface (NNI). The UNI connects ATM end systems (such as hosts and routers) to an ATM switch. The NNI connects two ATM switches.

In ATM the information to be transmitted is divided into short 53-byte packets or cells. There are reasons for such a short cell length. First, ATM must deliver real-time service at low bit rates. Thus the size allows ATM to carry multiple forms of traffic. Both time-sensitive traffic (voice) and time-insensitive traffic can be carried with the best possible balance between efficiency and minimal packetization delay. Second, using short, fixed-length cells allows for time-efficient and cost-effective hardware such as switches, multiplexers, etc. More information about ATM will be provided in the next chapter.

8.2.3 Broadband ISDN Protocol Architecture

Figure 8.9 depicts the architecture for the BISDN protocol. The figure also shows how ATM fits into BISDN. The ATM system is divided into three functional layers, namely the physical layer, the ATM layer, and the ATM adaptation layer. Thus, the protocol for BISDN also adopts a layered approach, made up of four layers:

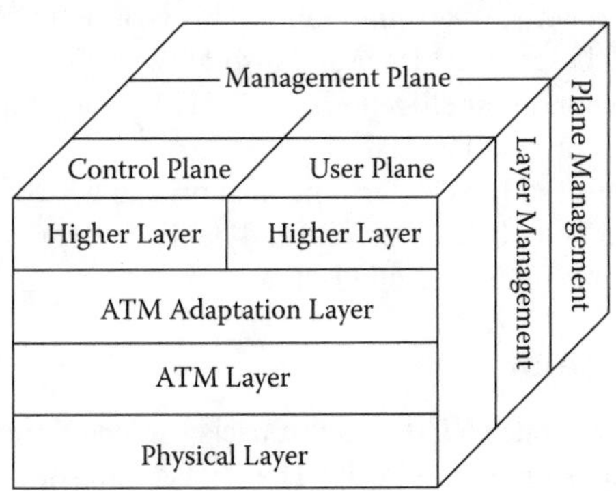

FIGURE 8.9 BISDN protocol reference model.

- Physical layer: The physical layer is responsible for basic physical layer activities such as information synchronization, bit timing, transmission frame generation and recovery, transmission frame adaptation, cell delineation, and cell header verification. The physical layer consists of two sublayers: the physical medium sublayer and the transmission convergence sublayer. The specification of the physical medium sublayer depends on the medium used. The sublayer is responsible for transmitting/receiving a continuous flow of bits with associated timing information to synchronize transmission and reception. The transmission convergence sublayer is responsible for transmission frame generation and recovery, transmission frame adaptation, and cell delineation.
- ATM layer: The ATM layer provides the packet transfer capabilities. It is independent of the physical medium. The ATM layer is responsible for cell multiplexing and demultiplexing. The cell multiplexing functions enable multiple logical connections across a single interface. The layer is also responsible for virtual path identifier (VPI) and virtual channel identifier (VCI) translation, cell header generation or extraction, and generic flow control.
- ATM adaptation layer (AAL): BISDN is required to be able to handle both packet and circuit-mode applications, thus the necessity for having the ATM adaptation layer (AAL). The AAL is needed to handle non-ATM protocol, such as link access protocol-D. The ATM adaptation layer consists of two sublayers: the segmentation and reassemble sublayer (SAR) and the convergence sublayer (CS). The convergent sublayer provides specific application support for applications using AAL. The segmentation reassemble sublayer packs the information received from CS into cells for transmission.
- Higher layers: The higher layers provide services for video, SMDS, frame relay, access, and network signaling. These will be discussed next.

The physical, ATM, and AAL layers form the lower layers of the BISDN protocol reference model. As shown in Figure 8.9, the higher BISDN layers follow a three-plane approach. These three separate planes are referenced as:

- user plane
- control plane
- management plane

The user plane (U-plane) is responsible for user information transfer from one point to another along with associated controls such as flow control, congestion control, and error control. The U-plane contains all of the ATM layers.

The control or signaling plane (C-plane) manages the call-control and connection-control functions. The C-plane shares the physical and ATM layers with the U-plane, and contains AAL functions dealing with signaling.

The management plane (M-plane) provides the management functions and the capability to transfer information between the C and U-planes. This plane performs two functions: plane management and layer management. The plane management deals with the management functions necessary for the whole system (end-to-end);

TABLE 8.2
BISDN Services

Service Categories		Example Services
Interactive Services	Conversational services	TV conference
	Messaging Services	Video Mail
	Retrieval Services	Videotex
Distribution Services	Without user presentation control	TV Broadcast
	With user presentation control	Videography

its duty is to provide coordination between all the different planes. The layer management performs layer-specific management functions such as managing resources and parameters within the protocol entities. For each layer, the layer management takes care of the specific operation and management information flow.

8.2.4 Applications for BISDN

The ITU-T has recommended the two service areas for application with BISDN: interactive services and distribution services. These services are summarized in Table 8.2.

Interactive Services

These include the following:

- Broadband video telephony services: Video telephony is the transfer of voice, moving pictures and scanned images, and documents between two points. Such technology is applied in sales, consulting, teaching, and legal services. However, the prohibitive costs of terminal equipment limit its widespread use.
- High-speed unrestricted data and information transmission services: This type of service will include LAN and WAN networking. Other applications include document transfers, fax, and multimedia documents including text, graphics, voice, and audiovisual information.
- Retrieval services: These involve retrieval of information stored at remote sites. Items transferred in this way could include medical information, share market information, and transmission of audio and video files. Unfortunately the amount of bandwidth required for transferring such resources restricts the number of transmissions that can occur at any one time.

Distribution Services

Distribution services are those in which information is broadcast to all subscribers of the service. They provide a continuous flow of information from a central source to any number of users. All of the users have access to the information. This type of service includes TV program distribution and document distribution.

Services that need BISDN include inter-LAN/PBX, multimedia, HDTV, and TV broadcast, which require bit rates of 100 Mbps, 80 Mbps, 30 Mbps, and 6 Mbps, respectively. Broadband ISDN will solve many disadvantages of networks today. The main problem with BISDN is the high price of implementation, because all copper wiring will have to be replaced with optical fibers. That is very expensive.

8.3 ISDN STANDARDS

Previously, ISDN was implemented in different ways in different countries. Thus, ISDN equipment that works in one country might not work correctly when installed in another country because of different national standards. The resulting incompatibility affected services and the interworking between equipment from different manufacturers.

By standardizing the interfaces to ISDN, all ISDN-compatible equipment (such as telephones, personal computers, etc.) will be able to attach to the network anywhere in the world and connect to any other attached system. However, ISDN standardization is an extensive task. It comprises signaling, switching, transmission, network architecture, operation and maintenance, and services. Standardization is a continuous process resulting in new designs and improvements to existing specifications.

ISDN standardization is primarily carried out by the International Telecommunication Union-Telecommunications Standardization Sector (ITU-T) and the European Telecommunications Standards Institute (ETSI).

The I-series (I stands for ISDN) of ITU recommendation for ISDN are as follows:

- I.100-141—General/terms/service attributes
- I.200-229—General service aspects
- I.230-239—Bearer services
- I.240-249—Teleservices
- I.250-259—Supplementary services - stage 1 descriptions
- I.300-349—Network functions/reference models/numbering
- I.350-359—Performance objectives
- I.360-369—Protocol layer requirements
- I.370-379—General network requirements
- I.410-429—User network interfaces - general
- I.430-459—User network interfaces (layer 1/2/3)
- I.460-469—Multiplexing and rate adaption
- I.500-599—Internetwork interfaces
- I.600-699—Maintenance principles

In addition, there are standards in Q-series, which include the following:

- Q.65-70—Methodology
- Q.80-89—Stage 2 descriptions for ISDN supplementary services
- Q.700-729—Signaling system number 7
- Q.760-769—ISDN user part

- Q.920-929—Data link layer protocol
- Q.930-939—Network layer protocol
- Q.940-949—User-network management
- Q.950-959—Stage 3 descriptions (DSS1) for ISDN supplementary services
- Q.2000-2099—Broadband ISDN
- Q.2100-2199—ATM adaptation layer
- Q.2600-2699—B-ISDN application protocols (common)
- Q.2700-2799—SS7 user part enhancements for B-ISDN
- Q.2900-2999—User-network protocol (DSS2) for B-ISDN

8.4 ISDN APPLICATIONS

With ISDN, you can connect with anyone, anywhere, including people using ordinary analog telephones, fax machines, and other terminal devices. Your ISDN line provides not one communications channel, but two. You can combine the channels to create a single information pipeline or conduct a voice call while sending a fax or surfing the net at 64 kbps. Table 8.3 provides the data rates required for different ISDN services.

Any telecommunication application can be handled by an ISDN, given bandwidth limitations of the ISDN channels. Applications of ISDN include file transfer, internet access, videoconferencing, internet service provider, PBXs, security, medicine, faxing and imaging, and LAN internetworking.

- Internet access: ISDN provides fast, reliable, low-cost connections to effectively use the Internet for business applications like providing company information, providing marketing material to customers, reducing operating costs, highlighting promotions or special deals, and selling products online.
- Videoconferencing: ISDN is exceptionally fitted to videoconferencing. Face-to-face meetings via videoconferencing can significantly reduce

TABLE 8.3
Data Rates for Various ISDN Services

Service	Data Rate (kbps)
Slow-scan video	10
Voice mail	16
Voice (telephony)/fax	64
High-fidelity sound	768
Videoconference	10,000–70,100
Television (TV)	30,100–70,100
High-definition TV (HDTV)	140,100–565,100

Source: M. L. Moore, *ISDN Strategies*, IDG Books, 1995, p. 167.

travel costs and promote faster consensus building and decision making. Using PRI or multiple BRI lines, channel bonding enables dynamic bandwidth allocation to support the high transmission rates (typically 64 kbps to 384 kbps) that video requires.

- High-speed transfers: This involves retrieving large amounts of information from distant locations. ISDN will link PBX to PBX, PBXs to LANs, LANs to WANs, and remote PCs to LANs. ISDN can be used as a backbone if one has low data transfer rates. For businesses already using high-speed data transfers such as T1, ISDN will offer the luxury of high-speed transfers without having to reverse bandwidth from a T1 system.
- Remote video surveillance: Using ISDN for remote surveillance allows temporary connections to cameras and other sensors at remote locations. The fast call setup time of ISDN allows a central operator to switch between various cameras. The technology can also be used for the remote monitoring of inaccessible machines.
- Telemedicine: With the application of ISDN in hospitals, health care providers and patients are now benefiting. Telemedicine applications range from radiology to psychiatry, even surgery. Medical images such as X-rays and computed tomography (CT) scans are now transmitted across ISDN lines from the hospital to the radiologist's office in a matter of minutes. This then allows two doctors to look at the same images on different computers and discuss primary diagnosis immediately.

SUMMARY

1. Integrated services digital network (ISDN) is the first network-based standard for simultaneous integration of voice, data, and video signal transmission.
2. There are two types of services associated with ISDN: BRI and PRI. The BRI service offers two B channels and one D channel (2B + D). The PRI service offers 23 B channels and 1 D channel in North America and Japan, yielding a total bit rate of 1.544 Mbps (the PRI D channel runs at 64 kbps).
3. Broadband ISDN uses fast packet switching ATM technology over optical fiber-based transmission circuits to provide high-speed data rates from 155 to 622 Mbps and beyond.
4. Asynchronous transfer mode (ATM) is a high-speed transmission technology for voice, data, and video that combines the best of circuit switching and packet switching.
5. The lower layers of BISDN protocol reference model consist of the physical, ATM, and AAL layers, while the higher layers are the management, user, and control planes.
6. Applications of ISDN include internet access, videoconferencing, high-speed transfers, remote video surveillance, and telemedicine.

REVIEW QUESTIONS

8.1 You buy a card that plugs into your computer and utilizes the device drivers for Ethernet, and the card would convert the Ethernet requests into an ISDN data stream. In this case, the card would be a TA, and your computer would be a TE2.
   a. True
   b. False

8.2 What type of device is commonly found between the S and T interfaces?
   a. TA
   b. LT
   c. NT1
   d. NT2

8.3 Which ISDN device is used to facilitate rate adaption, i.e., merging equipment with data rates lower than the 64 Kbps per B-channel to higher speed stream?
   a. TE1/TE2
   b. NT1/NT2
   c. U
   d. T/S
   e. TA

8.4 Which of the following messages is not used during the ISDN call setup phase?
   a. CONNECT
   b. RINGING
   c. SETUP
   d. CALL PROCEEDING

8.5 At which layer of the ISDN stack does LAPD protocol operate?
   a. Layer 1
   b. Layer 2
   c. Layer 3
   d. All layers

8.6 Broadband ISDN is simply an improved narrowband ISDN in that it replaces copper wire with optical fiber.
   a. True
   b. False

8.7 Which interface is between public BISDN and private BISDN?
   a. R
   b. S
   c. T
   d. U

8.8 Which of these functions is not the responsibility of the physical layer of BISDN protocol architecture?
   a. Transmission frame generation and recovery
   b. Transmission frame adaptation
   c. Cell delineation
   d. Generic flow control

8.9 The president of a company sends a video presentation to employees' video-equipped desktop computers. What kind of BISDN service is employed?
   a. Interactive service
   b. Distributive service
   c. Narrowband service
   d. Multicasting service

8.10 Which of the following is not ISDN interactive service?
   a. Video telephony
   b. TV program distribution
   c. LAN-WAN networking
   d. document transfer

Answer: 8.1a, 8.2d, 8.3e, 8.4b, 8.5b, 8.6b, 8.7c, 8.8d, 8.9b, 8.10b

PROBLEMS

8.1 What was the original goal of ISDN?

8.2 How fast is ISDN?

8.3 Describe the roles of TE1 and TE2 in ISDN.

8.4 Define the following terms: TA, NT1, NT2, LT, and ET.

8.5 (a) What type of information is carried by the ISDN B and D channels? (b) What is the purpose of using ISDN terminal adaptor (TA)?

8.6 What is the difference between BRI and PRI ISDN services?

8.7 A fax machine digitizes and represents image with 50 pixels per square centimeter and assigns 8 bits per pixel. If the fax is connected to an ISDN B channel, how long will it take to transmit a 27.5 cm by 21.5 cm (11 in by 8.5 in) image? How long will it take on a 155-Mbps ATM line?

8.8. For the primary rate interface used in North America, what is the percentage overhead?

8.9 Compare and contrast X.25 LAPB frame format and ISDN LAPD frame format.

8.10 Perform bit stuffing on the following data sequence:

11111111111011111101111111111111111101

8.11 A primary rate interface employs H12 + 3B + D. Determine the bit rate.

8.12 (a) Discuss some of the applications and services of ISDN. (b) Mention some of the limitations of ISDN.

8.13 Using web resources, discuss the format of the ISDN frame.

8.14 Explain the term "broadband" in BISDN.

8.15 What are the differences between NISDN and BISDN?

8.16 What are the functions of B-NT1 and B-NT2 in BISDN?

8.17 Discuss the three sublayers of the BISDN.

8.18 ATM is intended to be the "perfect transport." Describe what that means.

8.19 Describe the principal function(s) of the ATM layer.

8.20 Through research, find out what recommendations Q.931 and Q.933 are all about.

9 Asynchronous Transfer Mode

I cannot teach anybody anything. I can only make them think.

—Socrates

As mentioned in the previous chapter, asynchronous transfer mode (ATM) was selected in 1987 by ITU-T as the universal transfer mode for transferring information of the broadband integrated services digital networks (BISDN). It was originally conceived as a high-speed transfer technology for voice, video, and data over public networks. It combines the benefits of circuit switching with those of packet switching and provides a mechanism for integrating LANs and WANs. ATM allows the carriers to provide different classes of service over the network and to connect devices operating at different speeds.

The objective of this chapter is to provide a basic understanding of ATM networks and its services. The chapter provides summaries of ATM cells, protocols, services, and standards. It also discusses various applications of ATM, particularly how it applies in local area network (LAN) environments.

9.1 ATM TECHNOLOGY

The term "asynchronous" is used because ATM allows asynchronous operation between sender clock and receiver clock. A transfer mode usually refers to both the switching and multiplexing aspects of a communication network. The transfer mode for BISDN must be able to

- handle both narrowband and broadband rates
- handle both continuous and bursty traffic
- satisfy delay and/or loss sensitive quality requirements
- meet unforeseen demands

Neither circuit mode nor packet mode is suitable for meeting all these requirements. Asynchronous transfer mode (ATM), which is something between these two modes, has been selected as the target solution of BISDN. It is a specific packet-oriented transfer mode using an asynchronous time division multiplexing technique. This implies that the cells (units of information) of a specific application are not always in a fixed, periodic position in the channel and consequently, each cell must be identified by a header.

Asynchrous transfer mode (ATM) is a transfer mode in which information is organized into cells and is asynchronous in the sense that the recurrence of cells containing information from an individual user is not necessarily periodic.

ATM is a transfer technique with a special format that employs asynchronous time-division multiplexing. Information is transferred by way of a continuous flow of packets of fixed size and those packets are called ATM cells.

9.1.1 ATM Cell

Cells constitute the basic transport format for ATM. Each ATM cell consists of 48 bytes for the information field and 5 bytes for the header. (An advantage of using such a fixed, short cell is that it ensures that the time-sensitive information arrives at the destination in a timely manner.) The header is used to identify cells belonging to the same virtual channel and thus used in appropriate routing. The ATM cell structure is shown in Figure 9.1. ATM cell header fields are the following:

- Generic flow control (GFC): (4 bits) Default value 0000. The field is essentially used for controlling the flow of traffic on ATM connections from a terminal to the network.
- Virtual path identifier (VPI): (8 or 12 bits) This is used for physical layer routing, together with VCI.
- Virtual circuit identifier (VCI): (16 bits) This is used in conjunction with the VPI field to provide physical-level routing. Many cells with different VCIs may have the same VPI. Some VCI/VPI values are pre-assigned for special purposes. A VPI field of 8 to 12 bits allows 256 to 4096 virtual paths each of 64,000 VCIs. Both VPI and VCI provide information on the path that the cell will take in its transmission. VCI identifies a destination node, while VPI identifies the intermediate network path through interconnected networks.
- Payload type (PT): (3 bits) This field indicates whether the cell is user data or connection.
- Cell loss priority (CLP): (1 bit) A flag used for buffer management. This bit is set high (1) to indicate that the cell is low priority and may be discarded according to network conditions, i.e., it peforms congestion control. It is set low (0) to indicate high priority in which case the cell cannot be discarded.
- Header error control (HEC): (8 bits) This is used by the physical layer for detecting and correcting errors in the cell header. The HEC field is an 8-bit cyclic redundancy code (CRC) to check for errors in the header field.

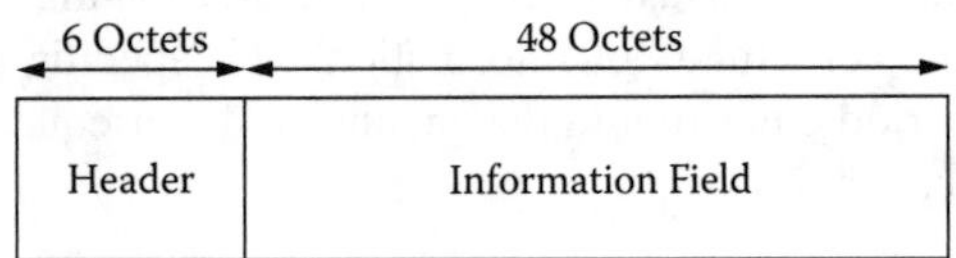

FIGURE 9.1 ATM cell structure.

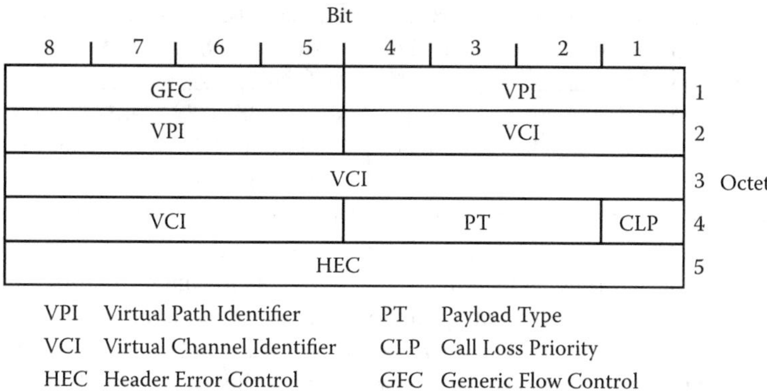

FIGURE 9.2 ATM cell header for UNI.

Notice that the cells do not contain a source or destination address because the route is prearranged and a virtual path is established before the cells are transmitted.

The cell header comes in two forms: the user-network interface (UNI) header and the network–network interface (NNI) header. The UNI is described as the point where the user enters the network. The NNI is the interface between networks. The typical header therefore looks like that shown in Figure 9.2 for the UNI. The header is slightly different for the NNI, as shown in Figure 9.3.

ATM is connection-oriented and connections are identified by the virtual channel identifier (VCI). ATM cell identifiers, i.e., virtual path identifier, virtual channel identifier, and payload type identifier (PTI), are used to recognize an ATM cell on a physical transmission medium. A virtual channel (VC) represents a given path between the user and the destination. A virtual path (VP) is created by multiple virtual channels heading to the same destination. The relationship between virtual channels and virtual paths is illustrated in Figure 9.4. A virtual channel is established at connection time and torn down at termination time. The establishment of the connections includes the

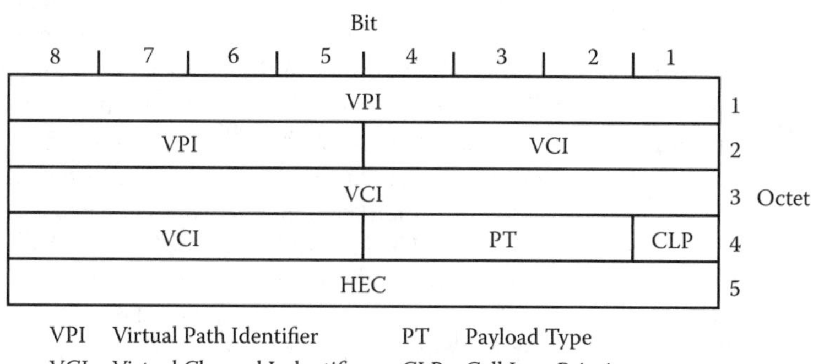

FIGURE 9.3 ATM cell header for NNI.

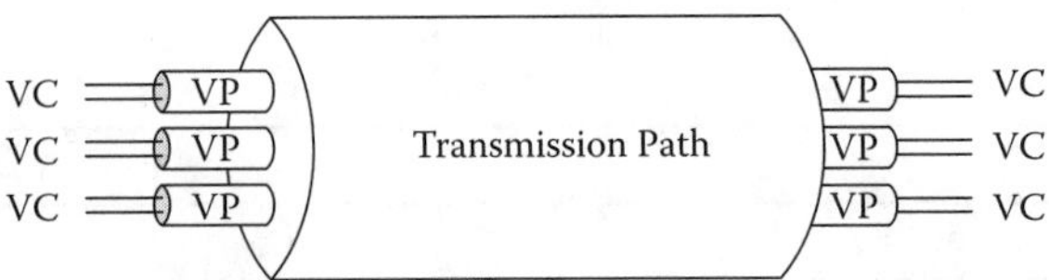

FIGURE 9.4 Relationship between virtual channel, virtual path, and transmission path.

allocation of a virtual channel identifier (VCI) and/or virtual path identifier (VPI) and also includes the allocation of the required resources on the user access and inside the network. These resources, expressed in terms of throughput and quality of service (QoS), can be negotiated between user and network either before the call setup or during the call. Having both virtual paths and channels make it easy for the switch to handle many connections with the same origin and destination. VPI and VCI are the same for cells belonging to the same virtual connection on a shared transmission medium.

Example 9.1

At a user-network interface (UNI), how many virtual paths can exist in a physical path? How many virtual channels can exist in a virtual path?

Solution:

The number of bits assigned to VPI for a UNI = 8.
The number of bits assigned to VCI for a UNI = 16.
Hence the number of virtual paths/physical paths = $2^8 = 256$.
The number of virtual channels = $2^{16} = 65{,}536$.

9.1.2 ATM Layered Architecture

The ATM is a four-layer architecture known as the ATM reference model and is shown in Figure 9.5. The ATM reference model is composed of the following planes, which span all layers:

- **Control:** This plane is responsible for generating and managing signaling requests.
- **User:** This plane is responsible for managing the transfer of data.
- **Management:** This plane contains two components:
  - Layer management manages layer-specific functions, such as the detection of failures and protocol problems.
  - Plane management manages and coordinates functions related to the complete system.

The ATM protocol reference model is composed of the following ATM layers:

- **Physical layer:** This is analogous to the physical layer of the OSI reference model. The ATM physical layer defines the electrical and/or optical characteristics and network interfaces and puts the bits on the wire.

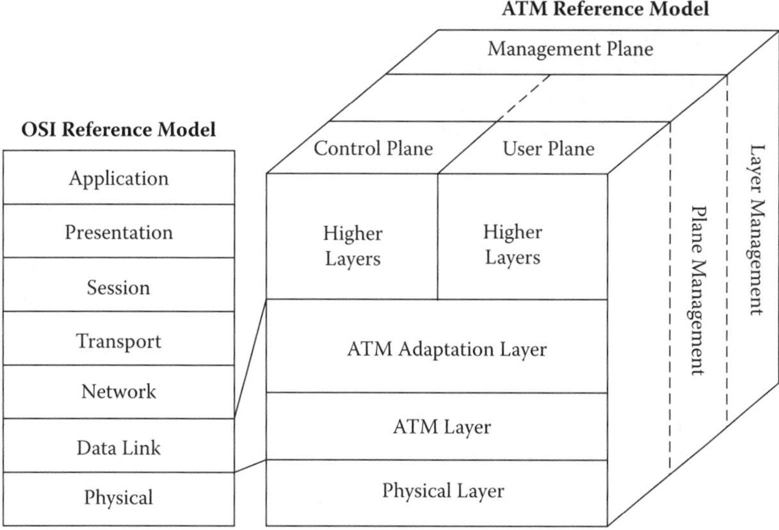

FIGURE 9.5 The ATM reference model relates to the lowest two layers of the OSI reference model.

The ATM physical layer is divided into two parts: the physical medium-dependent (PMD) sublayer and the transmission convergence (TC) sublayer. The PMD sublayer provides two key functions. First, it synchronizes transmission and reception by sending and receiving a continuous flow of bits with associated timing information. Second, it specifies the physical media for the physical medium used, including connector types and cable. The TC sublayer has four functions: cell delineation, header error control (HEC), sequence generation and verification, cell-rate decoupling, and transmission frame adaptation. The cell delineation function maintains ATM cell boundaries, allowing devices to locate cells within a stream of bits. HEC sequence generation and verification generates and checks the header error control code to ensure valid data. Cell-rate decoupling maintains synchronization and inserts or suppresses idle (unassigned) ATM cells to adapt the rate of valid ATM cells to the payload capacity of the transmission system. Transmission frame adaptation packages ATM cells into frames acceptable to the particular physical layer implementation. ATM was not designed around a specific physical medium. Currently, four physical layer specifications are approved: DS1 (T1), DS3 (T3), optical fiber using multimode transmission at 100 Mbps, and SONET OC-3.

- **ATM layer:** Combined with the ATM adaptation layer, the ATM layer is roughly analogous to the data link layer of the OSI reference model. It is at the ATM layer that cell construction and structuring occurs at the sender location. Cell reception and header validation take place at the destination end. The ATM layer determines how a cell is routed and ensures the quality of service of a circuit. It is responsible for the simultaneous sharing of virtual circuits over a physical link (cell multiplexing) and passing cells through the ATM network (cell relay). To do this, it uses the VPI and VCI information in the header of each ATM cell.

- **ATM adaptation layer (AAL):** Combined with the ATM layer, the AAL is roughly analogous to the data link layer of the OSI model. The AAL is responsible for isolating higher-layer protocols from the details of the ATM processes. It is also responsible for handling different requirements in different applications. The adaptation layer prepares user data for conversion into cells and segments the data into 48-byte cell payloads. Through the use of adaptation layers, ATM can handle both connection-oriented and connectionless traffic. The AAL is divided into two sublayers: convergence sublayer (CS) and segmentation and reassembly (SAR) sublayer. The CS divides very long packets into fixed-length packets known as service data units (SDUs). The SAR is responsible for converting the information received from the CS into ATM cells and for unpacking the information at the destination end.

9.2 ATM NETWORK

An *ATM network* consists of an *ATM switch* and *ATM endpoints*. The operation of an ATM switch is easy to understand. When a cell reaches a switch, the cell is received across a link on a known VCI or VPI value. The switch looks up the connection value in a local translation table to determine the outgoing port (or ports) of the connection and the new VPI/VCI value of the connection on that link. The switch then retransmits the cell on that outgoing link with the appropriate connection identifiers. An ATM endpoint (or end system) contains an ATM network interface adapter. Examples of ATM endpoints are workstations, routers, digital service units (DSUs), gateways, LAN switches, and video coder-decoders (CODECs). Figure 9.6 illustrates a typical ATM network made up of ATM switches and ATM endpoints.

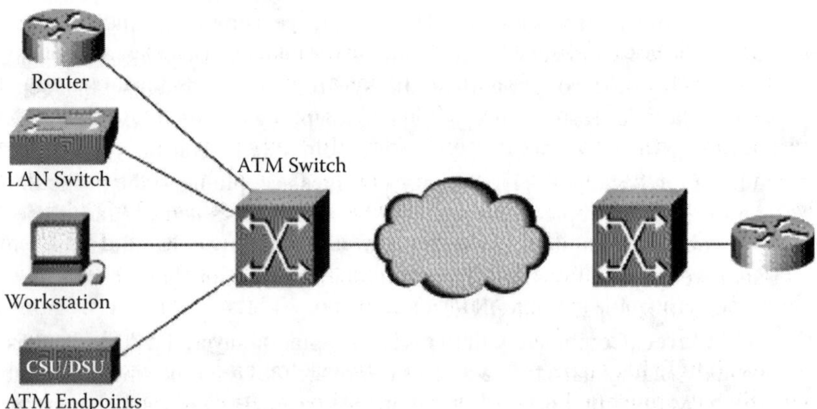

FIGURE 9.6 An ATM network consists of ATM switches and ATM endpoints.

An **ATM network** consists of a set of ATM switches and ATM endpoints interconnected by point-to-point ATM links or interfaces.

ATM networks are fundamentally connection oriented, which means that a virtual channel (VC) must be set up across the ATM network prior to any data transfer. (A virtual channel is roughly equivalent to a virtual circuit.)

Once configured, an ATM network will ensure that the end user enjoys the agreed quality of service (QoS). For ATM, the QoS of a connection is closely linked to the bandwidth it uses. In order to provide a guaranteed QoS, a traffic contract is usually established during connection setup. A traffic contract specifies one of the following classes of traffic:

- Constant bit rate (CBR): This is the simplest kind of traffic. It is used for emulating traditional circuit switching, where the bit rate is constant. For example, a connection between two PBXs will use CBR. Telephone traffic, television, video-on-demand, and videoconferencing may use this class.
- Available bit rate (ABR): This is meant for data applications that require reliable delivery of traffic but cannot specify the necessary data rate. The ABR is suited for applications that can endure longer delays. This applies to file transfer and email. ATM switches will flow control ABR connections to match the bandwidth available in the network.
- Variable bit rate (VBR): This is used for applications that have specific bandwidth requirements that vary between a minimum and a maximum. For example, a video encoder generates a VBR stream whose bandwidth depends on the images being encoded. The ATM Forum distinguishes between real-time VBR (VBR-RT) and nonreal-time VBR (VBR-NRT), thereby subdividing this class into two categories. VBR-RT is used for delay-sensitive traffic (e.g., packetized video), while VBR-NRT is used for delay-insensitive traffic. An example of VBR-RT is interactive compressed video, while that of VBR-NRT is multimedia email. Banking or investment transactions and airline reservations will use VBR-NRT.
- Unspecified bit rate (UBR): This service, specified only by the ATM Forum, is used by data applications that intend to use the ATM network as a "best-effort" network, i.e., applications that want to use any left-over capacity and are not sensitive to cell loss or delay. This service does not care if the cells are successfully delivered or not; it leaves things up to higher-level protocols such as the Internet protocol (IP) to handle. The ABR is suited for applications that can handle delay and latency. Internet access uses UBR.

These classes of ATM traffic are illustrated in Figure 9.7.

Example 9.2

Discuss the ATM traffic classes to which the following applications belong:

a. sending electronic mail (email)
b. a telephone conversation between two persons
c. downloading an image from the Internet

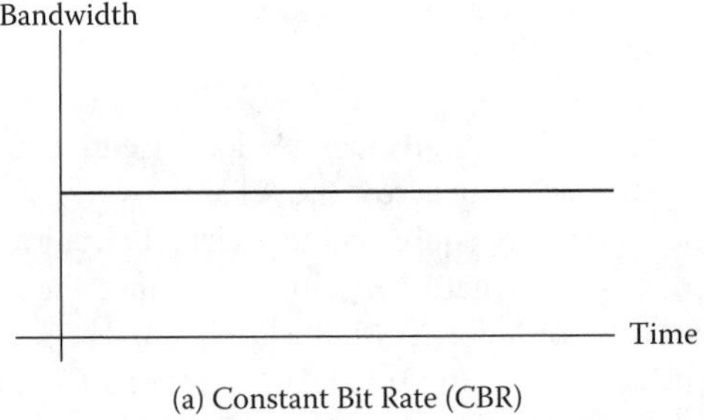

(a) Constant Bit Rate (CBR)

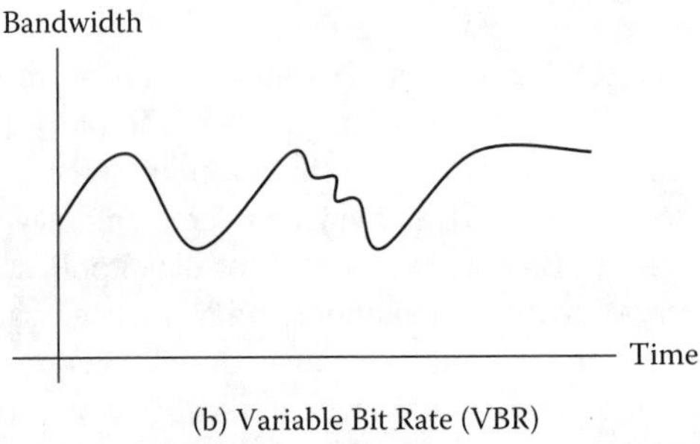

(b) Variable Bit Rate (VBR)

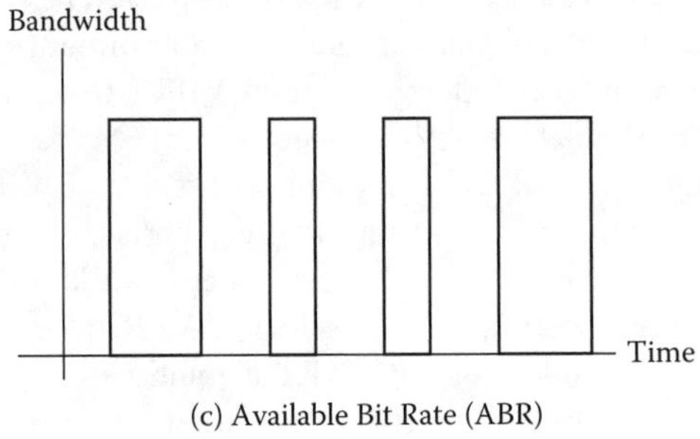

(c) Available Bit Rate (ABR)

FIGURE 9.7 Different types of ATM traffic.

Solution

a. For sending an email, either ABR or UBR can be used.
b. A telephone conversation is a real-time application. Either CBR or VBR-RT service can be used.
c. Downloading a document or image from the Internet requires a UBR service since it can accommodate variable delays and some cell loss.

9.3 ATM SERVICE CLASSES

In order to support different types of traffic, the ATM standards for AAL identified three parameters for defining service classification: (1) timing relation between source and destination (required or not required), (2) bit rate (constant or variable),

TABLE 9.1
Support Operations for AAL Classes

Class	A	B	C	D
Timing	Synchronous		Asynchronous	
Bit transfer	Constant	Variable		
Connection mode	Connection-oriented			Connectionless

and (3) connection mode (connection-oriented or connectionless). Based on these parameters, four different classes of services are specified as shown in Table 9.1 and explained as follows:

Class A is a connection-oriented and isochronous service. It maintains a constant bit rate (CBR) and timing relationship between the source and the destination. This can be used for voice or video transmission and circuit emulation.

Class B is a connection-oriented service with a variable bit rate (VBR). It maintains a timing relationship between the source and the destination. This service is intended for compressed audio and video transmission such as in teleconferencing.

Class C is a connection-oriented data service with a VBR. It maintains no timing relationship between the source and the destination. This service supports data transfer by a virtual circuit.

Class D is connectionless data service with a VBR. It maintains no timing relationship between the source and the destination. This service is intended to provide a connectionless data transfer such as LAN data.

AAL traffic are also classified into four traffic types as follows:

Type 1: In this case, we are dealing with a constant bit rate source, which allows ATM to emulate voice. By constant or fixed bit rate, we mean that the bit rate is (1) fixed and (2) synchronized between source and destination (class A service). As AAL is divided into CS and SAR, the CS is responsible for clocking and synchronization. The SAR has the task of packing the bits for transmission and unpacking them at the receiver. AAL1 requires timing synchronization between the source and the destination. For this reason, AAL1 depends on a medium, such as SONET, that supports clocking.

Type 2: The other traffic types (2, 3/4, 5) all deal with variable bit rate source. Type 2 is intended for VBR service where a timing relationship is required between source and destination (class B service). It is employed for analog applications such as audio and video.

Type 3/4: Originally, ATM standards designated type 3 for VBR connection-oriented service, while type 4 was for VBR connectionless service. Types 3 and 4 were combined because of their similarities. Type 3/4 is for VBR

TABLE 9.2
Service Classification for AAL

Higher Layers	Class A	Class B	Class C	Class D
ATM Adaptation layer	AAL1	AAL2	AAL3/4 or AAL5	
		ATM Layer		
		Physical Layer		

service without timing relationship between source and destination. It defines the procedure for providing class C and D services. It can multiplex different streams of data on the same ATM connection (VPI/VCI). It is employed in internetworking ATM with SMDS.

Type 5: Type 5 is an alternative protocol for providing class C and D service. It is less complex than type 3/4 and does not support multiplexing. It was introduced to reduce protocol processing overhead and transmission overhead, which are required in type 3/4. Type 5 has become popular in ATM LAN applications.

The relationship between the traffic types and classes is shown in Table 9.2.

9.4 ATM STANDARDS

There are three main standards bodies for ATM: International Telecommunications Union-Telecommunications (ITU-T), American National Standards Institute (ANSI), and European Telecommunications Standards Institute (ETSI). Industry forums also contribute to the standards process. There are four forums currently participating in the ATM standards process: ATM Forum, Internet Engineering Task Force (IETF), Frame Relay Forum, and Switched Multimegabit Data Service Interest Group.

For the most part, ATM is based on the efforts of the ITU-T BISDN standard. Standards already specified by the ITU-T include issues such as the definition of the BISDN protocol model and the function and sequence of each bit in the ATM cell.

The ATM Forum, which is comprised of 800 member companies worldwide, is an international organization formed in 1991, and it extended the ITU-T's vision of ATM for use over public and private networks. Through the ATM Forum, the proponents of ATM have banded together to develop technical standards and forward them to the International Telecommunications Union (ITU) to speed up the deployment of ATM.

The ATM Forum has released work on the following specifications:

- User-to-network interface (UNI) 2.0
- UNI 3.0
- UNI 3.1
- UNI 4.0
- Public-network node interface (P-NNI)
- ATM Security Specification Version 1.0

- LAN emulation (LANE) over ATM
- Multiprotocol over ATM
- Voice and telephony over ATM
- ATM trunking using AAL1 for narrowband services

The ATM Forum can be reached at

ATM Forum
2570 West El Camino Real, Suite 304
Mountain View, CA 94040-1313
Phone: (655) 949-6700
http://www.atmforum.com

9.5 ATM LAN EMULATION

As mentioned in the previous section, the ATM Forum is working on standards for ATM LAN emulation. The potential benefits of ATM technology have led to its acceptance as a local networking technology. Some believe that ATM gets its first significant usage in the LAN arena. This is based on a number of reasons. First, users need a higher-capacity LAN system. As personal computers and workstations increase in their capability, there is an attendant rise in the demand for LAN capacity. Second, data security in the LAN environment leaves much to be desired. In ATM, users cannot "listen" to what is going on and only a single end-user station receives the data intended for it. Third, there is a reduced need for routers and bridges. Fourth, ATM LAN is a multimedia LAN.

ATM LAN is one of the emerging ATM networks developed by the ATM Forum to be used as a high-speed LAN or backbone. Through a process known as LAN emulation (LANE), ATM can act as a backbone to interconnect multiple LANs. Why LANE? An enterprise migrating to ATM technology is faced with the fact that LAN and ATM technologies are different. For example, LAN is connectionless, while ATM is a point-to-point, connection-oriented protocol. ATM addresses are different from LAN addresses in more than just format. ATM is a nonbroadcast technology, whereas broadcast and multicast operations are easy on a LAN. To protect the huge investment made in traditional or legacy LANs and to avoid changing any components on the end-user station during migration, LANE is designed to provide transparency to the end-user applications and allow them to interact as though ATM did not exist. LANE is based on emulating LAN MAC service, as illustrated in Figure 9.8. It hides the complexity of call setup from the user. It makes an ATM interface appear like a LAN interface.

As LANE shields the layer 3 functions from ATM transport specifics, an emulated LAN appears to higher-layer functions as a legacy LAN. This means that an ATM network could be connected directly to traditional LANs without making any changes to application, transport, and network layers of the OSI model or to existing software applications.

In a network, LANE requires that an address resolution or translation take place before packets can be prepared for transmission over an ATM network. LANE is

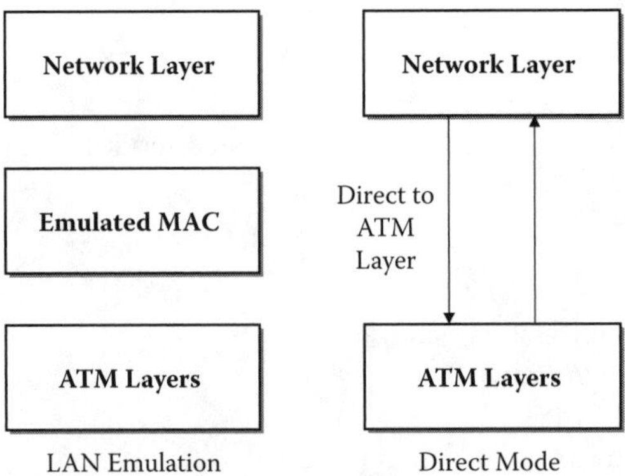

FIGURE 9.8 LANE compared with direct-ATM approach.

based on AAL5 adaptation protocol since ATM can carry multiple traffic types simultaneously at the adaptation layer.

Some have perceived LANE as the killer application that brings ATM into the mainstream. LANE is not part of ATM itself but a service developed by the ATM Forum. It is an effort in which ATM is made to perform the LAN function. In other words, LANE will enable legacy LANs to run on top of and coexist with ATM switched-based networks. LANE will emulate the characteristics and behaviors of conventional Ethernet and token ring. It must support a connectionless service and maintain the MAC address identity with each device attached to an emulated LAN.

Regardless of the specific LAN, an emulated LAN is implemented with the following four components:

- LAN emulation client (LEC) is an end system that provides a standard LAN service interface to higher-layer interfaces. It may be a workstation, bridge, or router. It performs data forwarding and control functions such as address resolution, establishment of virtual circuits, etc. It bridges LAN frames between LANs by using MAC addresses.
- LAN emulation server (LES) has the primary duty of providing directory and address resolution services to the LECs of the emulated LAN. It is used to initialize an LEC with information to the emulated LAN that the LEC is assigned. An LES implements the control function for an emulated LAN. A unique ATM address is used to identify the LES.
- Broadcast and unknown server (BUS) acts as a multicast server that floods unknown-destination address traffic and forwards broadcast and multicast traffic to clients within the emulated LAN. The BUS operates in a store-and-forward manner. Each LEC is assigned a BUS.
- LAN emulation configuration server (LECS) assigns LECs to different emulated LANs by providing each LEC with the ATM address of the LES for the emulated LAN assigned to it. Using an LECS to assign clients to different virtual LANs allows for central configuration and administration of many virtual LANs in an ATM network.

An emulated LAN consists of a LEC and the LANE services (LECS, LES, and BUS).

ATM in the local area has many forms. Shared-media ATM networks include DQDB, the Cambridge Fast Ring, the Cambridge Backbone Ring, and the Orwell Ring. For ATM LAN infrastructures, the ATM Forum has defined interfaces with transfer speeds of 52 Mbps and 155 Mbps for unshielded (UTP type 3) and shield copper twisted-pair cables, respectively. Since 1993, ATM LAN products and ATM LANs have been introduced. Vendors have developed ATM interfaces for workstations, terminals, bridges, and routers.

There are two important considerations that make ATM LAN different from the public BISDN. First, cost is the key consideration in the local network. Economics and deployment in the local and public networks are vastly different, even from timescale consideration alone. Overprovision of bandwidth may be the key to reducing the cost of local area ATM. Second, reliability in the local network is turning out to be an increasingly important consideration. General topology networks allow a continuum of tradeoffs between cost and fault tolerance.

The consideration of local wireless ATM networks highlights the need to consider and specify ATM service in a manner that is not based on a particular implementation.

9.6 ATM APPLICATIONS

ATM can be used in existing twisted pair, fiber optic, coaxial, and hybrid fiber/coax (HFC), SONET/SDH, T1, E1, T3, E3, E4, etc. for LAN and WAN communications. ATM is also compatible with wireless and satellite communications.

The ATM transport technique is ideal for connectivity networks because all kinds of service network can be built on top of them. ATM makes efficient use of common resources, functioning as both a transport network and as a connectivity network that supports on-demand switched connections. Commercial ATM deployment has been in two areas. First, in the customer premises ATM is cost effective for applications requiring higher dedicated bandwidth than can be provided by shared media alternatives such as FDDI. These applications are essentially data only. Second, ATM deployment is in the public network, both as a backbone technology and in the access network. In the network backbone, ATM is used to transport frame relay and SMDS traffic. In the access network, ATM is used to multiplex a combination of traffic onto a single access line, including CBR circuits at voice speeds, VBR data applications, and multimegabit video traffic.

ATM technologies and services are being applied in a variety of ways. These include

- Internet backbone: Internet service providers are deploying ATM backbones to meet their rapid growth. For the Internet, an ATM-based connectivity network provides an effective and flexible platform for interconnecting traditional IP routers. At the same time, ATM offers the foundation that the Internet needs to provide quality of service. ATM networks and the Internet are not two different types of network that compete with each other; rather, they complement one another.

- Interworking: There are two types of interworking for ATM networks. First, we have network interworking, which is the achievement of connectivity between other networks (e.g., X.25, ISDN, PSTN) and ATM. To achieve network interworking requires physical layer mapping and connection control signaling compatibility at the network interface. Second, we have service interworking, which builds on network interworking and requires interworking of services/applications between other networks and services provided by ATM networks. Typical examples of service interworking include voice over ATM, video over ATM, and IP over ATM.
- Frame relay-ATM interworking: Today's most predominant WAN solution is frame relay, although the backbone structure is rapidly migrating to ATM. Frame relay service providers are deploying ATM backbones to meet the rapid growth of their frame relay services and to enable frame relay-to-ATM interworking services.
- Voice over ATM: This refers to the transport of voice and voice-band data over ATM. It includes CBR using pulse code modulation (PCM) and VBR using adaptive differential PCM with silence detection. VBR techniques for voice exploit the inherently bursty nature of voice communications. Key application areas requiring voice over ATM include ATM to the desktop, distributed PBX, broadband computer telephony integration, cable company telephone service, cellular company access network, long-distance terrestrial transport, and ATM over satellite.
- Video over ATM: VBR video, packetized and transmitted over ATM, has a number of advantages over CBR video. However, packetized VBR video is likely to be one of the largest users of bandwidth among the various services offered by ATM. Packet loss is inevitable in the real-time transmission of VBR video due to stringent delay constraint. While circuit-based videoconferencing streams (such as JPEG running at 10 Mbps) can be handled by standard circuit emulation using AAL1, the ATM Forum has recommended the use of VBR using AAL5 for MPEF2 on ATM for video-on-demand, as this approach makes efficient use of resources.
- Multimedia services: The extensive class-of-service capabilities of ATM make it the technology of choice for multimedia communications. Service providers are building on their ATM networks to offer a broad range of services. Examples include LANE, voice over ATM, video over ATM, and full-service virtual private networking capabilities.
- Broadband services: These include LAN interconnection, video conferencing, multimedia database access, and high-speed multimedia document transfer. LAN interconnection is characterized by very bursty traffic. Videoconferencing requires a number of multicast (point-to-multipoint) broadband connections. Database access can also involve multicast connections with bursty traffic flows. High speed, multimedia document transfer exhibits a combination of the burstiness and variable bandwidth characteristics of the other applications.
- Broadband satellite network: Satellite can inherently provide wide geographic coverage to remote areas not connected to the terrestrial fabric.

Even in areas of existing terrestrial connectivity, satellite systems can serve as an alternative to fiber optic networks. A satellite network must support a wide variety of applications including data, voice, and video, both for unicast and multicast scenarios. ATM via satellite has been successfully demonstrated.

Although wired ATM is becoming the technology of choice for broadband service integration, the area of wireless ATM is an emerging technology where there are still many issues to be resolved. As the networking industry continues to develop, and new technologies and services are supplied to the market, ATM continues to stand out as the network backbone technology of choice.

9.7 TRAFFIC MANAGEMENT

An ATM network needs traffic management capabilities to cope with the various service classes and to cope with potential errors within the network at any time. Traffic management in ATM networks plays an important role in providing the capability to support a wide range of integrated services while utilizing network resources, such as buffer space and bandwidth, in an efficient manner.

Traffic management deals with the delivery of quality of service (QoS) to the end user and with the efficient use of network resources. The objectives of ATM traffic management are to deliver QoS guarantees for the multimedia applications and provide overall optimization of network resources. Meeting these objectives enables enhanced classes of service and offers the potential for service differentiation and increased revenues, while simplifying network operations and reducing network cost.

The ATM network defines QoS parameters based on speed, accuracy, and dependability required by cells being transmitted. When establishing a connection, an ATM network may use QoS information to determine if the network has sufficient resources to support the required connection. Since there are different types of information, there are different qualities of service parameters. QoS is defined on an end-to-end basis and in terms of the following parameters:

- Cell loss ratio (CLR), which is the ratio of lost or unsuccessfully transmitted cells within a period to the total number of cells transmitted during that period. It is required for CBR and VBR services. For a connection, CLR is defined as

$$\text{CLR} = \frac{\text{Lost cells}}{\text{Total transmitted cells}} \tag{9.1}$$

- Cell error ratio (CER), which is the ratio of cells with errors to the total number transmitted. For a connection,

$$\text{CER} = \frac{\text{Errored cells}}{\text{Successfully transmitted cells}} + \text{Errored cells} \tag{9.2}$$

An errored cell arrives at the destination with errors in the payload.

- Cell transfer delay (CTD), which is the arithmetic mean of the time taken for a cell to move from its source to its destination.
- Cell delay variation (CDV), which is the difference between the expected transfer delay and the actual transfer delay of a cell. It describes the extent of the variation in the time elapsing between the arrival of two cells on a given connection. This is primarily intended to determine the amount of jitter. It is required for CBR and real-time VBR services.
- Cell misinsertion rate (CMR), which is the number of misinserted cells (with invalid VPI/VCI due to header errors) transferred within a time period, divided by the time. It is the ratio of the cells that arrive from the wrong source to the total number of properly received cells per virtual connection. It can be defined as

$$\text{CMR} = \frac{\text{Misinserted cells}}{\text{Time interval}} \tag{9.3}$$

An ATM network must negotiate for these parameters depending on the types of traffic. The QoS is negotiated by the network before a connection is made. Once configured, the network will ensure that the end user enjoys the agreed QoS. For ATM, the QoS of a connection is closely linked to the bandwidth it uses. For example, using more bandwidth increases cell loss, delay, and delay variation and consequently decreases the QoS.

9.8 ADVANTAGES AND DISADVANTAGES

ATM advantages include

- Is a standards-based technology and this makes interoperability of ATM equipment possible.
- Supports voice, video, and data allowing multimedia and mixed services over a single network.
- Works with existing, legacy technologies. For example, it provides for legacy LAN emulation.
- Provides QoS classes and scalability.
- Supports both connection-oriented and connectionless traffic using AALs (ATM adaptation layers).
- Ability to use all common physical transmission paths (DS1, SONET/SDH, twisted-pair, coaxial or fiber optic, etc.).
- Efficient bandwidth use by statistical multiplexing.
- High data rate—speed in Mbps and possibly in Gbps.
- High flexibility—ATM operates independently of the type of transmission being generated at the upper layers and of the type and speed of the physical-layer medium below it.

ATM disadvantages include

- Flexible at efficiency's expense. At present, for any one application it is usually possible to find a more optimized technology.
- Cost, although it will decrease with time.
- New customer premises hardware and software are required.
- Competition from other technologies—100 Mbps FDDI, 100 Mbps Ethernet, and fast Ethernet.
- Presently the applications that can benefit from ATM such as multimedia are rare.

SUMMARY

1. Asynchronous transfer mode (ATM) is a high-speed transmission technology for voice, data, and video that combines the best of circuit switching and packet switching.
2. An ATM network comprises ATM switches and ATM endpoints.
3. ATM defines four categories (AAL 1, AAL 2, AAL 3/4, and AAL 5) for the ATM adaptation layer (AAL) to support the quality of service categories.
4. ATM is suitable to applications that cannot tolerate much time delay as well as for transporting LAN, frame relay, and IP traffic that are bursty.
5. ATM can be used in a LAN environment and has a LAN emulation (LANE) mode to cope with this. LANE software includes LAN emulation client server (LECS), LAN emulation configuration server (LECS), LAN emulation server (LES), and broadcast/unknown server (BUS) units.
6. The two major standards-making bodies for ATM are ITU-T and the ATM Forum.
7. ATM has some benefits and drawbacks.

REVIEW QUESTIONS

9.1 In ATM networks, all information is organized into cells of how many bytes?
   a. 40
   b. 53
   c. 64
   d. Variable

9.2 What is the maximum number of virtual paths an ATM network can have?
   a. 256
   b. 64,000
   c. 65,536
   d. Unlimited

9.3 Which field of an ATM cell header indicates whether the cell contains user data or layer management information?
a. GFC
b. VCI/VPI
c. PT
d. CLP
e. HEC

9.4 How many bits does the cell loss priority field of an ATM cell header have?
a. 1
b. 4
c. 8
d. 12
e. 16

9.5 This field is only used in the UNI:
a. Payload type
b. Virtual path identifier
c. Generic flow control
d. Virtual channel identifier
e. Cell loss priority

9.6 ATM cells do not contain a full address since the route is prearranged and a virtual path is established before transmission.
a. True
b. False

9.7 The basic connection unit in an ATM network is known as:
a. Virtual path connection
b. Virtual channel connection
c. Virtual path identifier
d. Virtual channel identifier

9.8 Which traffic type is used by Internet access?
a. Constant bit rate (CBR)
b. Variable bit rate (VBR)
c. Available bit rate (ABR)
d. Unspecified bit rate (UBR)

9.9 Which industry forum is responsible for ATM LAN emulation?
a. ATM Forum
b. Internet Engineering Task Force IETF
c. Frame Relay Forum
d. Switched Multimegabit Data Service Interest Group

9.10 This is not a benefit of ATM:
a. High performance via hardware switching
b. Class-of-service support for multimedia
c. Common LAN/WAN architecture
d. Flexibility at the expense of efficiency
e. International standards compliance

Answer: 9.1b, 9.2a, 9.3c, 9.4a, 9.5c, 9.6a, 9.7b, 9.8d, 9.9a, 9.10d

PROBLEMS

9.1 Discuss the ATM concepts of
   a. a virtual channel
   b. a virtual path

9.2 Describe an ATM cell. What is the percentage overhead (i.e., the percentage of the cell that is not user data)?

9.3 What is the function of the generic flow control?

9.4 Mention some of the advantages of using short fixed cells.

9.5 What is the purpose of the cell loss priority?

9.6 What are the four layers of the ATM architecture?

9.7 Explain the role of the "payload type."

9.8 What does a VPI identify?

9.9 Explain the difference between the header types for UNI and NNI.

9.10 Why is an ATM cell 53 bytes?

9.11 What is the purpose of an adaptation layer in an ATM protocol?

9.12 What is the function of the ATM layer?

9.13 Explain the concepts of
   a. Constant bit rate (CBR)
   b. Variable bit rate (VBR)
   c. Available bit rate (ABR)
   d. Unspecified bit rate (UBR)

9.14 Describe an ATM class of service.

9.15 In an ATM network, 6 out of 10,000 cells are lost during transmission, and 3 out of 10,000 cells contain errors. Calculate the cell loss ratio (CLR) and cell error ratio (CER).

9.16 How is an ATM LAN different from a legacy LAN?

9.17 (a) Why LAN emulation? (b) Discuss the entities an emulated LAN consists of.

9.18 What are the advantages and disadvantages of ATM?

9.19 Compare the bandwidth management capabilities provided by ATM virtual paths to the capabilities provided by SONET networks.

9.20 The probability P(x) that x errors occur in the ATM header (of length L = 40 bits) is given by

$$P(x) = \frac{L!}{x!(L-x)!}(1-p)^{L-x}p^x$$

where the error rate $p = 10^{-6}$. Determine the probability when the header has no errors, one error, and two errors.

Part 3

Access Networks

10 Digital Subscriber Line Systems

Education is a companion, which no misfortune can depress, no crime can destroy, no enemy can alienate, and no despotism can enslave...

—Joseph Addison

Digital subscriber line (DSL) service provides high-bit-rate digital service over ordinary phone lines, allowing from 100 kbps to tens of megabits per second to reach a telephone company customer. It is the phone company's alternative for broadband access. There is great interest in applying advanced new techniques to permit the use of existing copper wiring for access to high-rate digital services. Recent achievements include transmission of up to several Mbps over telephone subscriber lines, over 100 Mbps over short indoor wiring, and the use of coaxial cable television facilities for efficient two-way digital communications.

The objective of this chapter is to provide an overall discussion that will give an in-depth treatment of the basic principles, design, and achievable performance of DSL systems for multi-Mbps communication over the subscriber line, and for advanced LAN transmission. We will cover some of the varieties of DSL, which we call xDSL, where x stands for the different types of the DSL, such as asymmetric DSL (ADSL), high-bit data rate DSL (HDSL), single high-speed DSL (SHDSL), rate adaptive DSL (RADSL), consumer DSL (CDSL), single-line DSL (SDSL), ISDN DSL (IDSL), and very high data rate DSL (VDSL). Cable modem and its comparison with DSL will also be discussed for completeness.

10.1 DSL SYSTEM TECHNOLOGY

The evolution of digital subscriber line (DSL) system technologies is one of the greatest innovations of the past century in telecommunications. It has revolutionary opportunities in the wide area network (WAN), as well as the emerging residential broadband access. These xDSL technologies have the potential of enabling mega-fast access at speeds 300 times faster than what is presently achieved with most of the standard analog modems. Because xDSL technologies work on regular telephone (twisted-pair copper) lines, the DSL system can be considered as a key means of opening the bottleneck in the "last mile" of the existing telephone infrastructure. These xDSL technologies, therefore, have enormous telecommunications application potentials in areas such as video on demand, Internet, telecommute, electronic mail, broadcast feeds, and education.

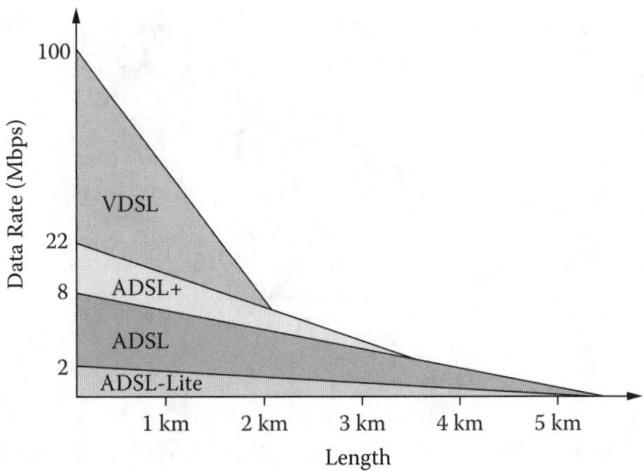

FIGURE 10.1 Data rate of different types of xDSL.

The DSL system technologies evolved from low-speed baseband dial-up modem to advanced discrete multitone (DMT) techniques such as ADSL and VDSL. Several types of the data rates for digital transmission are shown in Figure 10.1 for comparison as well as in Table 10.1. From ADSL-Lite to VDSL, the data rate increases while the service length decreases.

The most widely deployed asymmetric digital subscriber line (ADSL) service today can carry up to 8 Mbps downstream from a telephone company central office (CO) to a customer and up to 1.5 Mbps back upstream. Very high speed DSL (VDSL), the latest of the DSLs, can carry up to 60 Mbps on a single phone line. VDSL presumes some use of fiber to shorten phone-line lengths, consistent with eventual migration to fiber by phone companies.

DSL is touted as the next big push in Internet access technology. DSL is a generic term that describes a host of access technologies. It has x number of different types of DSLs that are classified based on their different lengths and speeds. For this reason, DSL is normally referred to as xDSL. Since DSL is not a "one-size-fits-all" technology, there are many "sizes" that are not available in certain areas. Table 10.1 shows the different types of DSL, their data rates, both downstream and upstream, distance, and applications. The two primary types of DSL are asymmetric DSL (ADSL) and symmetric DSL (SDSL). Of these two, ADSL is the most popular since it offers greater download speeds. A typical xDSL system can be applied to the broadband access network as shown in Figure 10.2. The analog front end (AFE) works with several xDSLs at the front end of the xDSL system.

ADSL technology is accomplished over a regular copper phone line using a special type of modem—a DSL modem. These devices can boost the typical, sluggish 56K dial-up connection to bit rates that exceed 1.54 Mb/s, which is T1 speed. In fact, DSL provides bandwidth that can be as much as four times faster than a T1 connection. To accomplish this feat, ADSL uses a frequency division multiplexing (FDM) scheme, by separating the upstream and downstream data flows as shown in Figure 10.3 and using discrete multitone (DMT) modulation for transmission as shown in Figure 10.4.

TABLE 10.1
Different Types of DSL Systems

DSL Type	Meaning	Data Rate Downstream Speed	Data Rate Upstream Speed	Distance	Applications
ADSL	Asymmetric DSL	1.5 to 8 Mbps	16 to 768 Kbps	Up to 4 km	Interactive multimedia, Internet access, remote office LAN residential and SOHO applications, video-on-demand
RADSL	Rate adaptive DSL	1.5 to 8 Mbps	16 to 768 Kbps	Up to 6 km	Interactive multimedia, Internet access, remote office LAN residential and SOHO applications, video-on-demand
HDSL	High data rate DSL	T1: 1.544 MbpsE1: 2.048 Mbps	T1: 1.544 MbpsE1: 2.048 Mbps	Up to 5 km; Up to 12 km with repeaters	Telco transport applications, cellular base stations connectivity, T1/E1 leased lines
HDSL2	High data rate DSL	T1: 1.544 MbpsE1: 2.048 Mbps	T1: 1.544 MbpsE1: 2.028 Mbps	Carrier Serving Area	Same as HDSL, and remote office LAN, Internet access, high-quality videoconferencing, residential and SOHO applications
CDSL	Consumer DSL	Up to 1 Mbps	16 to 128 Kbps	Around 4 km	Interactive multimedia, Internet access, remote office LAN residential and SOHO applications, video-on-demand
SDSL	Single line DSL	768 Kbps	768 Kbps	4 km	High-quality video conferencing, Internet access, residential and SOHO applications, remote office
SHDSL (G.SHDSL)	Single high-speed DSL	192 Kbps to 2.360 Mbps or 384 Kbps to 4.720 Mbps	192 Kbps to 2.360 Mbps or 384 Kbps to 4.720 Mbps	4 km	Remote LAN access, Web-hosting, application sharing, videoconferencing, etc.
IDSL	ISDN DSL	144 Kbps	144 Kbps	Up to 4 km	Internet, intranet applications, etc.
VDSL	Very high data rate DSL	13 to 52 Mbps	1.5 to 6.0 Mbps	Up to 1.5 km	Full service access network

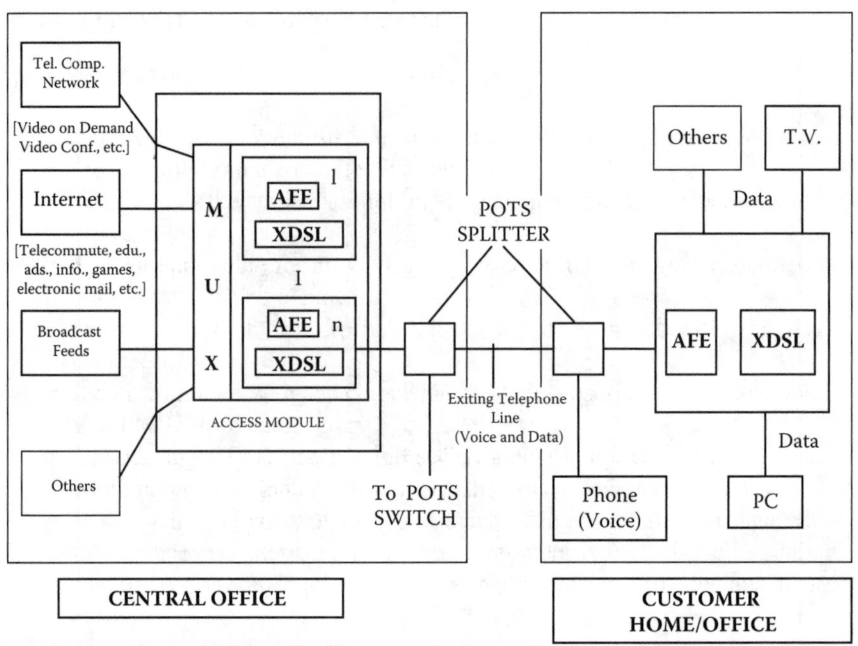

FIGURE 10.2 Basic XDSL system application block diagram.

One of the highest factors for the demand for ADSL is the simplicity in the implementation of the technology. The International Telecommunications Union (ITU) develops standards for the physical specifications of telecommunications technologies. Full ADSL (ITU G.922.1) requires a plain old telephone system (POTS) splitter, and ADSL lite (ANSI T1.419, ITU G.992.2), or G.lite, does not (see

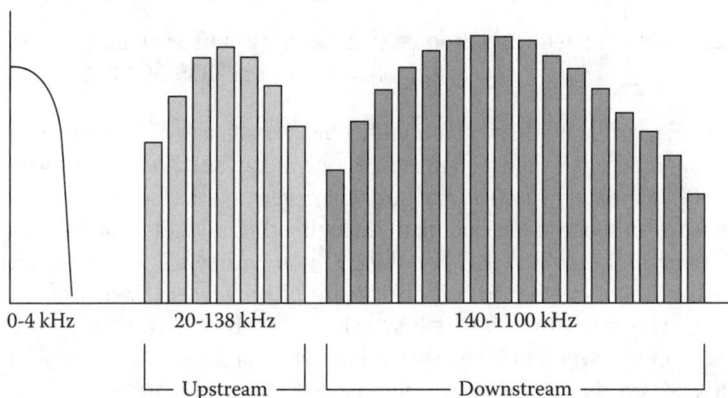

FIGURE 10.3 Upstream and downstream bandwidths for ADSL.

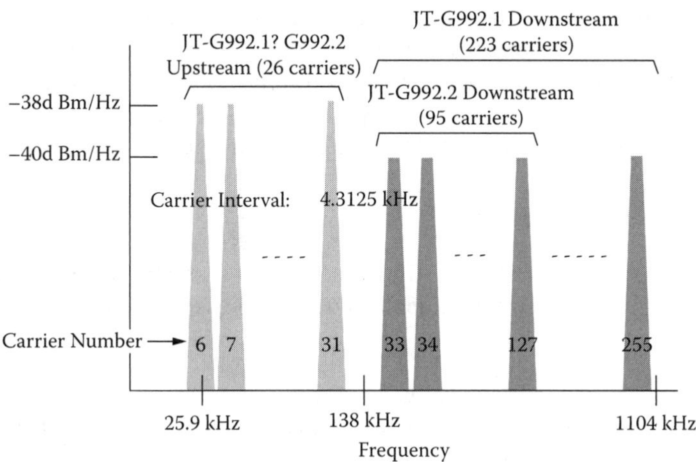

FIGURE 10.4 DMT modulation using QAM subchannel frequency separation for ADSL (and ADSL lite).

Figure 10.5). ADSL lite or "splitter-less" ADSL does have lower data rates, however. For ADSL lite, the maximum downstream data rate is up to 1.5 Mb/s and the maximum upstream data rate is up to 512 kb/s. At the customer's home, the POTS splitter and the DSL modem are known as the ADSL transmission unit-remote (ATU-R) or customer premises equipment (CPE). Most of the implementation of ADSL in the United States has been the G.lite (or splitter-less ADSL) variety.

Telephone companies (telcos) that provide ADSL service can do so for residences that are no more than 12,000 ft (approximately 2 miles) from their central office (CO). The telco's network DSL access multiplexer (DSLAM) at the CO handles the administration of the upstream and downstream traffic for the ADSL link. ADSL is available to many people because of this distance constraint, but not all. The demand for ADSL stems from the existence of an extensive phone system, the fast download speeds, which customers use most often, and the needs of business commuters. In addition, customers like the easy setup without needing to purchase and install a lot of equipment—only a DSL modem. Furthermore, the nature of ADSL makes it ideal for Internet/intranet surfing, video-on-demand and other multimedia services, and remote LAN access because ADSL users generally download more than they send. Therefore, industry pundits continue to predict a very bright future for this access technology.

The system model of ADSL network is shown in Figure 10.6. The aggregation of ADSL transceiver unit-central office (ATU-C) is implemented as the DSLAM resides at the telephone company side of the twisted pair. A splitter circuit may precede the DSLAM termination so that analog POTS signals can be passively separated from the DSL signals and conveyed to the voice telephone switches.

ADSL depends upon advanced digital signal processing and creative algorithms to squeeze so much information through twisted-pair telephone lines. ADSL modems divide the available bandwidth of a telephone line by the use of frequency

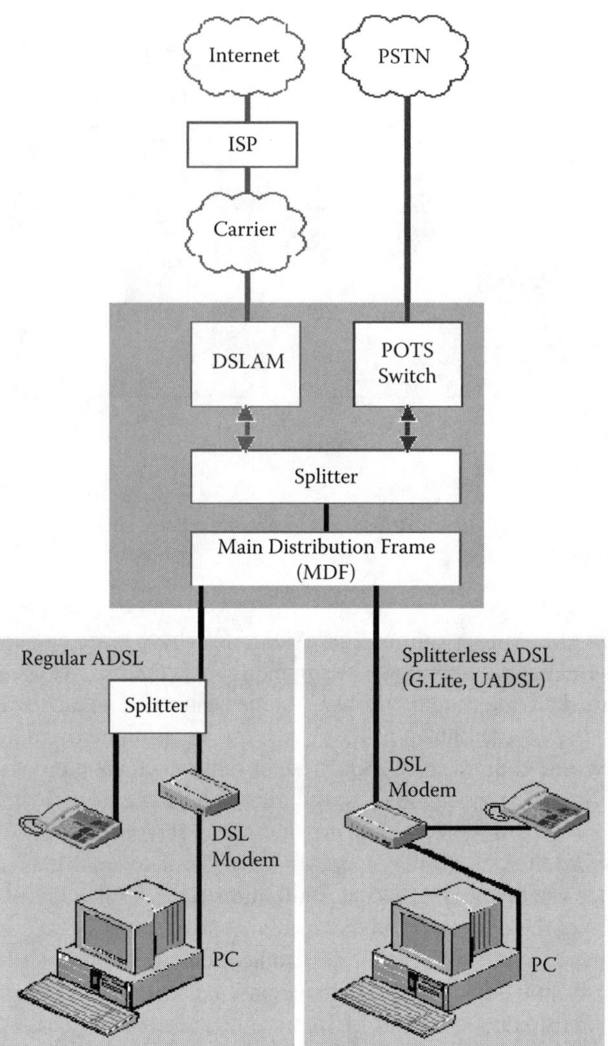

FIGURE 10.5 ADSL and ADSL lite models and connectivity to telco and the Internet.

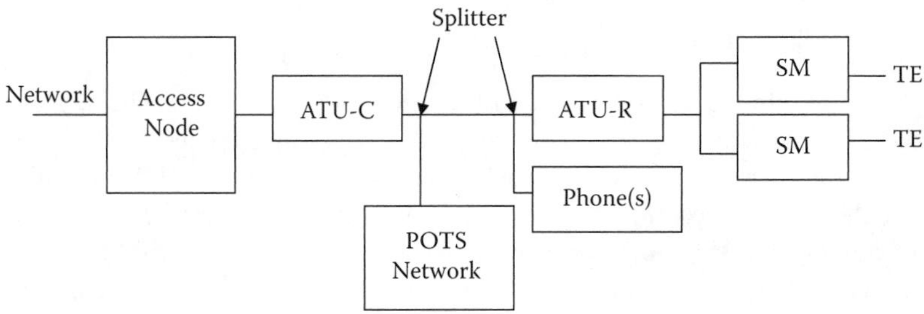

FIGURE 10.6 ADSL system model.

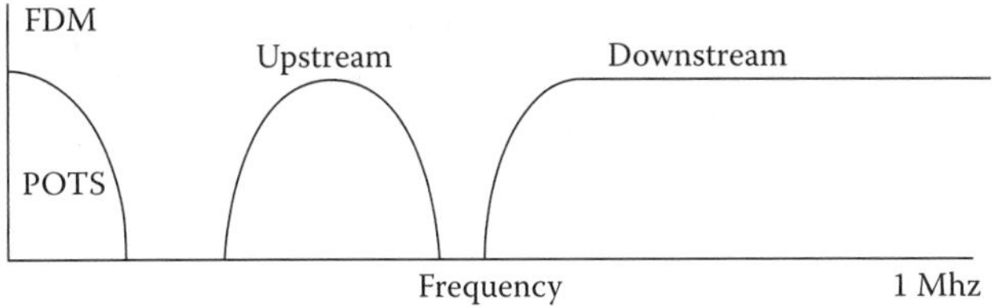

FIGURE 10.7 Frequency division diagram of ADSL.

division multiplexing (FDM). The low-band frequencies are used for upstream data and higher band for downstream data, shown in Figure 10.3 and Figure 10.7.

10.2 DMT LINE CODE FOR ADSL

DMT (discrete multitone) modulation is the standardized modulation technique for ADSL by the American National Standards Institute (ANSI). DMT ADSL (referred to as the ANSI T1.413 standard) utilizes quadrature amplitude modulation (QAM) to separate the upstream and downstream bandwidths into subchannels. The downstream bandwidth is separated into 256 subchannels that are 4.3125 kHz wide. In addition, the upstream bandwidth is separated into 32 4.3125 kHz-wide subchannels. For the downstream, the first subcarrier is at 4.1325 kHz, and the 256th subcarrier is at 1.104 MHz. Likewise, for upstream, the first subcarrier is at 4.3125 kHz and the 32nd subcarrier is at 138 kHz. Additionally, the standard limits the amount of data loading for the QAM subchannels to 15 bits (or 32,768 points) at a symbol rate of 4 kBaud (250 μs symbol period) (see Figure 11.4). Employing this FDM scheme, ADSL is capable of providing data rates for upstream from 64 kb/s to 1.5 Mb/s and data rates of 1.5 Mb/s to 9 Mb/s for downstream. This is one of the reasons for the increased demand for ADSL.

DMT is a transmission technique based upon the idea of frequency-division multiplexing (FDM). In FDM, the simultaneous transmission of multiple data signals in one frequency band is carried out on different carrier frequencies. Modulation is the process of varying a carrier signal in order to use that signal to convey information, and demodulation is the inverse operation of modulation. The processes are implemented via the discrete Fourier transform (DFT), using the fast Fourier transform (FFT). A fast Fourier transform is an efficient algorithm to compute the discrete Fourier transform and its inverse.

Let $x_0, \ldots, x_{N-1}$ be complex numbers. The DFT is defined by the formula shown in Equation 10.1.

$$X_k = \sum_{n=0}^{N-1} \chi_n e^{-\frac{2\pi i}{N}nk} \quad k = 0, \ldots, N-1. \tag{10.1}$$

In DMT, a single transmitter transmits on many different orthogonal frequencies closely spaced; each one only has room for a narrowband signal. This modulation technique, coupled with the use of advanced modulation techniques on each component, results in a signal with high resistance to interference. The fact that DMT does not interfere easily with other signals is the main reason it is frequently used in many communications systems.

10.3 DMT SCHEME

The DMT scheme makes use of variations of the QAM/CAP coding methods. However, the primary technical difference, which makes DMT unique, is the splitting of the frequency spectrum into equally spaced subchannels. Splitting the subchannel does not add any information handling capability. What it does is allow for interference and "automatic" rate adaptation by eliminating subchannels from a data path (upstream or downstream). Like other common modulation techniques used in our everyday communications, DMT has two major entities, which is to transmit and receive data as shown in Figure 10.8.

In the transmitter, a number of carriers, which hold vectors containing digital signals, are input into an IDFT (which transforms an N-point complex vector to another N-point complex vector). The data that come out of the IDFT are treated as upstream data and each carrier is delayed sequentially allowing each carrier to be converted from digital to analog one at a time. The analog signals are then sent through a low pass filter, clearing the distortion. Once they are filtered, they are sent to the channel. In the receiver the signals are sent through another low pass filter, sampled and held/delayed, and then converted to digital from analog. The digital signals are then treated as downstream data where they will be put in vector form and then input to the DFT to be transformed to the correct data.

Based on measured signal-to-noise ratio (SNR) of each subcarrier, the transmission throughput of a DMT ADSL system can be estimated. Using this SNR, we can

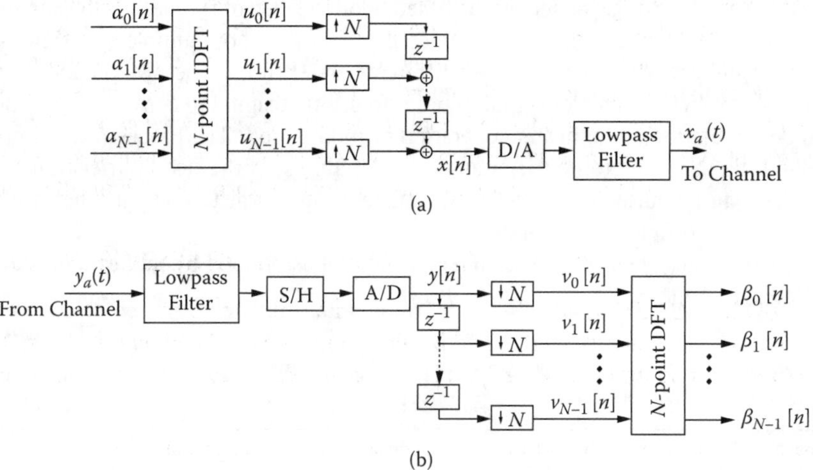

FIGURE 10.8 Block diagram of the DMT scheme. (a) transmitter, (b) receiver.

calculate the total transmission throughput, R. The product of the prefix-modified baud rate and the number of bits each baud carries is known as R. This can be represented as shown in Equation 10.2.

$$R = \frac{1}{T}\sum_{i=1}^{256}\log_2\left(1+\frac{SNR_j}{\Gamma}\right) \tag{10.2}$$

where $\Gamma = 9.55$, the value corresponds to an error rate of 10^{-7} without any coding gain, SNR_j is the signal-to-noise-power ratio of the subcarrier, and $\frac{1}{T}$ is the modified baud rate.

Note: We calculate SNR_j according to loop attenuation, noise power density, and transmit power density.

10.4 THE BENEFITS AND DISADVANTAGES OF USING DMT

Some of the benefits of using DMT are

- high spectrum efficiency
- resistance against multipath interference
- ease of filtering out noise
- duration of each symbol is relatively long

Some of the disadvantages of using DMT are

- time-variations in the channel or presence of a carrier frequency offset
- circuitry must be linear

Note: Even though there are a few disadvantages in using DMT, the advantages of using DMT far outweigh the disadvantages.

10.5 OFDM ENCODER

Figure 10.9 shows the diagram of an OFDM encoder. The source S generates a flow of binary symbols. These binary symbols are split up into several channels, thus creating a short sequence of binary symbols. In Figure 10.9, M_i represents the binary symbols. They can have different length generally. Then each sequence is represented by a complex number a_i according to any kind of modulation [QAM, phase shift keying (PSK), etc.]. The spectrum of the signal transmitted is demonstrated as the sequence of a_i. It means that all the complex values are sent to a block that calculates the inverse Fourier transform using the fast Fourier transform (FFT) algorithm. Some zeros can be added at the beginning and at the end of the sequence. Then a header H is appended to the code at the output of the FFT.

The obtained sequence will be, in general, complex-valued. The real and imaginary parts are sent separately through the channel by QAM modulation. Before modulation, they are alternatively multiplied by (–1) in order to have a null mean value.

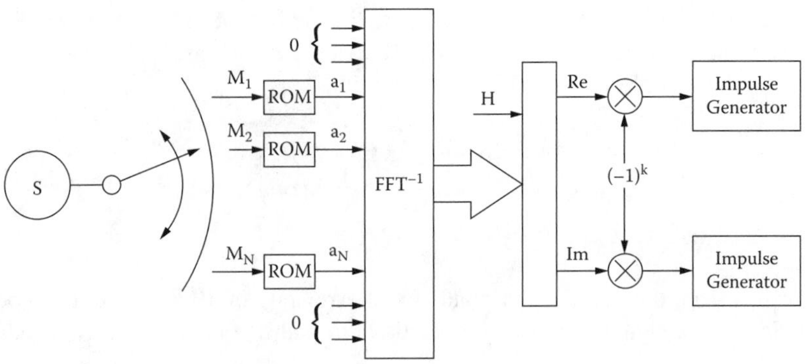

FIGURE 10.9 Diagram of an OFDM encoder.

In general, there can be a different modulating technique on each virtual channel. Since different symbols are sent on different samples of the spectrum of the signal to be sent, if the channel has a linear behavior, there cannot be any interference between the different frequencies, so any possibility of intersymbol interference is removed.

The low-pass equivalent of the OFDM signal $v(t)$ is expressed as shown in Equation 10.3:

$$v(t) = \sum_{k=0}^{N-1} I_k e^{i2\pi kt/T}, \quad 0 \le t < T, \tag{10.3}$$

where $\{I_k\}$ are the data symbols, N is the number of subcarriers, and T is the OFDM block time. The subcarrier spacing of $\frac{1}{T}$ Hz makes the subcarriers orthogonal; this property is expressed as shown in Equation 10.4.

$$\frac{1}{T}\int_0^T (e^{i2\pi k_1 t/T})^* (e^{i2\pi k_2 t/T})dt = \frac{1}{T}\int_0^T e^{i2\pi(k_2-k_1)t/T}dt = \begin{cases} 1, & k_1 = k_{2,} \\ 0, & k_1 \neq k_{2,} \end{cases} \tag{10.4}$$

where (•)* denotes the complex conjugate operator.

To avoid intersymbol interference in multipath fading channels, a guard interval $-T_g \le t < 0$, where T_g is the guard period, is inserted prior to the OFDM block. During this interval, a cyclic prefix is transmitted. The cyclic prefix is equal to the last T_g of the OFDM block. The OFDM signal $v(t)$ with cyclic prefix is thus shown in Equation 10.5.

$$v(t) = \sum_{k=0}^{N-1} I_k e^{i2\pi kt/T}, \quad -T_g \le t < T. \tag{10.5}$$

where $\{I_k\}$ are the data symbols, N is the number of subcarriers, and T is the OFDM block time.

The low-pass signal above can be either real or complex-valued. Real-valued low-pass equivalent signals are typically transmitted at baseband—wireline applications such as DSL use this approach. For wireless applications, the low-pass signal is typically complex-valued, in which case the transmitted signal is up-converted to a carrier frequency f_c. In general, the transmitted signal *s(t)* can be represented as shown in Equation 10.6.

$$s(t) = \Re\{v(t)e^{i2\pi f_c t}\}. \tag{10.6}$$

For a wireless application, *s(t)* is as shown in Equation 10.7.

$$s(t) = \sum_{k=0}^{N-1} |I_k| \cos(2\pi[fc + k/T]t + \arg[I_k]). \tag{10.7}$$

10.6 CAP LINE CODE FOR ADSL

Carrierless amplitude and phase modulation (CAP) is another line code for ADSL. However, the one adopted by ANSI and ETSI for ADSL is the DMT line code. CAP is a coding technique that is based on quadrature amplitude modulation used in DSL transmission. Figure 10.10 shows the transceiver architecture. Note it differs from the QAM passband architecture by the removal of the phase rotators in the modulator and demodulator. Therefore, the primary difference between QAM and CAP is that CAP utilizes quadrature passband filtering. This is one of the advantages that are put forward by CAP proponents. Passband operation avoids centering on DC and thus the impact of noise interfering sources near DC normal in twisted pair signaling applications is lessened.

Figure 10.11 and Figure 10.12 show the transmitter and receiver functional elements. The inphase and quadrature sequences comprising the complex signal, c_k, can be independently filtered and then summed to form an input to the D/A. Low-pass filtering produces the analog signal out. In Figure 10.10, the channel between the transmitter and receiver is implied by the break in the arrows between them. Let a complex baseband signal *c(t)* be transmitted through a linear complex baseband channel described by the filter *H(f)*, then the received baseband signal can be written as:

$$z(t) = \int_0^{\infty} c(t-\tau)h^*(\tau)d\tau = \tilde{a}(t) + j\tilde{b}(t) \tag{10.8}$$

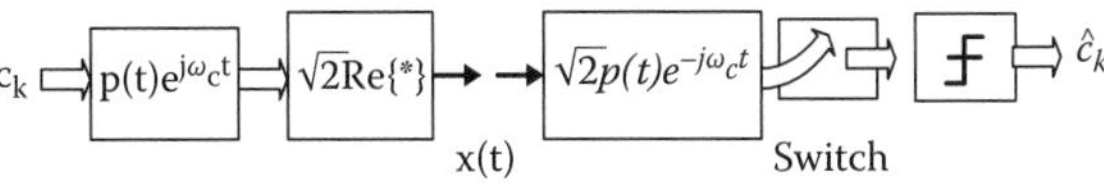

FIGURE 10.10 CAP basic transceiver.

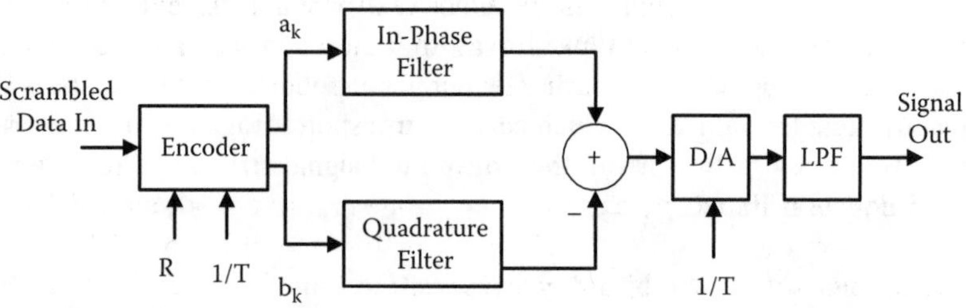

FIGURE 10.11 CAP transmitter.

where *h(t)* is the filter's complex impulse response, * denotes complex conjugation, and ~ indicates the channel's filtering effect on the data. It follows that the received modulated signal is then as shown in Equation 10.9. The filtered signals are of the same form as *a(t)* and *b(t)* given in Equation 10.8, but with the concatenated filtering effect of the transmitter's passband filter and the channel's impulse response.

$$r(t) = RE\{z(t)e^{jw_c t}\} = \tilde{a}(t)\cos(\omega_c t) - \tilde{b}(t)\sin(\omega_c t) \tag{10.9}$$

In Figure 10.12, demodulation begins processing *r(t)* with analog-to-digital conversion to convert to discrete operation. Equation 10.8 and Figure 10.12 demonstrate how the matched filtering operation will recover the signal. In general, CAP architects suggest the use of combined fractionally spaced equalizers and decision feedback equalizer (DFE) structures to accomplish not only the matched filtering, but also the channel equalization. Optimum sampling of the equalized signal followed by the decision and decoding process produces the output data.

10.7 QAM BASIC EQUATIONS

M-QAM is quadrature amplitude modulation by M signal vectors defined in a two-dimensional I-Q signal constellation plane. For square signal sets (constellations) M-QAM is the complex form of √M-PAM (pulse amplitude modulation). Let a and b take on √M amplitude levels from the set {±d, ±3d, …, ±(√M – 1)d}, where d is a

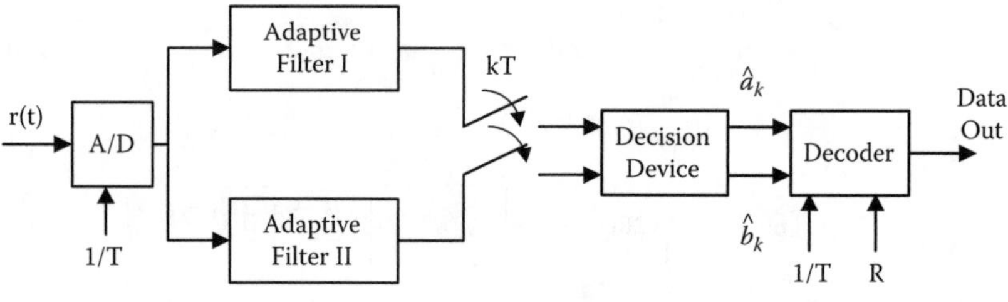

FIGURE 10.12 CAP receiver.

normalization constant. Then $c = a + jb$ takes on M complex values with magnitude $|c| = \sqrt{a^2 + b^2}$ and phase $\arg(c) = \tan^{-1}(b/a)$ in the constellation plane. Let the complex symbol sequence $c_k = a_k + jb_k$ serve as the M-QAM modulation source with each symbol occurring every T_s seconds (i.e., at rate $R_s = 1/T_s$ symbols per second).

Thus, the symbols remain constant over the symbol interval. Direct modulation of this sequence by the inphase and quadrature sinusoids would produce an unshaped waveform with its spectrum following the $\sin^2 x/x^2$ curve. Shaping the modulation can reduce the spectral occupation of the waveform. Let $p(t)$ be a pulse such that the convolution of $p(t)$ with itself is a Nyquist pulse. Such a pulse will be called a root-Nyquist pulse. The conventional passband QAM transmitted signal, $x(t)$, modulated to the carrier frequency f_c is now given by Equation 10.10. Note that $x(t)$ is a real signal.

$$x(t) = \sqrt{2}\mathrm{Re}\{c(t)e^{j2\pi f_c t}\} \tag{10.10}$$

where $c(t)$ is the complex baseband signal with baseband transmit pulse shape $p(t)$

$$c(t) = \sum_{k=0}^{\infty} (a_k + jb_k)p[t - kT_s] = a(t) + jb(t) \tag{10.11}$$

$a(t)$ and $b(t)$ are inphase and quadrature baseband signals

$$a(t) = \sum_{k=0}^{\infty} a_k p[t - kT_s] \tag{10.12}$$

$$b(t) = \sum_{k=0}^{\infty} b_k p[t - kT_s] \tag{10.13}$$

The complex data symbol sequence, $c_k = a_k + jb_k$, consists of a_k and b_k inphase and quadrature data symbols, transmitted at the rate $R_s = 1/T_s$, where T_s is a symbol period.

Equation 10.10 is the QAM basic modulation expression. Figure 10.13 shows one possible functional form of a QAM transceiver suggested by Equation 10.10. The demodulator portion is simply the inverse of the modulator. The two-rail structure, with the real and imaginary filtering having the same basic pulse shape, is shown in Figure 10.13.

Figure 10.13 can be redrawn compactly in its complex processing form and is shown in Figure 10.14. It is equivalent and represents the first line of Equation 10.10. The receiver first demodulates $x(t)$. Then a complex matched filter operation precedes the sampler and decision process, which recovers the complex sequence. An alternative,

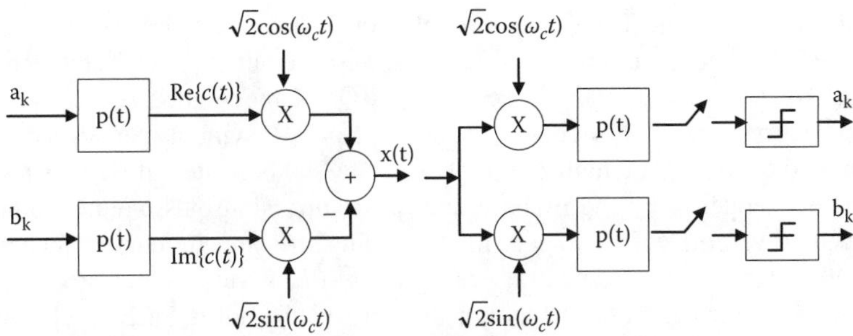

FIGURE 10.13 QAM transceiver with baseband filtering: two rail architecture.

QAM passband filtering structure can be obtained by rewriting Equation 10.10 again as shown in Equations 10.14a and 10.14b.

$$x(t) = \sqrt{2}\mathrm{Re}\left\{\sum_{k=0}^{\infty} c_k p[t-kT_s]e^{j2\pi f_c t}\right\}$$

$$= \sqrt{2}\mathrm{Re}\left\{\sum_{k=0}^{\infty} c_k e^{j2\pi f_c kT} p[t-kT]e^{j2\pi f_c(t-kT)}\right\} \tag{10.14a}$$

$$= \sqrt{2}\sum_{k=0}^{\infty}(a_k\cos(2\pi f_c kT) - b_k\sin(2\pi f_c kT))p[t-kT_s]\cos(2\pi f_c(t-kT))$$

$$-\sqrt{2}\sum_{k=0}^{\infty}(a_k\sin(2\pi f_c kT) + b_k\cos(2\pi f_c kT))p[t-kT_s]\sin(2\pi f_c(t-kT)) \tag{10.14b}$$

Note in Equation 10.14a, that the complex symbol sequence, c_k, is first phase rotated by $e^{j2\pi f_c kT}$. This simply modulates the signal to the passband centered around f_c. It is then shaped by the passband complex filter $p[t-kT_s]e^{j2\pi f_c(t-kT)}$. Equation 10.14b shows the two rails structure, with the real and imaginary filtering having the same basic pulse shape. Sampling t at $n\Delta$ provides the sampled form of Equation

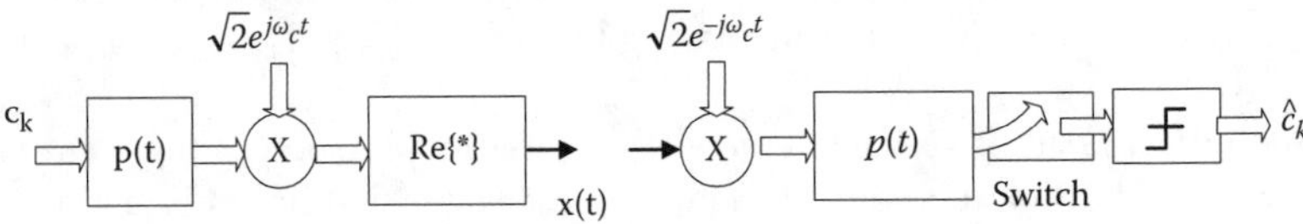

FIGURE 10.14 QAM transceiver with baseband filtering: complex symbol architecture.

10.14, which can be recognized as a linear discrete convolution, with the passband filter folded then shifted. This sampled form is shown in Equation 10.15.

$$x(n\Delta) = \sqrt{2}\,\mathrm{Re}\left\{\sum_{k=0}^{\infty} c_k e^{j2\pi f_c kT} p[n\Delta - kT_s] e^{j2\pi f_c}(n\Delta - kT)\right\} \tag{10.15}$$

Recalling that discrete convolution is commutative will allow simplification of the filtering process, that is

$$y(n) = \sum_{k=-\infty}^{\infty} x(k)h(n-k) = \sum_{k=-\infty}^{\infty} x(n-k)h(k) \tag{10.16}$$

The complex sequence, c_k, has only nonzero values at $\Delta = T$, by construction. Using this fact and dropping Δ in the notation, Equation 10.15 can be rewritten in sampled form with the input folded and shifted as (thus, the passband filter, $p[k]e^{j2\pi f_c kT}$, can easily be stored in memory with static values).

$$x(n) = \mathrm{Re}\left\{\sum_{k=-N}^{N} c_k(n-k)e^{j2\pi f_c(n-k)T} p[k] e^{j2\pi f_c kT}\right\} \tag{10.17}$$

Figure 10.15 shows the architecture that implements Equation 10.17. The complex sequence is first modulated to passband. The complex passband filtering has real and imaginary parts formed simply by taking the baseband root Nyquist pulse shaping filter and multiplying by cos() and sin() respectively. Taking the real part forms the real signal. The demodulator forms the inverse processing. First, complex quadrature passband filtering is applied. Derotation is followed by sampling and the decision process to form the estimated complex sequence.

EXAMPLE 10.1: CONSTELLATION FOR 32-QAM

Figure 10.16 shows the plot of a QAM constellation having 32 points and a peak power of 1 W. The example also illustrates how to label the plot with the numbers that form the input to the modulator. The constellation of the signal is the modulation

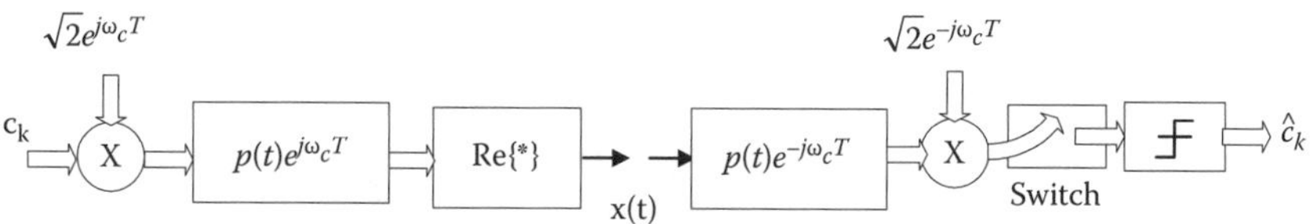

FIGURE 10.15 QAM transceiver using passband filtering.

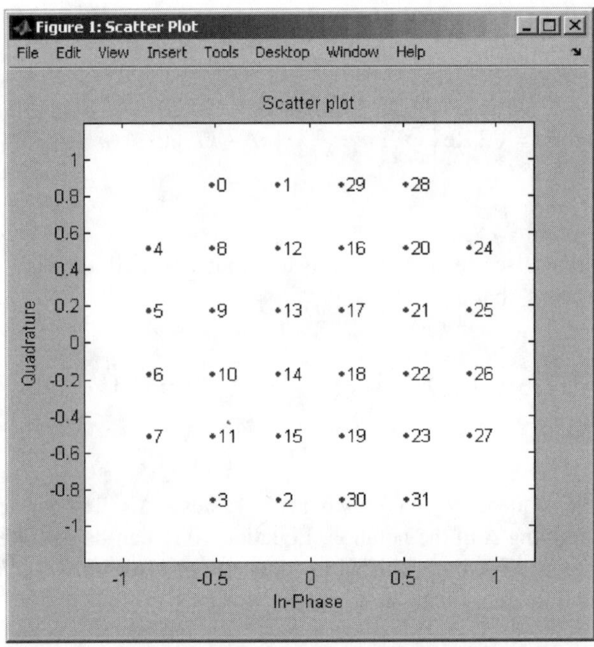

FIGURE 10.16 A QAM constellation having 32 points and a peak power of 1 W.

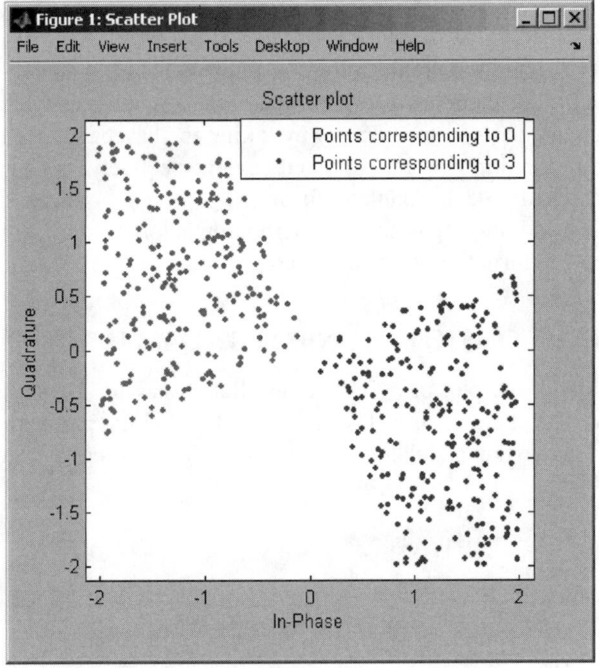

FIGURE 10.17 QAM signal demodulation.

of a signal onto a sequence of complex numbers that lie on a lattice of points in the complex plane. Figure 10.17 shows which regions in the complex plane are associated with different digits that can form the output of the demodulator. Random points are demodulated. It looks for points that were demapped to the digits 0 and 3. Those points in red and blue are plotted respectively. The regions reflect a rotation of the signal constellation by pi/8.

10.8 CAP BASIC EQUATIONS

In Section 10.6, most of the basic equations were discussed. However, the CAP transceiver system is a variation of the passband QAM system, resulting by simply removing the phase rotator in the modulator and derotator in the receiver of Figure 10.14 or Figure 10.15. That is, the complex data sequence, c_k, is directly applied to the QAM transceiver's passband filtering structure. From the second line of Equation 10.14a, we have the basic CAP modulation equation shown in Equation 10.18. The second line results from the commutative property of convolution and suggests the implementation of shifting the data into the passband filter.

$$\begin{aligned} x(t) &= \sqrt{2}\,\mathrm{Re}\left\{\sum_{k=0}^{\infty} c_k p[t-kT]e^{j2\pi f_c(t-kT)}\right\} \\ \text{or,}\quad x(t) &= \sqrt{2}\,\mathrm{Re}\left\{\sum_{k=0}^{\infty} c_k[t-kT]p(t)e^{j2\pi f_c t}\right\} \end{aligned} \tag{10.18}$$

Figure 10.18 shows the basic CAP transceiver architecture. Note it differs from the QAM passband architecture of Figure 10.15 only by the removal of the phase rotators in the modulator and demodulator. Therefore, the primary difference between QAM and CAP is that CAP utilizes quadrature passband filtering. This is one of the advantages that is put forward by CAP proponents. Passband operation avoids centering around DC and thus the impact of noise interfering sources near DC normal in twisted pair signaling applications is lessened.

Figure 10.11 and Figure 10.12 show the transmitter and receiver functional elements. As Equation 10.10 and Equation 10.13 suggest, the inphase and quadrature sequences comprising the complex signal, c_k, can be independently filtered and then summed to form an input to the D/A. Low-pass filtering produces the analog signal out. Note that the overview of the CAP demodulation process above did not take into account echo cancellation (NEXT or FEXT as referred to in the literature). This will be addressed later in this chapter.

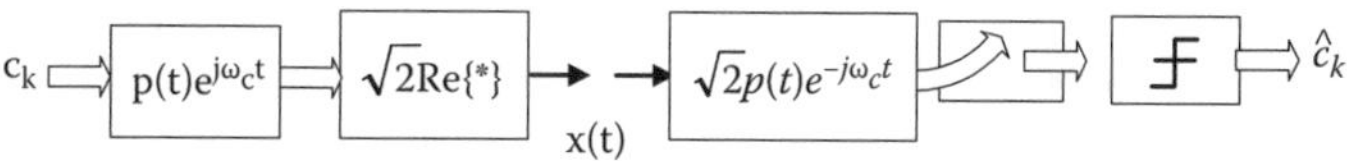

FIGURE 10.18 CAP basic transceiver.

10.9 CAP PASSBAND FILTERING

The primary difference between CAP and QAM is the passband filtering. We will fully discuss the passband filtering in the remainder of this section.

10.9.1 Root Raised Cosine Pulse Shaping

Having an ideal Nyquist pulse to the decision device of the receiver assures zero intersymbol interference (ISI) at the decision instants. Using matched root-Nyquist pulse shaping filters at both the transmitter and receiver and assuming an ideal channel with no distortion accomplishes this goal. A typical root-Nyquist pulse is based on the raised cosine family. The root-Nyquist pulse for the CAP RADSL standard for example is given in Equation 10.19. Note this is the same pulse shape used in cable and satellite applications.

$$p(t,\alpha,R_s) := \frac{\sin[\pi \cdot t \cdot R_s \cdot (1-\alpha)] + 4\alpha \cdot t \cdot R_s \cdot \cos[\pi \cdot t \cdot R_s \cdot (1+\alpha)]}{\pi \cdot t \cdot R_s \cdot [1-(4 \cdot \alpha \cdot t \cdot R_s)^2]} \tag{10.19}$$

In Equation 10.19, $R_s = 1/T$, which is again the symbol rate, α is the raised cosine rolloff factor, and t is the time argument. α normally varies from 0 to 100% in QAM applications and represents the excess bandwidth relative to the sinx/x pulse shape (raised cosine shape with $\alpha = 0$). Some CAP applications recommend use of an α greater than 100%. Figure 10.19 shows the baseband pulse response for $\alpha = 0$ and 15%. The time scale is normalized by Rs and thus represents symbol time. For

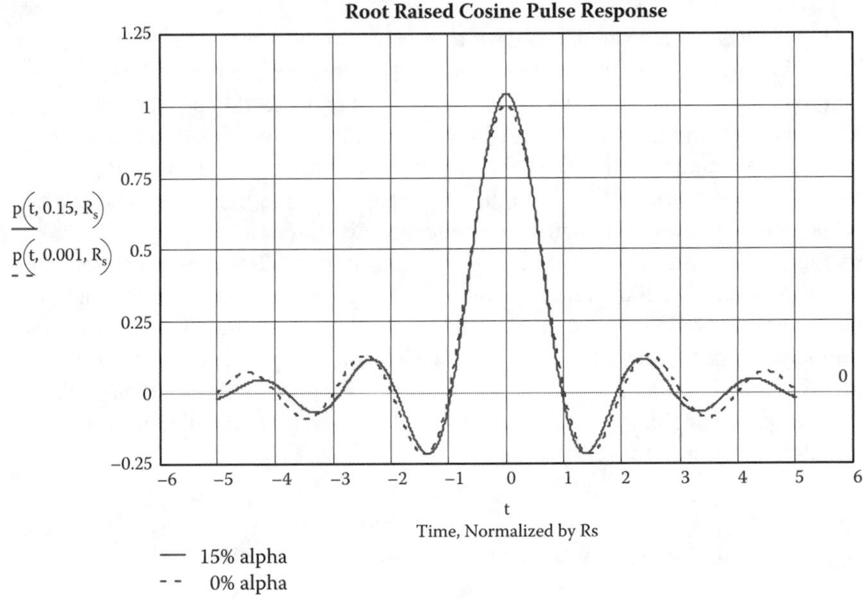

FIGURE 10.19 Baseband root-raised cosine pulse response.

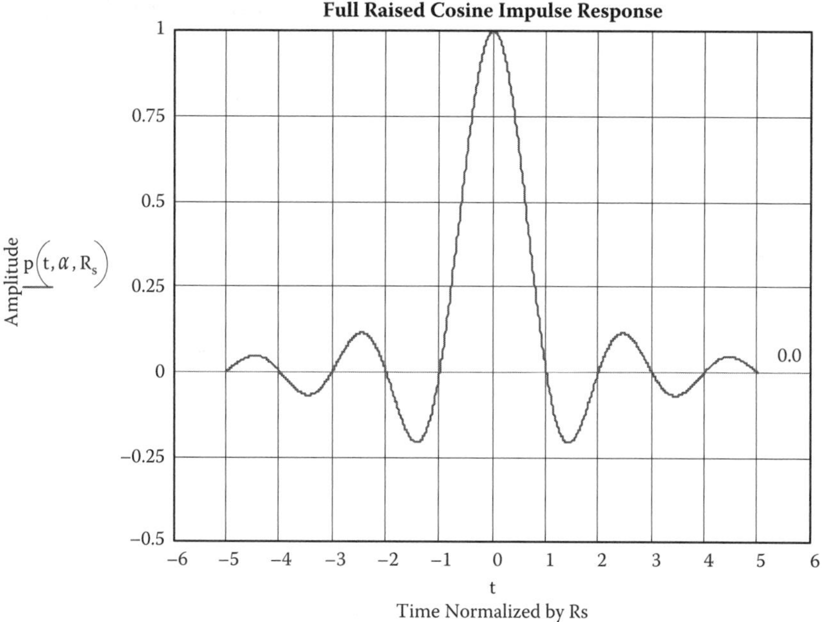

FIGURE 10.20 Baseband full-raised cosine pulse.

comparison, the full raised cosine pulse is plotted in Figure 10.20 for $\alpha = 15\%$. Note that zero crossings occur at symbol times, as desired for zero ISI at the decision device in the receiver.

10.9.2 Orthogonal Passband Filtering Providing Separation and Data Recovery

The orthogonal passband filters in Figure 10.12 are formed as described by Equation 10.20 and Equation 10.21. When matched filtering is applied with optimum sampling, the filtering convolution becomes the integral of the product of the two responses. If the integral equals zero, as it does with the passband root raised cosine pulses, then orthogonality is established. To show data recovery, consider the following. Assume a_k had been formed on the inphase rail at the modulator. The kth transmitted signal pulse is $a_k R_I(t, f_c, \alpha, R_s)$. With optimum sampling ($t = t_{opt}$) and zero ISI (equalization provided raised cosine pulses, which at the optimum sampling instants exhibit zero ISI), convolution with the adaptive inphase filter ideally produces the received data symbol a_k times the energy of the matched filtering process (Equation 10.23). Note the received signal, *r(t)*, is directly separated into its inphase and quadrature components by the passband filtering in Figure 10.18. That is, the signal separation is accomplished without explicit complex derotation.

$$R_I(t, f_c, \alpha, R_s) := p(t, \alpha, R_s) \cdot \cos(2 \cdot \pi \cdot f_c \cdot t) \tag{10.20}$$

$$R_q(t, f_c, \alpha, R_s) := p(t, \alpha, R_s) \cdot \sin(2 \cdot \pi \cdot f_c \cdot t) \tag{10.21}$$

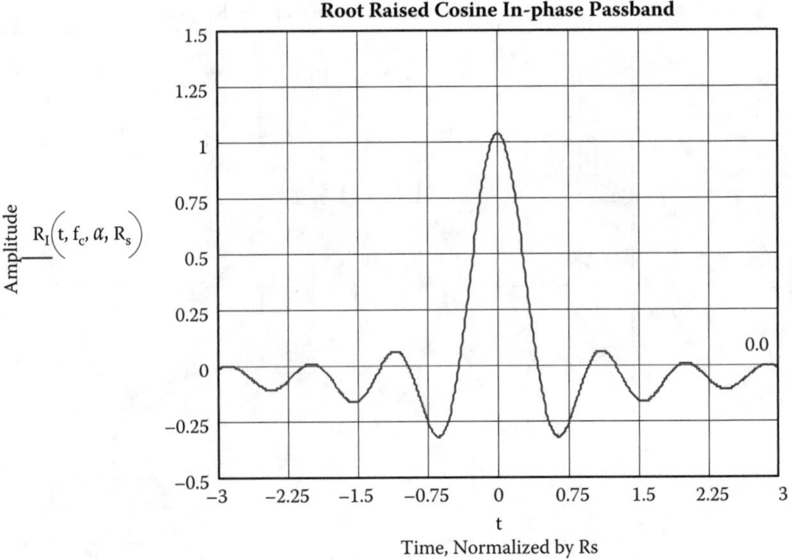

FIGURE 10.21 Inphase passband root-raised cosine response.

$$\int_{-3}^{3} R_I(t, f_c, \alpha, R_s) \cdot R_Q(t, f_c, \alpha, R_s)dt = 0 \tag{10.22}$$

$$\int_{-3}^{3} a_k R_I(t_{opt}, f_c, \alpha, R_s) R_I(t_{opt}, f_c, \alpha, R_s)dt = a\xi_{signal} \tag{10.23}$$

10.9.3 Inphase Passband Response

The inphase passband root raised cosine pulse of Figure 10.11 is plotted in Figure 10.21. It is even about zero. Note that it does not exhibit the zero crossing property of the full raised cosine pulse.

10.9.4 Quadrature Passband Response

The quadrature passband root raised cosine pulse of Figure 10.11 is plotted in Figure 10.22. Note that it is odd about zero and also does not exhibit the zero crossing property of the full raised cosine pulse. The odd symmetry and the fact that the inphase and quadrature passband filters are not equivalent will need to be planned for in an implementation.

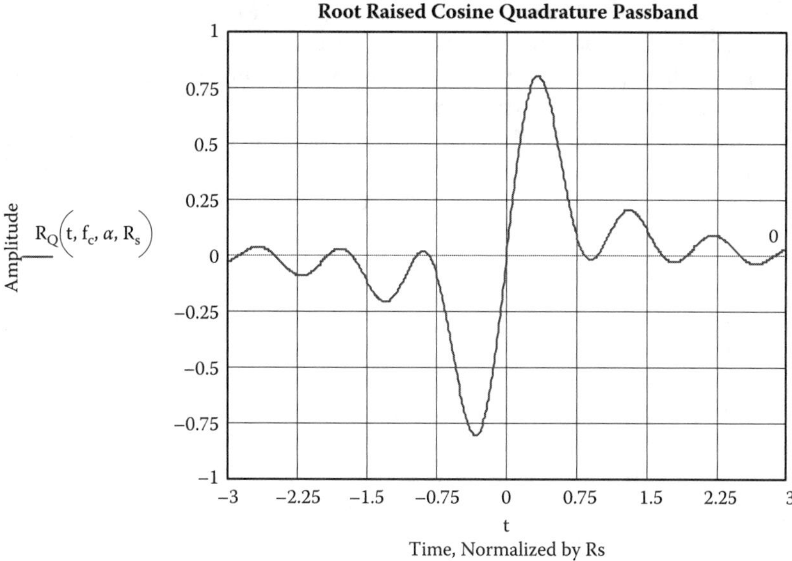

FIGURE 10.22 Quadrature passband root-raised cosine response.

10.10 CAP TRANSCEIVER ARCHITECTURAL CONSIDERATIONS

Figure 10.23 shows the CAP transceiver functional architecture. The CAP transmitter and receiver were shown earlier in Figure 10.11 and Figure 10.12. The transmitter and receiver adaptive filtering, echo canceller/equalization, precoding, QAM size, carrier, and symbol rate synchronization are described in later sections.

10.10.1 Transmitter Filtering

FIR filtering is used for the quadrature filters of the transmitter. Since there are two, this is a doubling of the filtering requirement. The implementation must take into

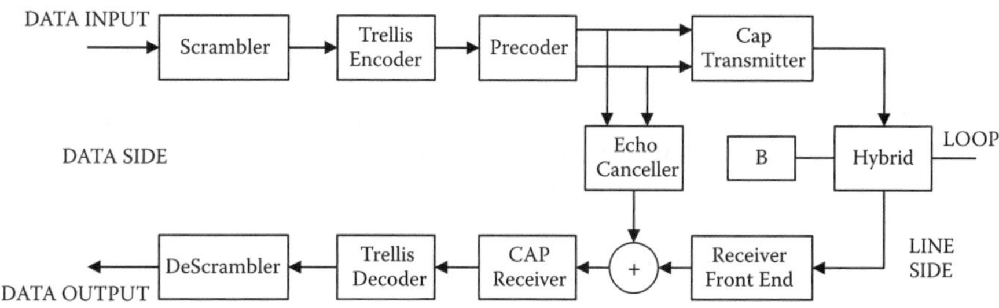

FIGURE 10.23 CAP transceiver functional architecture.

account the modulation function, not required by QAM systems, as indicated by Equation 10.20 and Equation 10.21. Note the quadrature phase filter is an odd function. Hence at the very least, sign inversions are needed. In most cases, the designers ask whether there will be an intrinsic raised cosine coefficient generation function for different α_s, such as in the development system, or whether it will be the user's responsibility in an offline calculation. The answer depends on the level of simplification of the design. It also depends on how much the potential customer wants to pay.

10.10.2 Receiver Adaptive Filtering

In general, CAP architects suggest the use of fractionally spaced equalizers and this could drive the number of taps that potential DSL systems need to provide. The use of fractionally spaced equalizers is not uncommon. However, there is a penalty. To maintain the same span as a symbol spaced equalizer, the fractionally spaced equalizer must grow in number of taps. For example, a $T/2$ spaced equalizer will have twice as many as the T spaced one. A doubling sounds significant. However, it could be more. For instance, in some cases, there could be a need for potential use of $T/6$ equalizers. The other important consideration to recall about fractionally spaced equalizers is that they clock at multiples of the incoming symbol rate.

10.10.3 Echo Canceller/Equalization

This supports cancellation of the next signal or near-end crosstalk. The number of taps required can be large. A span of between $16T$ and $32T$ may be useful. Most CAP designers use adaptive LMS operation for both the cancellation and equalization functions.

10.10.4 Precoding

Tomlinson precoding has been mentioned in some DSL-based standard specifications. Use of this precoding technique in the transmitter lessens the impact of the equalizer's decision feedback error (DFE) propagation. In addition to lessening the impact of the DFE, it also indicates that Trellis coding, required by the key DSL standards, makes precoding a necessity. The receiver design must include modulo operation, as this is part of the precoding functionality.

10.10.5 QAM Size

The use of up to 256 QAM constellations has been specified for some of the variations of the DSL systems such as the rate adaptive digital subscriber line (RADSL). It is important to note that for larger constellations, the performance requirements become increasingly more difficult for processing to satisfy. For example, synchronization (both carrier and symbol timing) and word widths (e.g., precision) become especially sensitive parameters.

10.10.6 Carrier and Symbol Rate Synchronization

Digital implementations of the CAP transceiver rely on synchronization of the center (carrier) frequency with the signaling baud rate. This should simplify the passband filtering. It also provides the added benefit that only a timing recovery circuit is necessary for the receiver to synchronize to the CAP transmitter signal. That is, a carrier recovery circuit is unnecessary.

10.11 COMPARING CAP AND QAM

This section focuses on a top-down complexity comparison between CAP and QAM functionalities to help in forming early observations on processing requirements. This is achieved by examination of the QAM signal processing flow and extracting the requirements that drive each functional step. In digital video broadcasting-cable (DVB-C) QAM applications, the following requirements drive the design: equalization, QAM modes, filtering, supported rates, and synchronization (frequency and timing). Other functionalities, such as coding (Reed Solomon and Trellis), interleaving, and scrambling, are considered as backend processing. Thus, we did not consider them in this comparison.

RADSL, like ANSI T1.413 CAP, is representative of a more demanding CAP application. For instance, both utilize 256-point constellation mappings and operate at similar symbol rates in excess of 1 Msps. A T1E1 Ad Hoc Working group released a draft specification proposal for CAP/QAM xDSL applications. This supports speculation of the continuing interest in the deployment of CAP systems. Therefore, we are using RADSL in this case as the basis for this comparison of CAP and QAM.

10.12 OBSERVATIONS

From the top-down comparison of QAM signal processing, the following observations can be formed:

Rate requirements for CAP are less:

- QAM symbol rates are about a magnitude higher than CAP (12 Msps versus 1 Msps).
- Loops, such as the carrier or symbol timing loops, are symbol-rate proportional. Thus, time criticality of this type of loop processing versus processor raw speed is relaxed.

Additional functionality is not required by QAM processing:

- Echo cancellation and passband filtering are additional front-end requirements for CAP, but are not required by QAM.
- Development of these requirements, and potentially test codings, are required in the feasibility assessment of any design.

Equalization processing requirements should be similar:

- In the driving processing function for QAM, since it is an adaptive FIR operation, the general FIR requirements also apply.
- Feed forward and decision feedback architecture prevails.

- Employs LMS as the kernel adaptation algorithm, also used for the echo canceller required by CAP. The equalization task is assumed at this point to be slightly more demanding than the echo cancellation requirement. Inherently, echo cancellation relies on cooperative processing with the transmitter (example training) whereas the equalizer must operate in the blind and adverse channel effects.
- Some of the new DSL system product capabilities will be influenced by the design choice of fractional versus symbol spaced taps for equalization or cancellation operation. CAP proponents call for fractionally spaced operation.
- Update rate on order of symbol rate is normally required.

Constellation map requirements similar:

- Both QAM and CAP require support for 256-point constellations.
- This drives precision requirements, as well as amplitude and phase control processing.

Passband filtering, which is not included in QAM, is the primary filtering difference for CAP:

- Must implement the implicit sine and cosine quadrature modulation functionality to achieve passband.
- Root raised cosine filtering with similar $\alpha = 15\%$ rolloff.
- May be combined with the equalization function.

10.13 HDSL

The earlier version of TI lines at the rate of 1.536 Mbps needed repeaters strategically placed at different locations for it to function properly. This presented problems with cost and the deployment of TI access lines. However, there was a need to have a TI line without the repeaters. Bellcore engineers worked on this issue for years before developing the repeaterless TI line called the high-bit-rate DSL (HDSL). The line code is called 2 binary, 1 quaternary (2B1Q). This line code made it possible for the encoding of 2 bits/baud rather than 1 bit/baud. It is the same line code used in the basic rate interface (BRI) of integrated service digital network (ISDN).

HDSL used two copper wire pairs. There was no need for repeaters. Each of the pairs of HDSL carried 784 kbps, which is half the total T1 rate of 1.544 Mbps. This new DSL technology required the use of HDSL termination unit-central (HTU-C) and HDSL termination unit-remote (HTU-R) at the central office (CO) and customer premises ends respectively. HDSL supports 1.544 Mbps T1 line (North American Standard; ANSI) and 2.048 Mbps El line (European Standard; ETSI) at both the upstream and downstream rates at a distance of up to 5 km repeaterless and up to 12 km with repeaters. The key applications are in the areas of telco transports, cellular base station connectivity, and T1/E1 leased lines. Figure 10.24 shows HDSL T1 basic network architecture.

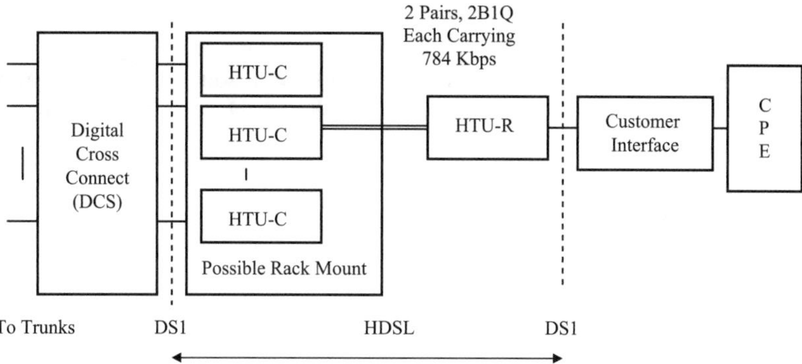

FIGURE 10.24 HDSL T1 basic network architecture.

The HDSL T1 basic network architecture is based on copper loops between the central office and the subscriber premises. There are some differences between the HDSL and the ADSL architectures. Simultaneous POTS and service are assumed in ADSL. In the case of HDSL, it does not include support for POTS. It uses frequencies that do overlap the voice passband. This is the reason why the splitterless techniques are used in HDSL architecture. HDSL uses two copper pairs and in some implementations, up to three pairs. ADSL uses only one copper pair. It should be noted, however, that HDSL 2 operates over a single pair.

10.13.1 Basic HDSL Characteristics

It is true that different vendors make different types of HDSL systems. Therefore, each design may be different. However, they all share the following characteristics:

- Full-duplex, DS1-framed transport at 1.544 Mb/s.
- Repeaterless operation for full carrier serving area (CSA). It should be noted that CSA permits loop lengths up to 12,000 ft of 24 AWG cable, or 9,000 ft of 26 AWG cable.
- Transmission over existing distribution wire pairs.
- Specified bit error rate ten times better than T1 carrier.
- Spectral compatibility to permit coexistence in the same cable-sheath with voice, basic rate ISDN, T1 carrier, and many other transmission systems.
- Operation in the presence of bridge taps and gauge changes.

10.13.2 HDSL Applications

- Videoconferencing: Multiple H0 channels of 384 × 384 kbps that can deliver high-quality videoconferencing are supported by HDSL and HDSL2.
- Private T1 line: These scenarios can be supported:
  - voice trunk to wireless cell site
  - ATM line from business site to data switch or frame relay

  - videoconference line to business site
  - X.25 link between business sites or frame relay
- LAN-LAN interconnect: Files and other resources of companies can be shared in their LANs as a result of the symmetric nature of HDSL/HDSL2.
- Web hosting: HDSL/HDSL2 allows multiple users access simultaneously because it can make available symmetric bandwidth necessary for hosting Web home pages on a server at the company's premises.
- Telecommuting: HDSL/HDSL2 has the potential of providing high-speed bidirectional communications like bulk file transfers, unlike ADSL telecommuting.
- Integrated services to businesses: In integrating voice and data access over a single copper loop to a business using multiline derived from digital voice lines, HDSL2 provides an economical way to do it.
- Shared Internet access: Bidirectional high-speed bandwidth can be shared in a business by PCs on a LAN.

10.14 SDSL

One single wire loop is what single line DSL (SDSL) or the single-pair high-bit-rate DSL (SHDSL) uses. SDSL has a reach of about 10,000 feet. It has a speed that is about the same as that of the HDSL. It is intended to provide a basic DSL option delivering 1.544 Mbps data rates in both the upstream and downstream directions over existing copper twisted-pair wireline connections. Like ADSL and HDSL, SDSL is symmetric, hence the name symmetric DSL (SDSL) found in the literature.

The line code for SDSL is 2B1Q. It is that type of signal that has 2 bits per baud that are arranged as one quaternary (four levels). This type of modulation scheme is a baseband four-level pulse amplitude modulation scheme. Data in this type of scheme are transmitted at twice the frequency of the signal. In the full duplex scheme, these types of systems transmit the same bit rate in upstream and downstream directions. Since the transceivers in the upstream end are colocated in the local exchange carrier (LEC) central office (CO), SDSL receivers suffer much from its self-NEXT (near end-crosstalk). It causes severe degradation on its data rates. Vector equalization for the SDSL upstream receiver has the capabilities of combating its crosstalk effectively instead of the conventional scalar equalizer by about 15 dB gain in mean squared error (MSE).

10.15 SHDSL

The symmetric high bit rate DSL (SHDSL) is one of those varieties of DSL that is within the xDSL family. It leverages on the capabilities of some of the earlier versions of DSL like HDSL, HDSL2, SDSL, IDSL, ISDN e-1, and T-1. It provides services such as delivering voice and data. It also provides high symmetric data rates, which are one of the advantages with guaranteed bandwidth and low interference with other services. There is up to 2.3 Mbps available on a single copper wire in a typical installation of SHDSL. A four-wire model is also available for future greater bandwidth needs of up to 4.6 Mbps.

Since SHDSL is symmetric, it has a better fit for the following needs:

- video conferencing
- Web-hosting
- remote LAN access
- application sharing

The small business markets are prime targets of the SHDSL. Some of the major characteristics include remote LAN access, data channels, videoconferencing, leased lines with customer-specific rates, and multiple telephones.

10.16 RADSL

The rate-adaptive digital subscriber line (RADSL) uses one wire pair, but it can adapt its data rates to line conditions. It is a means to optimize the throughput of data communications in a DSL service. It does it by adjusting the connection to compensate for variations in the line characteristics of the local loop. In most cases, the desired bit rate may be manually or automatically initialized at startup. In other cases, once there is a way to monitor the line conditions, it may be adjusted as needed, based upon the performance characteristics of the line. This can be done in addition to the normal rate ADSL service adaptation capabilities. RADSL uses the DMT line code.

The signal of the RADSL can have greater spectral occupancy than normal ISDN systems. Therefore, crosstalk seems to be a major concern. We have to consider crosstalk in RADSL from two points of view: (1) effects of an RADSL on traditional systems and (2) increased susceptibility of an RADSL system to crosstalk. It is possible that RADSL systems may not punish existing fixed-rate systems more than crosstalk from systems of the same type because RADSL does not increase the power spectral density of the signal in the bands admitted by receivers of traditional systems. An RADSL system with the same total power as a standard design may result in lower than normal crosstalk because power densities are lower than normal in the crosstalk passband of the standard systems.

In the receive direction most RADSL hardware may have the following:

- an adaptive equalizer capable of functioning over a range of clock rates
- a wide-acquisition-range, ambiguity-free clock recovery
- programmable demultiplexer or stat-mux interface
- a programmable attenuator for use during adaptation
- framing and fast BER evaluation circuits
- AGC circuit with means to produce "signal present" indication
- a frequency analyzer for determining the currently received symbol rate

While in the transmit direction, most RADSL hardware may have the following:

- a programmable output attenuator
- a transmit encoding logic and line driving circuits operable over a range of frequencies

- programmable input mux or stat mux input buffer
- a programmable transmit frequency synthesizer and pattern generator

Some of the application areas of the RADSL loop are

- reduced provisioning strategies
- bandwidth-on-demand applications
- PBX-CO trunk groups over a single pair
- multiline POTS to apartment buildings
- stat mux applications
- long reach 1B + D or 0.5B + D minimal service

10.17 IDSL

IDSL stands for ISDN DSL where, ISDN stands for the integrated service digital network. The basic idea behind IDSL is to provide basic rate ISDN (BRI ISDN) service to potential customers, making it cost effective to them by not requiring access to the public switch telephone network (PSTN) at the central office. In this type of DSL, the normal 2B + D ISDN basic rate interface (BRI) channels, which run at 144 kbps, that is, 64 kbps B-channels, and one 16 kbps D-channel, transmit digital data and then runs the BRI into the xDSL equipment instead of the ISDN voice switch. IDSL can extend its reach up to 18,000 ft, just like an ISDN DSL. It uses one pair of wires. Internet and intranet are some of the application areas of IDSL.

10.18 CDSL

CDSL is the consumer digital subscriber line. Generally, the speeds of ADSL technology may be more than the average consumer may require. As a result of that, with the lower-speed communications capability in mind, CDSL as a model was developed. It has a data rate downstream speed of about 1 Mbps. The data rate upstream speed is about 16 to 128 kbps. Its reach can be extended to about 4 km. In general, it has a very modest speed and distance when compared to ADSL and RADSL. CDSL does not need a splitter that can separate the incoming signal to analog and data so that the consumer can continue to receive analog telephone voice without interruption while still receiving faxes and computer information data. Some of the application areas of the CDSL are

- interactive multimedia
- Internet access
- remote office LAN residential and SOHO applications
- video-on-demand

10.19 VDSL

High-speed digital subscriber line (VDSL) technology provides up to 52 Mbps of digital bandwidth over a single pair of copper wires. This allows several channels of high-quality video streams to be simultaneously transmitted to a user's location without rewiring. Some enhanced versions of VDSL have already been deployed in several countries around the world.

VDSL-based products have been developed for the evolving applications that will take our civilization into the next millennium. These applications allow us to increase the speed at which we communicate and learn:

- electronic commerce
- Internet/intranet applications
- video on demand
- telemedicine
- videoconferencing
- distance learning
- electronic publishing

Many of these applications will require more bandwidth than IDSL, SDSL, and even ADSL can supply. VDSL has been developed to enhance our ability to communicate using multimedia at high bandwidths.

Distance learning: The ability to have interactive video to communicate with remote students with a qualified teacher is just what some locations need to educate their young population. This also works well with additional training for professionals.

Telemedicine: Doctors with specialties can communicate procedures remotely after viewing the patient's medical video images transmitted from smaller medical facilities. This allows better treatment for more people.

Electronic commerce: Companies are already purchasing billions of dollars of goods and services electronically and it is expected that consumers will soon jump into electronic purchasing through video and data applications.

Video on demand: Many hotels have been successfully offering near video on demand services to their guests. Pay-per-view on satellite and cable services only offer a few movies at certain presubscribed times. Subscribers want to watch a movie when they want to watch and control play, pause, rewind, forward, and stop.

Internet access: Easy and fast access is made available with the VDSL products. Data, pictures, and graphics from the Internet will rapidly be displayed at speeds of over 100 times faster than today's modems. VDSL's two-way communication offers the fast downstream bandwidth without having to use a different medium for the upstream like Web TV has to.

Some of the facilities that can benefit from VDSL systems today include universities, hospitals, hotels, apartment buildings, condominiums, government buildings, and office buildings for small businesses. These facilities service many disparate organizations that each alone cannot afford their own private high-bandwidth network access, but together they can share a network through VDSL connections over their existing copper telephone wire.

Universities: They require the ability to have interactive video to communicate with students at remote locations or for training professionals. Their teachers and students need high-speed Internet from their dorms. Students need to have access to educational videos when they want the information.

Hospitals: Their specialists need to communicate procedures remotely after viewing the patient's medical video images. Their patients need to see their medical images in the comfort of their rooms. Patients should have access to the Internet to communicate with loved ones. Doctors can have patients watch medical videos on treatments or special diets. This allows hospitals to offer better treatment for more people.

Small business building: Many companies are already purchasing billions of dollars of goods and services electronically and it is expected that a majority of small businesses will receive most of their new business through the Internet. Consumers will soon use electronic purchasing through video and data applications for most of their shopping requirements.

Hotels: Many hotels have been somewhat successful offering near video-on-demand services to their guests. Their guests really want to watch a movie when they want to watch and control play, pause, rewind, forward, and stop using full video-on-demand services. Guests also want to have high-speed Internet access from their hotel rooms.

Residential buildings: Owners of multidwelling units want to add multimedia services to their facilities to attract more tenants and earn more revenue. They want to provide their tenants with CATV or DBS video services, pay-per-view or video-on-demand services, high-speed Internet access, and complete voice services.

10.20 FUTURE OF DSL SYSTEMS AND BROADBAND SYSTEMS

As the communications industry advances, broadband is beginning to take on an entirely different meaning from its initial concept. Both DSL and cable are technologies that are forming the broadband landscape for tomorrow's latest devices and applications.

DSL can literally transform the existing information network from one limited to voice, text, and grainy pictures to a powerful universal system capable of bringing new multimedia services (including live streaming video) to everyone's desktop. DSL (or ADSL) is slated to be a key player in the next decade as telcos offer new services for video and other multimedia. It will take years to reach all customers with phone lines that provide quality service, but their strategy is to reach as many subscribers as possible until the build-out is complete. The goal of telcos is to develop a system around DSL that will bring movies, television, video catalogs, remote CDs, corporate LANs, and the Internet to both consumer and corporate users. DSL will make markets that are profitable for telcos and application suppliers.

Broadband also has a promising future with cable Internet access. Data over cable service interface specification (DOCSIS) is a cable access standard and an initiative backed by many companies to answer issues of security and performance limitations due to device standards interoperability. One of the goals of DOCSIS is to develop a set of standards to deal with cable modem-to-telephone return interfaces (CMTRI). CMTRI is an important standard because it offers CATV companies with another way to transmit upstream, and this standard will allow CATV companies to compete with local exchange carriers for telephone service. The cable industry is working on standards (as part of the DOCSIS protocol) that will address key issues to successfully provide business connectivity over cable networks. The companies leading the protocol initiative are Comcast Corporation, Cox Communications, and Time-Warner Cable along with Rogers Communications, Continental Cablevision, and Cable Laboratories. Collectively this initiative group is known as the Multimedia Cable Network System (MCNS). These key companies are sure to influence the cable broadband industry to adhere to standards that provide legitimate security and sound interoperability.

As ADSL continues its steady growth in the competition between ADSL and cable, the price of both services will go down in the future. At the same time, broadband wireless will also play a more important role in the broadband access market. Satellites will also extend high-speed access to even the remotest locations—a particular benefit for businesses in areas that cannot afford to build traditional telecommunications networks from scratch, such as most developing countries.

10.21 XDSL STANDARDS

There are a number of standards-making bodies and consortiums who are responsible for evaluating issues relating to multimedia service definition and multimedia systems, including the associated modems, protocols, and signal processing. To name a few: ANSI and ITU-T Study Group 16 are involved in studies on multimedia standards; Cable Labs, who is looking at, among other things, interoperability specs for devices connected to cable networks; the IETF, who is looking at IP over cable networks; and the IEEE 802.14 and the ETSI who are promoting standards for interactivity. The main focus of these standards-making bodies is on the physical and MAC layer issues, looking at the different modulation schemes. This research reflects network requirements for delivery of video and data, or interactive multimedia to the home, via cable modems and DSL wireline technologies.

With DSL applications, technical and market trials are being conducted, and standards are still under development. The ANSI organization is reviewing EDI standards and the ITU-T Study Group 16 is looking at the following:

- audiovisual/multimedia services
- interactive multimedia information retrieval services (MIRS)
- data protocols for multimedia conferencing
- modems for switched telephone networks and telephone-type leased circuits
- accessibility to multimedia for people with disabilities
- advanced video coding
- harmonization of multimedia systems, applications, and services
- audio and wideband coding in public telecommunications networks
- software and hardware tools for signal processing standardization activities

Associated video encoding standards under consideration are H.320, MPEG, and MPEG2. H.320 is a standard for interoperability. MPEG is a worldwide standard for image compression for full motion video. MPEG2 covers digital cable TV signals or more of a generic coding of moving pictures and associated audio information that is considered to be HDTV/LD quality. DSL needs a true international standard with the flexibility to address country-specific requirements and to provide for interoperability and portability.

There is an ever-essential need for open and coherent standards for interconnectivity and interoperability, as well as ease of use and global consumer protection. The IEEE and ETSI are two of the technical and scientific organizations that play an important role in promoting standards for interactivity. Because end users for data services are still mostly corporations, while voice services are used by corporations and individuals alike, emerging multimedia services that require real-time or near-real-time speech and audio interactivity in addition to video and/or data communications may require a new breed of standards that combine compatibility and some similarity.

Multimedia standards under development by the ANSI group IISP are

- reliability
- network-to-network interface for users, interface between local access serving offices, interface between LAPs and IXCs
- application-to-application requirements; application integrity
- application-to-appliance software installation services and application software
- portability
- security
- quality of service
- survivability

It must be noted that while discrete multitone (DMT) is the standard modulation scheme for DSL, CAP (carrierless AM/PM) equally can be used. This may be a hindrance to DSL gaining widespread acceptance in some areas. It may not be a significant problem since each technology works equally well as long as the same type of modem is used at both ends of the link and many vendors are building equipment that supports both modulation schemes.

10.22 XDSL STANDARDS COMPLIANCE

10.22.1 ADSL

- ANSI T1.413, 1995: Telecommunications-network and customer installation interfaces (ADSL).
- ETSI ETR 328: Transmission and multiplexing (TM); asymmetric digital subscriber line (ADSL); requirements and performance.
- T1E1.4/97-007R6 (ANSI T1.413 Issue 2), 1997: Standards project for interfaces relating to carrier-to-customer connection of asymmetrical digital subscriber line (ADSL) equipment.

10.22.2 HDSL

- ETSI ETR 152: Transmission and multiplexing (TM); high-bit-rate digital subscriber line (HDSL) transmission system on metallic local lines; HDSL core specification and applications for 2.048 kbps access digital sections.
- TA-NWT-001210, Issue 1, 1991: Generic requirements for high-bit-rate digital subscriber lines.
- ANSI TR No.28: High-bit-rate digital subscriber line (HDSL).

10.22.3 VDSL

- T1E1.4/98-054, 1998: VDSL draft specification.
- ANSI T1E1.4/98-043, 1998: Very-high-speed digital subscriber lines system requirements, draft technical document, revision 13.
- T1E1.4/98-036, 1998: SDMT VDSL draft standard proposal.
- ETSI T1E1.4/98-032 1998: Transmission and multiplexing (TM); access transmission systems on metallic access cables; very-high-speed digital subscriber line (VDSL); part 1; functional requirements.

10.22.4 RADSL

- T1E1/97-104R2a, 1997: Draft physical specification for CAP/QAM-based rate adaptive digital subscriber line (RADSL).

10.22.5 ISDN

- ANSI T1.601-1992: Integrated services digital network (ISDN)-basic access interface for use on metallic loops for application on the network side of the NT (layer 1 specification).
- ANSI T1.602, 1992: Telecommunications-integrated services digital network (ISDN) - basic access interface for use on metallic loops for application on network side of the NT.
- ANSI T1.231, 1993: Telecommunications-in-service layer 1 digital transmission performance monitoring.
- ANSI T1.401, 1993: Telecommunications-Interface between carriers and customer installations-Analog voice-grade switched access using loop-start and ground-start signaling.
- ETSI ETR 080, 1996, 2nd ed.: Transmission and multiplexing (TM); integrated services digital network (ISDN) basic rate access; digital transmission system on metallic local lines.
- ITU-T Q.920, 03/93: Digital subscriber signaling system no. 1 (DSS1), ISDN user-network interface data layer-general aspects.
- ITU-T Q.921, 03/93: Digital subscriber signaling system no. 1, ISDN user-network interface-data link layer specification.

- ITU-T Q.931, 03/93: Digital subscriber signaling system no. 1 (DSS1), ISDN user-network interface layer 3 - specification for basic call control.
- ITU-T I.430, 11/95: Integrated services digital network (ISDN), ISDN user-network interfaces, and basic user-network interface - layer 1 specification.

10.22.6 Wireless Channel

- IEEE Standard 802.11-1997: Wireless LAN medium access control (MAC) and physical layer (PHY) specifications.
- ITU[8-1/29]: Spectrum requirements for terrestrial third generation wireless IMT-2000 in the year 2010.

10.22.7 LAN

- IEEE Standard 802.11-1997: Wireless LAN medium access control (MAC) and physical layer (PHY) specifications.
- ISO/IEC 15802-1: 1995, Information technology-telecommunications and information exchange between systems-local and metropolitan area networks-common specifications, part 1: medium access control (MAC) service definition.

10.22.8 Others

- CSA - Carrier Serving area defined as all customer premises within 12 kft on 24 AWG (1260 Ω d.c. loop resistance) or 9 kft on 26 AWG (1502Ω) of a central office. CSA includes about 75% of all customers approximately.
- ECSA - Extended carrier serving area is defined as customer premises within 18 kft on 24 AWG (1890Ω) or 13.5 kft on 26 AWG (2254Ω) of a central office. ECSA includes about 90% of all customers approximately.

SUMMARY

1. The invention of the telephone was radical. This invention subsequently led to the telephone system public service telephone network (PSTN). In turn, the PSTN in place today allows for public access to the Internet.
2. Although broadband is a term that refers to high-speed Internet access, broadband is evolving into more than just fast downloads. Internet access is possible using DSL, T1, T3, cable, or satellite. There are many different ways to access the Internet, but the two leading access technologies today are DSL and cable. Due to these two access technologies, the broadband industry is experiencing tremendous growth.
3. DSL is in major deployment in North America, and there is a major ramp up in parts of Europe. All related industries are playing their part, as well. Major telcos and local exchange carriers are aggressively advertising DSL service, and they continue to increase their DSL customer base. Also, chip makers are designing chipsets that can be used right off the shelf for broadband applications.

4. The quick adoption of these technologies by consumers and the limited use of the technologies by business indicate that broadband usage will increase.
5. DSL provides very high data speeds of up to 1.5 Mb/s for upstream and up to 9 Mb/s for downstream.
6. The most promising factor for broadband access technology is the major efforts from the business sector to make improvements and to exceed previous expectations of the capabilities of broadband.
7. Major companies such as Intel, Microsoft, Comcast Corporation, and Time-Warner Cable are actively participating in the progression of broadband. Indeed, the future for broadband is very good.
8. The demand for broadband communication is growing at a fast pace. We have given an introduction of the concept of broadband communications and the necessary technologies and techniques required for telecommunication infrastructure developments. In the immediate future, a particular benefit for businesses and individuals that cannot afford to build traditional telecommunications networks from scratch is to consider satellites, which can extend high-speed access to even the remotest locations while gradually moving to xDSL and cable modem, as two typical broadband communication systems.
9. We have provided a top-level description of the CAP modulation/demodulation process to support any development of DSL systems requirements for CAP operation. Its relationship to the more familiar QAM signaling was established via examination of the key modulation/demodulation mathematical representations. From these expressions, functional architectures were also presented. The use of quadrature passband filtering by CAP was highlighted as a primary difference. The formation of these filters, their shapes, and preliminary observations of the processing implications were discussed.
10. A review of the key processing steps for QAM receiver technology as it relates to CAP processing on a DSL system processor was provided. The RADSL requirements for CAP were used as baseline for this comparison. Observations from this exercise indicate that CAP processing requirements are very similar to QAM, with the exception of rate.

REVIEW QUESTIONS

10.1 ADSL is a symmetric digital subscriber line.

- a. True
- b. False

10.2 The DSL that has the greatest data rate upstream and downstream speed is:

- a. ADSL
- b. RADSL
- c. HDSL
- d. HDSL2

e. CDSL
f. SDSL
g. IDSL
h. VDSL

10.3 Which of the following can be classified as a line code?
a. DMT
b. CAP
c. 2B1Q
d. All of the above

10.4 DSL is one of the leading access technologies in broadband communications.
a. True
b. False

10.5 The data rate upstream and downstream speed of 1.544 Mbps is for which type of DSL?
a. ADSL
b. RADSL
c. HDSL
d. HDSL2
e. CDSL
f. SDSL
g. IDSL
h. VDSL

10.6 The DSL that has a distance of up to 6 km without repeaters is:
a. ADSL
b. RADSL
c. HDSL
d. HDSL2
e. CDSL
f. SDSL
g. IDSL
h. VDSL

10.7 Which DSL has a single wire loop with a reach of about 10,000 feet?
a. ADSL
b. RADSL
c. HDSL
d. HDSL2
e. CDSL
f. SDSL
g. IDSL
h. VDSL

10.8 Time Warner Corporation has both DSL technology and cable technology modems.
a. True
b. False

10.9 CSA and ECSA can each reach customer premises within 12 kft on 24 AWG.
   a. True
   b. False

10.10 Which of these organizations are responsible for making XDSL standards?
   a. NSI
   b. ETSI
   c. ITU-T
   d. TIEI.4
   e. All of the above

Answer: 10.1b, 10.2h, 10.3d, 10.4a, 10.5c,d, 10.6b, 10.7f, 10.8d, 10.9a, 10.10e

PROBLEMS

10.1 What is DSL? What is the role of DSL in broadband communication systems?

10.2 Describe the various types of DSL broadband communication systems.

10.3 What are the differences between the various DSL broadband communication systems?

10.4 What are the differences between DSL and cable modem technologies?

10.5 What is DMT? Describe the role of DMT techniques in an ADSL system.

10.6 What is CAP? Describe role of CAP techniques in an ADSL system.

10.7 Describe the differences between DMT and CAP in an ADSL system.

10.8 Why is DMT the preferred standard line code in an ADSL system?

10.9 Compare the DMT, CAP, and 2B1Q line codes used in DSL technologies.

10.10 Describe the various applications of the different types of DSL modems.

10.11 Describe the various application areas of VDSL.

10.12 A DMT ADSL system has the value of $\frac{1}{T}$ to be 4044 Hz and the SNR_j is calculated to be 40 dB. Estimate the total throughput of this DMT transmission system.

10.13 Write a MATLAB code that can modulate a QAM constellation having 32 points and a peak power of 1 W. Label the plot with the numbers that form the input to the modulator.

10.14 Write a MATLAB code that can demodulate a QAM constellation having 32 points and a peak power of 1 W. Label the plot with the numbers that form the input to the demodulator.

10.15 If the downstream bandwidth is separated into 256 subchannels that are 4.3125 kHz wide and the upstream bandwidth is separated into 32 4.3125-kHz-wide subchannels, what is the total bandwidth at the 256th subcarrier?

10.16 Using MATLAB code, design a raised cosine pulse that has $\alpha = 20\%$. How does it compare with the raised cosine pulse of Figure 11.19?

10.17 What is the usefulness of raised cosine pulse filters in DSL systems? Are there other types of filters that perform the same type of functions in a DSL system?

10.18 Why are standards very crucial in the design and specifications of DSL systems?

10.19 Is it possible for systems such as DSL systems to survive in the marketplace without conforming to xDSL standards? Why?

10.20 Compare and contrast CAP and QAM line codes using any of the DSL systems.

11 Cable Modem Systems

No one can make you feel inferior without your consent.

— Eleanor Roosevelt

Cable modem systems are modems designed to operate over cable television (cable TV) lines. Accessing the Internet using a cable network (and a cable modem) is becoming increasingly popular due to the much greater speed than is available through telephone-modem access. A cable modem uses coaxial cable, which has the capability to provide greater bandwidth than regular telephone lines. This makes it possible for cable modem systems to achieve extremely fast access to the World Wide Web (WWW). The fact that millions of homes are already wired for cable TV has provided a tremendous boost to many homes to have access to Internet. However, there are a number of technical difficulties. The fact that the cable TV infrastructure is designed to broadcast TV signals in just one direction—from the cable TV company to people's homes—presents one problem. This is because the Internet is a two-way system where data also needs to flow from the client to the server. It is debatable as to whether the cable TV networks can handle the amount of traffic that could result if millions of users began using the cable modem system for Internet access. Cable modem system organizations have made significant progress in accommodating the two problems mentioned.

11.1 CABLE MODEM TECHNOLOGY

Cable modem (CM) is another way to provide broadband communication services. It is designed to operate over the hybrid fiber coax (HFC) network as shown in Figures 11.1, 11.2, and 11.3. Future HFC cable networks will need fiber out to homes to deliver emerging services.

CM frequency bandwidth is up to 862 MHz. This design is different from the traditional cable network, which did not have the HFC connections as shown in Figure 11.4. The major part of the distance between the headend and the subscriber is covered by one or several cascaded optical transmission systems. At the receiving end of an optical link, a relatively simple photo-detector is used to convert the optical signals back into an RF spectrum, essentially identical to the one presented at the input (transmitting) end. With the aid of optical amplifiers, the transmission range can be relatively disturbance-free in length. Coaxial network technology is only used for a small part of the distance involved. The data rate can be as high as 10 Mbps, much faster than all but VDSL technology.

A single downstream 6-MHz television channel may support up to 27 Mbps of downstream data throughput from the cable headend using 64 QAM (quadrature amplitude modulation) transmission technologies. Speeds can be boosted to

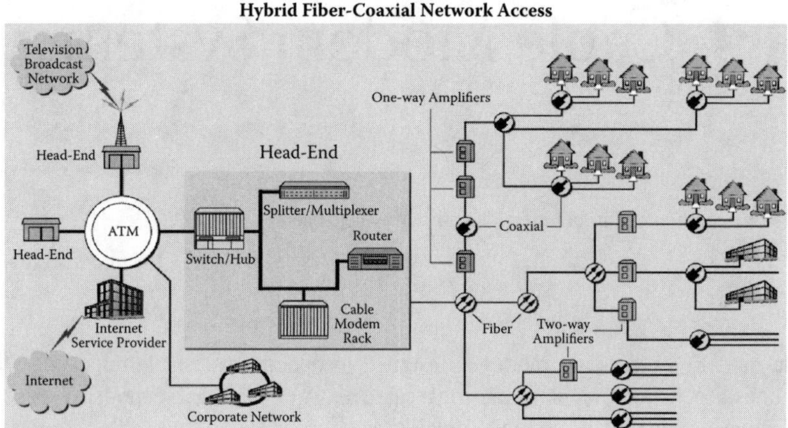

FIGURE 11.1 Future HFC cable network services.

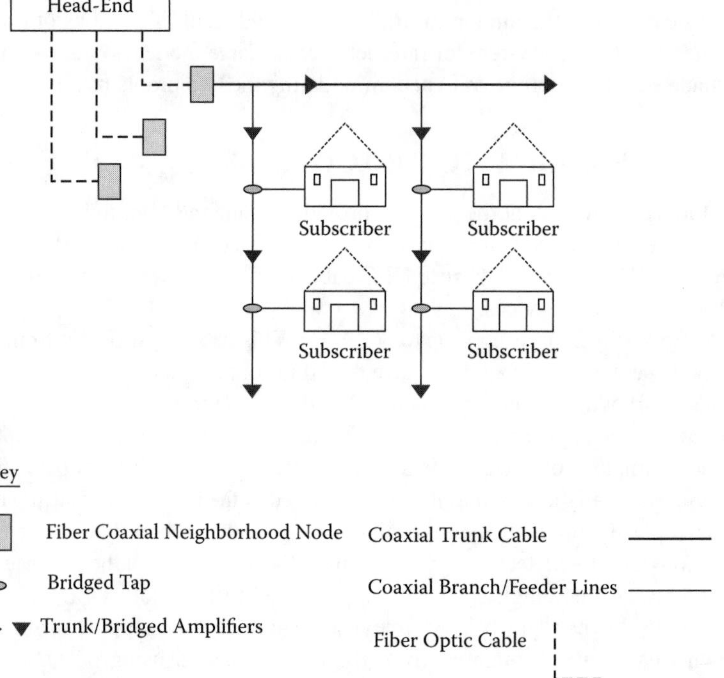

FIGURE 11.2 Hybrid/fiber coax cable network.

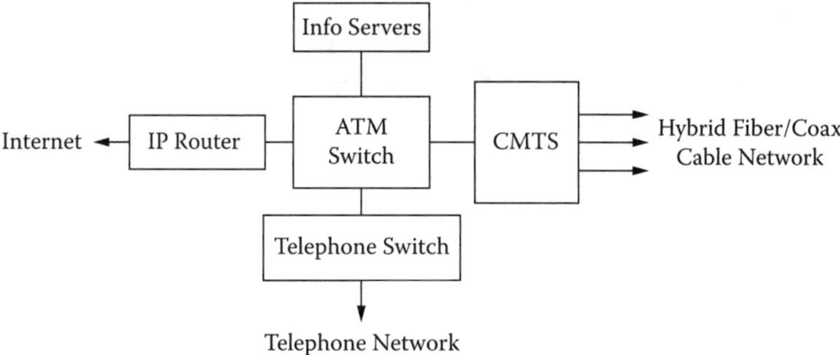

FIGURE 11.3 Type of equipment located at the headend of the cable modem systems network.

36 Mbps using 256 QAM. Upstream channels may deliver 500 kbps to 10 Mbps from homes using 16QAM or QPSK (quadrature phase shift key) modulation techniques, depending on the amount of spectrum allocated for service. To deliver data services over a cable network, one television channel (in the 50–750 MHz range) is typically allocated for downstream traffic to homes and another channel (in the 5–42 MHz band) is used to carry upstream signals. This upstream and downstream bandwidth is shared by the active data subscribers connected to a given coaxial cable distribution network segment, typically 500 to 2000 homes on a modern HFC network.

11.1.1 Structure of a Cable Modem System

The structure of a cable modem is shown in Figure 11.5. The tuner that connects directly to the cable television (CATV) outlet is a frequency-selective duplex filter that allows simultaneous transmission of information in opposite directions. During the modem boot process, the modem tuner scans the downstream spectrum until a

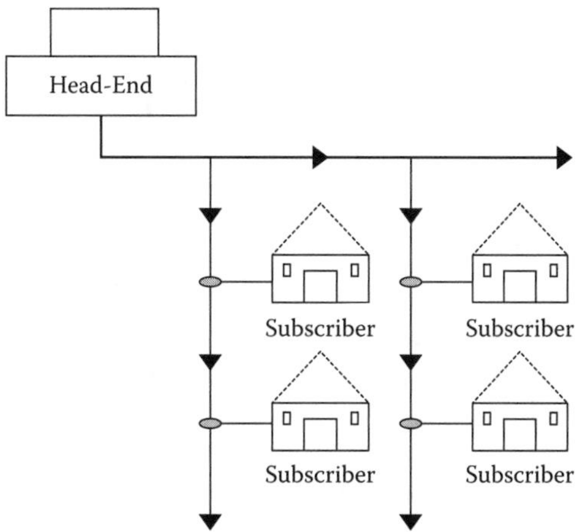

FIGURE 11.4 Traditional cable network.

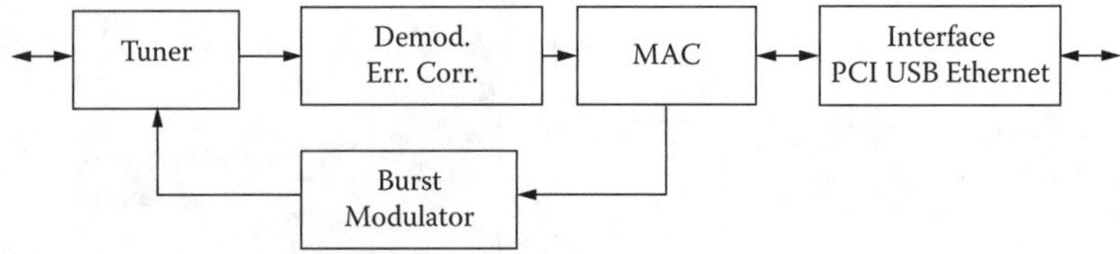

FIGURE 11.5 Inside the cable modem systems.

digital QAM modulated signal is encountered. The demodulator in the receive direction normally consists of analog-to-digital (A/D) converter, QAM-64/256 demodulator, MPEG frame synchronization, and Reed Solomon error correction.

The burst modulator in the transmit direction does Reed Solomon encoding of each burst modulation of the QPSK/QAM-16 on the selected frequency and digital-to-analog (D/A) conversion. The media access control (MAC) mechanism sits between the transmit and receive paths. The data that pass through the MAC go into the computer interface of the cable modem, be it Ethernet, USB, or PCI bus.

11.1.2 Cable Modem Systems Connection

In cable technology, the coax cable is connected to the splitter as shown in Figure 11.6 and Figure 11.7 from the cable television (CATV), which is the noninterrupted service part of the network architecture. The splitter is also connected to the cable modem. The cable modem is connected to the computer by Ethernet cable for high-speed Internet access. To deliver data services over a cable network, one television channel (in the 50–750 MHz range) is typically allocated for downstream traffic to homes and another channel (in the 5–42 MHz band) is used to carry upstream signals.

Cable operators had already begun upgrading their networks to handle more channels and offer more services before the Internet and the Web became household

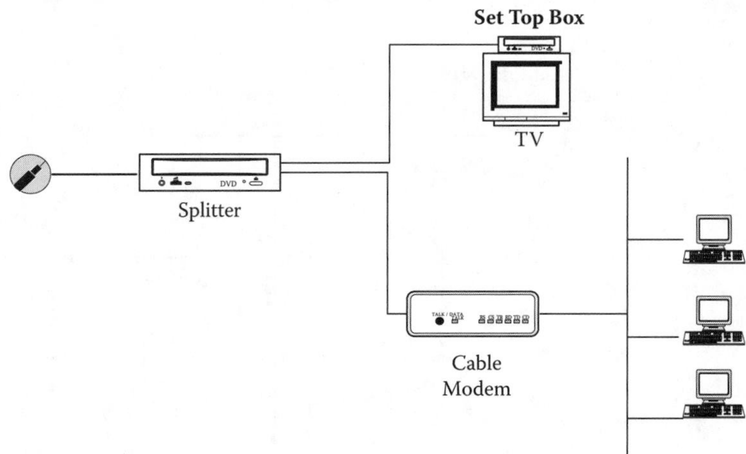

FIGURE 11.6 Cable modem systems connection at subscriber location.

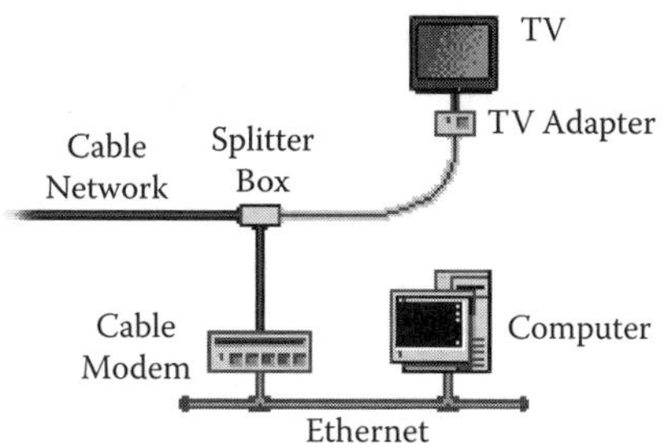

FIGURE 11.7 How cable modems work.

items. In recent years, the cable industry is at the undisputed heart of broadband. They are positioned to deliver video, voice, data, text, graphics, and more to both the PC and the TV in an inherent two-way capability. Three types of cable modem are possible: external modem, internal modem, and interactive set-top cable box as shown in Figures 11.6, 11.8, and 11.9.

11.2 EXTERNAL CABLE MODEM

The external cable modem is the small external box that connects to your computer normally through an ordinary Ethernet connection. The downside is that you need to add a (cheap) Ethernet card to your computer before you can connect the cable modem. A plus is that you can connect more computers to the Ethernet. In addition, the cable modem works with most operating systems and hardware platforms, including Mac, UNIX, laptop computers, etc.

Another interface for external cable modems is USB, which has the advantage of installing much faster (something that matters, because the cable operators are normally sending technicians out to install each cable modem). The downside is that you can only connect one PC to a USB-based cable modem.

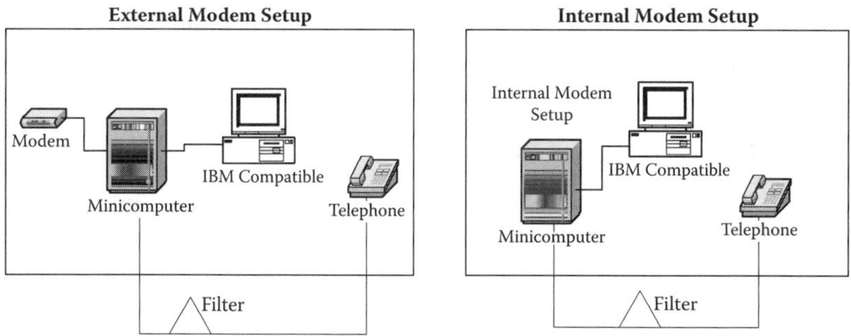

FIGURE 11.8 Typical setup in a home for an external DSL modem and an internal DSL modem.

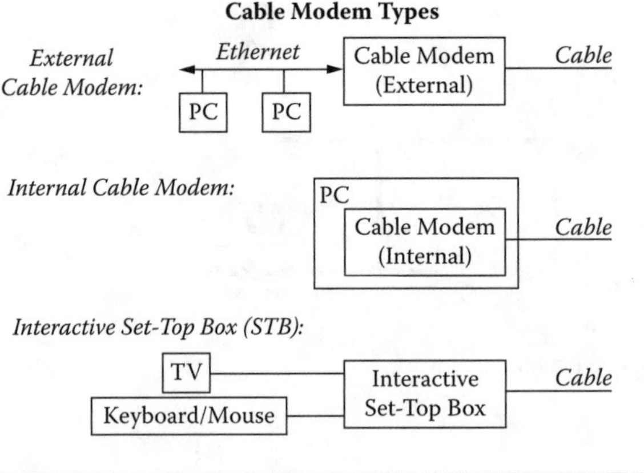

FIGURE 11.9 Different types of cable modem configurations.

11.3 INTERNAL CABLE MODEM

The internal cable modem is typically a PCI bus add-in card for a PC. This might be the cheapest implementation possible, but it has a number of drawbacks. The first problem is that it can only be used in desktop PCs. Macs and laptops are possible, but require a different design. The second problem is that the cable connector is not galvanic isolated from AC mains. This may pose a problem in some CATV networks, requiring a more expensive upgrade of the network installations. Some countries and/or CATV networks may not be able to use internal cable modems at all for technical and/or regulatory reasons.

11.4 INTERACTIVE SET-TOP CABLE BOX

The interactive set-top box is really a cable modem in disguise. The primary function of the set-top box is to provide more TV channels on the same limited number of frequencies. This is possible with the use of digital television encoding (DVB). An interactive set-top box provides a return channel—often through the ordinary plain old telephone system (POTS)—that allows the user access to Web browsing, email, etc. directly on the TV screen.

11.5 CABLE INTERNET ACCESS

The popularity of ADSL has not gone unchallenged. Cable television companies also wanted to meet the demand for broadband by offering Internet access through cable networks. Some of the restrictions for ADSL availability make it possible for CATV companies to offer Internet access to the public. Almost half of the households in the United States are still using dial-up connections to access the Internet. By contrast, in 2002 Korea had a market penetration of about 57% of all households using cable Internet access.

One of the strengths of the cable system is the structure of the network. Cable networks were built with coaxial cable, which carries analog signals for several miles. To reach remote areas, signal amplifiers are placed along cable paths as shown in Figure 11.1. Moreover, cable companies discovered that using fiber-optic cable as shown in Figures 11.1, 11.2, and 11.3 in place of coaxial cable in the network allows the signal to travel even further distances.

While most of the network is still coaxial, the current cable network is a hybrid of both fiber and coaxial mediums. The benefit of the HFC network is a system capable of full duplex communications—not just broadcast. For all the available 750-MHz bandwidth, only a small portion is used for voice, data, and other services. Upstream is in the range of 5 MHz to 42 MHz, and downstream covers 50 MHz to 750 MHz. Data rates for upstream and downstream for cable are comparable to DSL.

However, speeds are expected to climb to 3 Mb/s for upstream and as much as 30 Mb/s for downstream. In order to reach these extreme speeds, CATV companies will need to place fiber completely out to the customer premises, not just to local headends as shown in Figures 11.1 and 11.2. More network rebuilding is necessary for efficient duplex communication. Some of the present strengths of the cable network serve as problems that must be solved in order for cable companies to realize future communication strategies.

In spite of some pitfalls with the cable network architecture, CATV companies are competing with telcos to provide Internet access by utilizing a modulation technique that is much different from DMT for ADSL. At the CATV company's headend, the cable modem termination system (CMTS) operates in much the same way as the DSLAM for ADSL. The CMTS as shown in Figure 11.3 handles the administration of the upstream and downstream traffic by using allocated servers for assigning Internet packet (IP) addresses, by routing traffic directly to the Internet, or routing traffic to a third party Internet service provider (ISP) for connection to the Internet.

The CMTS can enable up to 1000 users to connect to the Internet using only a 6-MHz channel—the same bandwidth used for cable TV. In addition, at the headend, the cable provider will have dedicated control servers that operate on the Cable Labs certified cable modem protocol (or better known as data over cable service interface specifications; DOCSIS) for data transmission.

DOCSIS is the standardized protocol for Internet access using cable. The DOCSIS standard specifies the protocol and type of modulation techniques used for duplex communication over cable. It supports upstream data rates of 10 Mb/s and downstream data rates of up 27 Mb/s. Normally, however, due to throughput limitations of T1 lines to the Internet and interoperability standards, the average user will see speeds of 500 kb/s to 2.5 Mb/s for upstream and 1.5 Mb/s to 3 Mb/s for downstream.

Many customers like cable Internet access because it is easy to subscribe to the service. Performance with cable does not depend on the distance of the customer's residence from the headend. Also, cable Internet users can watch cable television and surf the Internet at the same time. The download speeds users experience rival those of DSL, and the connection is always on. Cable access only requires a cable modem and sometimes a cable splitter as shown in Figure 11.7. In addition, many of the individuals that cannot sign up for ADSL because of central office (CO) proximity can get Internet access through a CATV company. Thus, the market for cable Internet access is steadily climbing.

11.6 USING QPSK AND QAM IN CABLE MODEM SYSTEMS

Several modulation techniques are available and can readily be used in many communication systems depending on the design and structure. In the case of cable modem systems, quadrature phase shift keying (QPSK) and quadrature amplitude modulation (QAM) are two different forms of modulation techniques used with cable modem systems technology. QPSK can be implemented easily and tends to resist noise fairly well. It is used primarily for sending data from the cable modem system subscriber upstream to the headend, but sometimes it can also be used to send data downstream.

QAM is primarily used for sending data downstream. QAM is very efficient, but QAM's susceptibility to interfering signals makes it ill suited to noisy upstream transmissions from the subscriber to the headend. There are three variations, 16QAM, 64QAM, and 256QAM. A single downstream 6-MHz channel may support up to 27 Mbps of data throughput from the headend using 64QAM, while speeds can be boosted to 36 Mbps using 256QAM. A single 6-MHz upstream channel may support between 500 kbps and 10 Mbps of data throughput from cable modem system subscribers' homes using 16QAM or QPSK, depending on the amount of frequency spectrum allocated for the service.

11.7 COMPARISON BETWEEN BROADBAND DSL AND CABLE MODEM TECHNOLOGIES

Both DSL technology and cable modem technology have their specific pros and cons. For example, once the cable plant has been upgraded to two-way broadband, adding high-speed access is extremely cheap. This is important for telecommunications broadband infrastructure developments. Adding new subscribers on a traditional wireline network is expensive, as it is on a wireless network. Cable modem services offer shared bandwidth among all the neighbors. The actual speed will vary with how many people are on the cable modem network, which may be a disadvantage. With DSL service, every user has a dedicated connection.

The impedance of widespread installation of DSL is the physical constraints of the local loop (number of load coils and bridge taps as well as its distance from the central office), and the interoperability problems between products, and spectrum compatibility issues. Phone lines endure a lot of noise, which obscures or disguises the bit. The noise is typically electromagnetically coupled into phone lines. The external sources of energy that contribute to noise can include signals on other phone lines (known as crosstalk), radio, and ham broadcasts, and virtually any type of electrical or mechanical equipment within close proximity to the phone line.

While crosstalk was not much of a problem in the analog world, it has the potential to become a severe problem for DSL users. DSL requires much higher frequencies than analog; consequently, the number of lines subject to crosstalk or other interference increases significantly. Performance will be severely impaired if the DSL line is tied to a T1 line or other high-speed lines.

The problem with deploying DSL is the use of bridge taps, which are branches that are attached to circuits to reroute the line to another location. In most cases, 56 percent of the loop population has bridge taps. A bridge tap is any portion of the loop that is not in the direct path between the central office (CO) and the end user. Telephone companies have used bridge taps so they could cheaply build in extra capacity in a neighborhood without knowing in advance whether the demand existed. The number of load coils and bridge taps on the lines and the telecom companies' inability to locate them quickly without costly analysis and mapping of all lines is a real problem without an easy solution.

The other main obstacle to a successful deployment of DSL to the mass market is loop testing and qualification issues. Prequalification is the testing of loops to determine if the loop is capable of supporting DSL transmission prior to attempting to provide the service to the customer and is critical because DSL deployment is dependent on the design and quality of the copper loop.

Operation of the network in the return path demands that the cable modem HFC network be designed with particular care. Usually the frequency range of 5 to a maximum of 65 MHz is used. Problems may occur due to interfering signals (harmonics and other RF signals generated from household appliances) often noted in this low frequency range. When building up or upgrading a system it is therefore absolutely necessary to use components with high screening factors equipped with filtering to suppress any ingress noise.

The real commonality between DSL modems and cable modems is that they both are broadband communication systems that allow high-speed connection to the Internet by increasing the bandwidth available to the user. The two technologies access the broadband networks through different mediums. The DSL offers high-speed connection using the existing copper twisted wires of the phone line while cable modems offer high-speed connection by using the cable lines going to the user televisions. The DSL has a specific bandwidth dedicated to each user so that the user can expect a certain speed at all times during the day or night, while in the case of the cable, the technology is a shared technology where in times of high occupancy and usage, one can experience slower than normal speeds.

Some of the different types of DSL modems are asymmetric DSL (ADSL), symmetric DSL (SDSL), ISDN DSL (IDSL), high-bit-rate DSL (HDSL), very high-bit-rate DSL (VDSL), and rate adaptive DSL (RADSL). The cable technology has three different modems available: external modem, internal modem, and interactive set-top cable box as shown in Figures 11.6, 11.8, and 11.9. In Figure 11.8, Bell South Fast Access shows a typical DSL setup for an external and internal DSL modem.

When deciding whether or not to invest in DSL or cable access broadband communication modems, you need to study the comparisons of the two technologies so that an intelligent decision can be made. The DSL is more than 300 times faster than the 56-kbps dial-up modem. It is always on and has better security than cable modem. The performance of the DSL depends on the location of the modem. The setup is usually difficult and for business users, higher speeds can get pricey. In the case of cable modem, it is potentially faster than DSL. There may be price breaks if you get cable television service. The cable line can be shared with others in the neighborhood and speeds vary accordingly.

The installation fee for DSL is about \$100.00 to \$200.00 while in the case of cable, it is from \$75 to \$200. The monthly rate for DSL is from \$40 to \$80 for home users, and \$80 to \$320 for business users. In cable, the monthly rate is from \$39.95 to \$49.95. The

TABLE 11.1
Comparison of Data Rates for ADSL and Cable Modem Systems

	Cable Modem	ADSL
Downstream Rate	27 Mb/s (64QAM)–38 Mb/s(256 QAM)	0–8 Mb/s (DMT)0–1.5 Mb/s (G.lite)
Upstream Rate	320 kb/s (QPSK)–10 Mb/s (16QAM)	0–1 Mb/s (DMT and G.lite)
Monthly Price	~$45/month	~$50/month
Availability	75% of USA HH 100% of cable modem-ready HH	~80% of USA HH

connection speed ranges are ADSL: 384 kbps to 9 Mbps download/128 kbps upload; SDSL: 1.5 Mbps download/upload; ISDL: up to 144 kbps download/upload. The cable modem connection speed ranges are 1 to 2 Mbps download/128 to 384 kbps upload.

Recently, Internet access via cable has become available in many cities. When configured correctly and under good conditions, cable can transmit data just as fast as DSL. Actually, cable has the potential to surpass DSL speeds (see Table 11.1), but this capability is contingent on resolving certain problems with the cable network itself. However, on the corporate side, cable cannot compete with DSL because cable networks usually do not reach many commercial districts. From its inception, cable networks were designed for residential use. Thus, the real competition is for the consumer Internet dollar.

Looking at both access technologies there are benefits and limitations that cable and DSL have. For example, cable does offer speeds that are comparable to DSL. However, cable access is provided over a shard network (public). During heavy usage of cable network, a user will experience a significant drop in performance; on the other hand, DSL is provided on a dedicated line—it is not public—so the performance does not degrade. Since cable networks are public, there are security issues that must be addressed. This is the reason that telecommuters who connect to the Internet using cable are limited or even prohibited from accessing the company LAN.

Both technologies pose for a situation of the "best of both worlds." The telephone network is global, yet proximity to the CO can disallow ADSL service. Cable TV networks, though not as extensive as the public switched telephone network (PSTN), are also vast global networks. Still, cable Internet users experience major performance degradation—even a halt in service—with too many users on any one segment of the CATV network simultaneously. In order to meet the needs of a growing demand for Internet access from the general public, both telcos and CATV providers will need to address their particular network shortcomings. Both industries are planning for the future, and the outcome for both industries appears to benefit the consumer and business alike.

Cable is great for home or teleworker use, since the upgraded digital cable plant passes most homes in North America. However, for small, medium, and distributed enterprises, cable Internet availability is a challenge. While DSL is broadly available in most business districts, the cable Internet plant is not generally available

TABLE 11.2
Comparison of DSL and Cable Modem Systems

	DSL Systems	Cable Modem System
Always-on	Yes	Yes
Speed	8 Mbps down	27–36 Mbps down
	1.5 Mbps up	500 kbps–10 Mbps up
	Fixed speed, but depends on location	Speed varies as number of active users
Bandwidth	25 k–120 kHz down	50–750 MHz down
	138 k–1.1 MHz up	5–42 MHz up
Security	Dedicated line, better security	Shared line, easier to snoop
Wire	Unshielded twisted-pair line	Screening coax cable
Price	High	Price break if you get cable TV service too
Interference	Crosstalk	RF interference at upstream frequency range
Coverage	Steady growth	Wider user coverage

in commercial areas. DSL is a better choice for business-class services. The lessons therefore, are that they should adopt, as much as possible, the current North American model where cable and DSL technologies are used in specific application areas. This is also the case for most of the technologically developed countries of the world. Table 11.2 lists the comparison of DSL and cable modem system for several criteria.

11.8 CABLE MODEM SYSTEMS STANDARDS COMPLIANCE

There are a number of standards-making bodies and consortiums who are responsible for evaluating issues relating to multimedia service definition and multimedia systems, including the associated modems, protocols, and signal processing. To name a few, ANSI and the ITU-T Study Group 16 are involved in studies on multimedia standards; Cable Labs is looking at, among other things, interoperability specs for devices connected to cable networks; the IETF is looking at IP over cable networks; and the IEEE 802.14 and the ETSI are promoting standards for interactivity. The main focus of these standards-making bodies is on the physical and MAC layer issues, looking at the different modulation schemes.

With reference to cable modem systems, there are standard interfaces and protocols associated with cable. Currently, a majority of cable modem system vendors are applying the Cable Labs' specifications. The Cable Labs' standard defines three layers for delivering IP traffic over the network:

- Network layer uses the Internet protocol (IP)
- Data link layer consists of three sublayers:
  - logical link control conforms to Ethernet standards
  - link security sublayer supports authorization and authentication
  - media access control sublayer is suitable for cable system operation that supports variable-length PDUs. This is the most critical layer in the

protocol since it provides for bandwidth efficiency through variable-length PDUs, extensions for future support of ATM, and support for multiple grades of service and data rates.

- Physical layer provides support for both downstream and upstream traffic.

Other specifications are defined that enable the cable modem systems to "discover" the network and provide reliable transmission. An interesting aspect of using Ethernet-type protocols is the media access control for HFC. Standard LAN techniques, such as CSMA/CD or token ring, are inadequate due to the extreme distances that cable networks span. Instead of using those standard LAN techniques, vendors are converging to credit-based schemes that grant access to the medium for a specified period of time. Collision resolution algorithms to address collisions on grant requests are also an important consideration. There are few existing cable modem standards developed today, but there are many standards under review.

11.8.1 Cable Modem Systems

- Multimedia cable network systems - MCNS
- IEEE 802.14: Cable-TV access method and physical layer specification
- SP-RFI-I02-971008: Data-over-cable service interface specifications: radio frequency interface specification
- SP-CMCII01-960702: Data-over-cable service interface specifications: cable modem to customer premise equipment interface specification.
- SP-CMTS-NSII01-960702: Data-over-cable service interface specifications: cable modem termination system-network side interface specification.
- SP-CMTRI-I01-970804: Data-over-cable service interface specifications: radio frequency interface specification.
- Digital Audio-Visual Council (DAVIC)
- ITU-T J83/ANNEX A and J83/ANNEX B
- Digital video broadcasting (DVB) cable

11.8.2 Wireless Channel

- IEEE Standard 802.11-1997: Wireless LAN medium access control (MAC) and physical layer (PHY) specifications.
- ITU[8-1/29]: Spectrum requirements for terrestrial third-generation wireless IMT-2000 in the year 2010.

11.8.3 LAN

- IEEE Standard 802.11-1997: Wireless LAN medium access control (MAC) and physical layer (PHY) specifications.
- ISO/IEC 15802-1: 1995, Information technology-telecommunications and information exchange between systems-local and metropolitan area networks-common specifications-part 1: medium access control (MAC) service definition.

11.8.4 Set-Top Boxes

- MCNS: Multimedia Cable Network System Partners Ltd, March 1997.
- IEEE 802.14: Standard protocol for cable-TV based broadband communication network.
- IEEE 802.14/a Draft 2 Revision2: Cable-TV access method and physical layer specification, 1997.
- MPEG-2, MPEG-4

11.8.5 Digital Broadcast Systems

- Digital Audio-Visual Council (DAVIC).
- DAVIC 1.0 Rev. 3.1, 1995(T1E1.4): Lower layer protocols and physical interfaces, Part 8.
- Digital video broadcasting (DVB).
- ATSC digital television standard - DOCS. A/52, A/54, A/56, A/58.

11.8.6 Others

- CSA - Carrier serving area defined as all customer premises within 12 kft on 24 AWG (1260 Ω d.c. loop resistance) or 9 kft on 26 AWG (1502Ω) of a central office. CSA includes about 75% of all customers approximately.
- ECSA - Extended carrier serving area is defined as customer premises within 18 kft on 24 AWG (1890Ω) or 13.5 kft on 26 AWG (2254Ω) of a central office. ECSA includes about 90% of all customers approximately.

SUMMARY

1. Cable modem systems are modems designed to operate over cable television (cable TV) lines.
2. CATV companies are currently laying more fiber optic cable in order to deliver potential data rates and comply with future standards. Moreover, many large communications companies, multimedia cable network system (MCNS), are working to resolve security and other issues in the cable industry.
3. To deliver data services over a cable network, one television channel (in the 50–750 MHz range) is typically allocated for downstream traffic to homes and another channel (in the 5–42 MHz band) is used to carry upstream signals.
4. The quick adoption of these technologies by consumers and the limited use of the technologies by business indicate that broadband usage will increase.
5. Cable also provides high data rates of up to 2.5 Mb/s for upstream and up to 3 Mb/s for downstream. DSL provides very high data speeds of up to 1.5 Mb/s for upstream and up to 9 Mb/s for downstream.
6. Cable modem services offer shared bandwidth among all the neighbors.

REVIEW QUESTIONS

11.1 Which one is a type of cable modem system?
   a. Set-top box
   b. Diplex filter
   c. Local headend
   d. Fiber node

11.2 The data that pass through the MAC go into the computer interface of the cable modem, be it
   a. Ethernet
   b. USB
   c. PCI bus
   d. All of the above
   e. None of the above

11.3 During the modem boot process, how does the modem acquire the downstream channel?
   a. The modem is commanded by the CMTS to set to the specific channel.
   b. The modem uses the default value in the Broadcom chipset.
   c. The modem tuner sets a level as defined by the DOCSIS specification.
   d. The modem tuner scans the downstream spectrum until digital QAM modulated signal is encountered.

11.4 QPSK and QAM are two types of modulation techniques used in cable modem systems.
   a. True
   b. False

11.5 Cable modem systems have wider user coverage than DSL systems.
   a. True
   b. False

11.6 Cable operators can allow signals to flow in both directions.
   a. True
   b. False

11.7 Cable modem systems provide high data rates of up to 2.5 Mb/s for upstream and up to 3 Mb/s for downstream.
   a. True
   b. False

11.8 Cable modem can be classified as another way to provide broadband communication services.
   a. True
   b. False

11.9 Cable technology has which other type(s) of modems?
   a. External modem
   b. Internal modem
   c. Interactive set-top box
   d. All of the above

11.10 Which of these organizations are responsible for making cable standards?
   a. Cable Labs
   b. DAVIC

c. MCNS
d. DVB
e. All of the above

Answer: 11.1a, 11.2d, 11.3d, 11.4a, 11.5a, 11.6a, 11.7a, 11.8a, 11.9d, 11.10e

PROBLEMS

11.1 Describe the meaning of a cable modem system.
11.2 Describe the different types of cable modem systems. Illustrate with appropriate diagrams.
11.3 What are the differences between DSL systems and cable modem systems technologies?
11.4 What are the traditional roles of the cable systems network?
11.5 Compare the traditional roles of the cable systems network and the current changes that have been made to the traditional roles.
11.6 Describe the various applications of cable modem system technology.
11.7 Describe how cable modem system operators allow signal to flow in both directions.
11.8 Describe the traditional cable modem system network and compare to that of today's cable network.
11.9 In cable modem systems, what proportion of the bandwidth capacity is assigned to carry signals upstream and downstream?
11.10 Describe a cable modem termination system (CMTS).
11.11 What is the significance of some cable operators changing from coaxial cables to fiber optic cables?
11.12 What is the bandwidth capacity of the cable modem systems?
11.13 Is it possible for a cable modem system to communicate directly with another cable modem system and why?
11.14 Describe the major difference between upstream and downstream signals in cable modem systems.
11.15 Describe a media access control layer (MAC) and logical link control layer (LLC) used in cable modem systems.
11.16 What is a fiber coaxial neighborhood node in cable modem systems?
11.17 What is the relationship between a cable modem system and the Internet?
11.18 Explain why it is not possible for everyone to have access to cable modem systems.
11.19 Briefly describe how a cable modem system works. Illustrate using a diagram.
11.20 Using a table, illustrate the real world performance of a number of connection methods. Use as a base factor for this table the standard "33.6 modem." This is a device that can transmit data at a rate of 33,600 bits per second, which is equivalent to 4200 characters per second.

12 Passive Optical Networks

It is every man's obligation to put back into the world at least the equivalent of what he takes out of it.

— Albert Einstein

One of the emerging new technologies in broadband communication systems is the passive optical networks (PONs). The generic topology of a PON is as shown in Figure 12.1. The word "passive" implies that optical transmission has no power requirements or active electronic parts once the signal is going through the network. In fiber-to-the-x (FTTx) deployments, PON is the leading technology used. The "x" in FTTx stands for premises or home or curb or neighborhood, etc. For example, depending on where the PON terminates, FTTx can stand for fiber-to-the-premises (FTTP), fiber-to-the-home (FTTH), fiber-to-the-curb (FTTC), fiber-to-the-neighborhood, fiber-to-the-village (FTTV), or fiber-to-the-business (FTTB). They are some of the hottest topics in optics. Figure 12.2 shows some of the multiple services supported by PON, especially, the wavelength division multiplexed PON (WDM-PON). The $\lambda_{(i)}$ represent the various wavelength signals. The difference between PON and the other optical network topologies is that the former is a point-to-multipoint (P2MP) topology while the latter is not. The PON current standards have defined downstream data rates up to 2.5 Gbps.

In PON, the enabling technology is laser, which is used in the optical line terminal (OLT) and the optical network unit (ONU) components for transmission and reception. It is used also in wave division multiplexing (WDM) with only one wavelength currently being used for each upstream and downstream direction. In Japan, the PON market is well established and growing at a rate of 300% a year, while in North America, the PON market is just starting to emerge. When compared to the DSL broadband access technology, DSL is growing at 150% a year. In the United States, Verizon, SBC, and Bell South have all announced some type of PON deployment.

The objective of this chapter is to provide an in-depth discussion of PON, the basic principles, design, advantages, and disadvantages of this new emerging broadband technology. As part of the objectives of this chapter, we will discuss the security and the key standard issues facing the PONs.

12.1 BASIC OPERATIONAL PRINCIPLES OF PON

Definition 1: A passive optical network (PON) is a system that brings optical fiber cabling and signals all or most of the way to the end user.

Definition 2: Optical fiber, also known as "fiber optic," is defined as the medium and the technology that is associated with the transmission of information as light pulses along a glass or plastic wire or fiber.

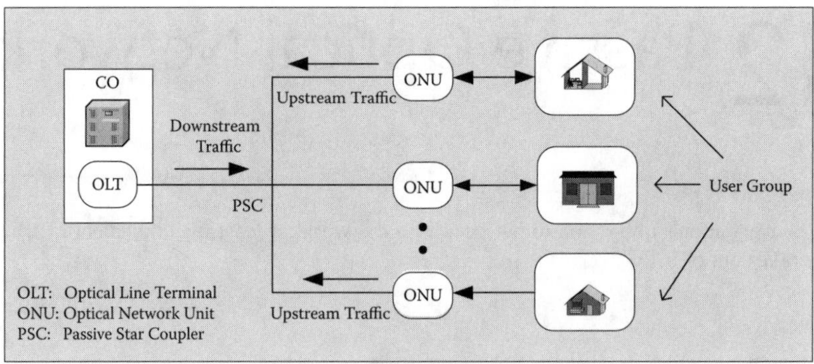

FIGURE 12.1 Passive optical network (PON) topology.

Optical fiber carries much more information than conventional copper wire and is in general not subject to electromagnetic interference. Carriers always want to connect each of their customer's sites with a wavelength of light. However, the carriers would like to avoid having to dedicate a fiber to every wavelength. They use PONs to address this issue by bundling together multiple wavelengths (up to 32 at present) so they can be carried over a single access line from the carrier's central office (CO) to a manhole or controlled environmental vault close to a cluster of customer sites.

PON operates from the central office's (CO) optical line terminal (OLT) where a single strand of fiber goes out to a passive optical splitter where its signal is multiplied to 32 different lines. The transmit and receive signals operate on different wavelengths allowing bidirectional operation over a single fiber. From the OLT,

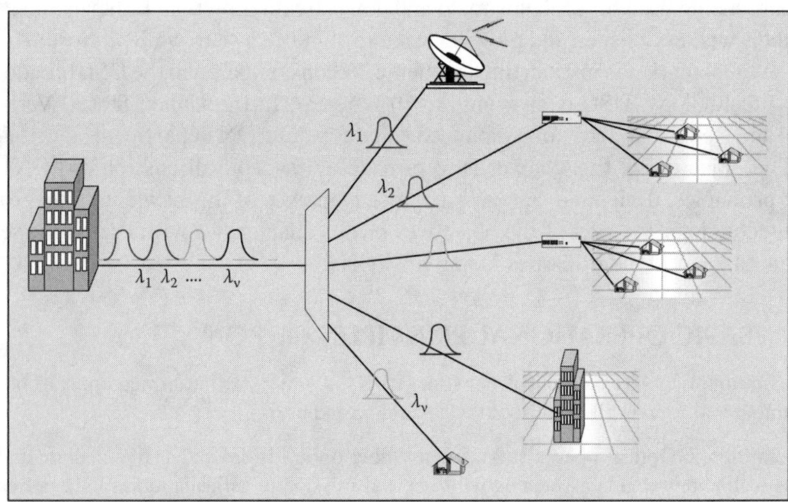

FIGURE 12.2 Multiple services supported by PON, especially WDM-PON.

anything transmitted is transmitted to all 32 lines. It is up to the customer's system optical network unit (ONU) to determine what packets are for that customer. The other packets are all discarded.

In a stand-alone system, a PON could deliver up to 622 Mbps downstream to the user and up to 155 Mbps upstream. Multiple users of a PON could be allocated portions of this bandwidth. A PON could also serve as a trunk between a larger system, such as a CATV system, and a neighborhood, building, or home Ethernet network on coaxial cable. In Japan, Nippon Telephone and Telegraph (NTT) is planning to install passive optical networks throughout its system.

Because all 32 lines are multiplexed to a single fiber, only one ONU may transmit at a time. To solve this problem, each ONU is given a time slot where it can transmit a burst of data. The order of transmission is determined by OLT. From a component point of view:

- OLT to ONU: Exactly like in an ordinary point-to-point link (P2P), the link operates in a continuous mode. Standard post amplifier parts may be used in the ONU.
- ONU to OLT: The link must operate in a bursty mode. Each ONU will send a burst of data. If the ONU is not transmitting, the laser must be in an off state. Therefore, the laser driver needs to be able to turn on and off very quickly.

The promise of high-speed data services as well as emerging triple play products delivered over a low-cost network structure make PON systems a very interesting option to all network operators, especially the ones with aging DSL lines in need of upgrade/overhaul.

12.2 THE VARIOUS TYPES OF PON ARCHITECTURES

There are various types of PON architectures. The general technology is the same. The differences are in the specifications and the upper layer protocols. The main PON architectures are ATM PON (APON), broadband PON (BPON), Ethernet-based PON (EPON), gigabit PON (GPON), and wavelength-division-multiplexed passive optical network (WDM-PON).

12.2.1 APON—ATM PON Architecture

APON is a point-to-multipoint architecture. The downstream traffic transport is broadcasting. The upstream traffic transport uses a TDMA protocol. The architecture is packet-based access technology. APON is based on the ATM framework with a maximum reach of 20 km and is designed to be a cost-sensitive access solution. It is standardized through the ITU-T 983.1 recommendations. Figure 12.3 shows the basic architecture of APON while Figure 12.4 shows the APON solution design to support voice, video, and data. Its systems are for voice, video, and data services. APON systems carry Ethernet traffic seamlessly. The highly developed, mature, and standardized suite of ATM traffic management techniques and off-the-shelf ATM

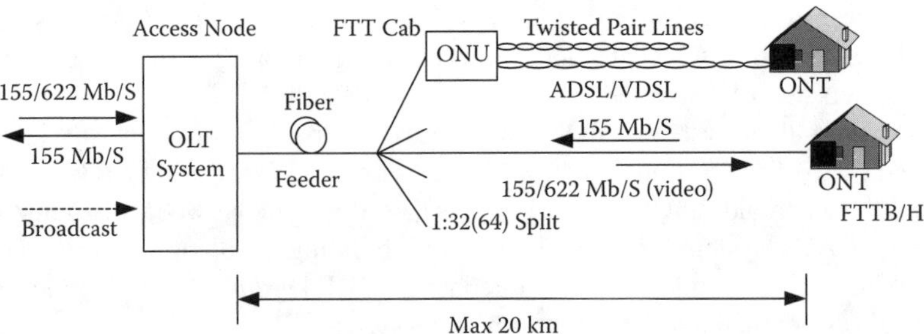

FIGURE 12.3 Basic APON architecture (FTTCabinet/Curb/Cab, FTTBuilding/Home).

components gives APON a solid technological base. Figure 12.5 shows the APON access techniques in the broadband communication networks.

The synchronization of the frames enables the proper alignment of time slots for upstream time division multiplexer access (TDMA) traffic. In order to avoid cell collisions, the OLT sends "grants" to the ONT that assign time slots for upstream transmission. The OLT controls the flow of information from multiple OLTs. APON can guarantee QoS for diverse network applications cost effectively and is thus attractive to major service providers such as incumbent local exchange carriers (ILECs). APON systems support downlink performance of 155 to 622 Mbps and uplink performance of 155 Mbps.

12.2.2 BPON—Broadband PON Architecture

Broadband passive optical networking (BPON) is a point-to-multipoint architecture for delivering last-mile connectivity without any active components in the

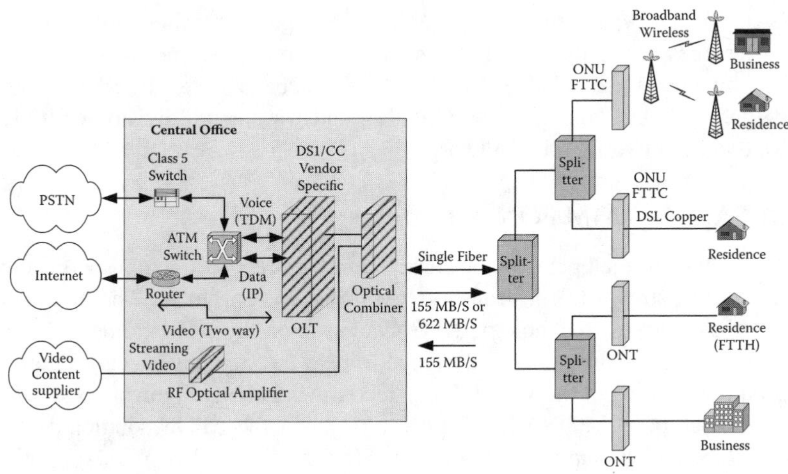

FIGURE 12.4 APON solution designed to support voice, data, and video.

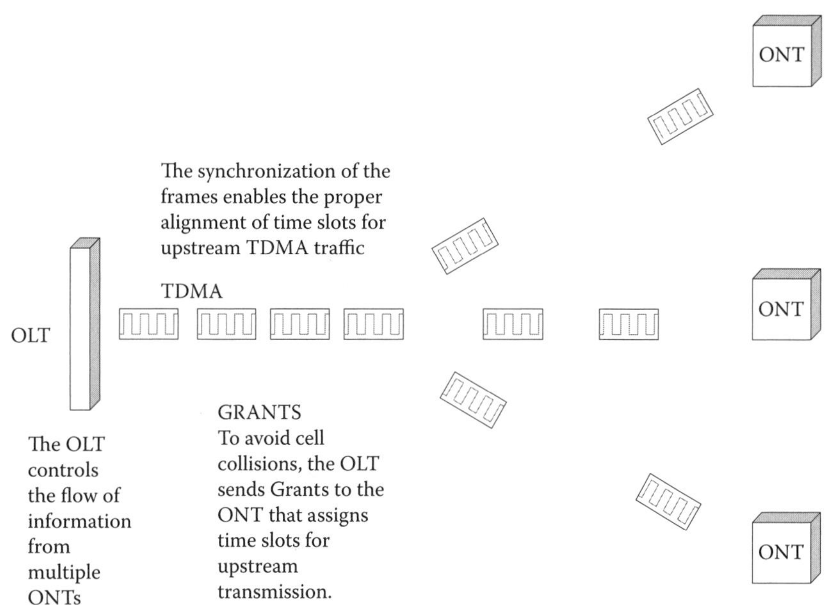

FIGURE 12.5 APON access.

distribution network. Figure 12.6 shows the typical BPON architecture while Figure 12.7 shows a typical BPON architecture with video application service. The fact that there is no powered equipment means that there is less cost, less network management, longer reach, and no need for an upgrade. BPON results in a 10x increase in bandwidth over existing broadband technologies, such as xDSL and broadband cable.

BPON systems support downlink performance of 155 to 622 Mbps and uplink performance of 155 Mbps. BPON provides efficient convergence of voice, data, telemetry, ATM, and TDM private line circuit services, with full QoS and service level agreement (SLA) management.

12.2.3 EPON—Ethernet-Based PON Architecture

There is a promising future for EPON as a low-cost solution for the bottleneck access. It is more efficient and cost effective than APONs. Figure 12.8 shows the EPON configuration. It has the headend (OLT), the 1:N optical splitter, the single fiber optics, and the subscriber (ONU). The subscribers see traffic only from headend, not from each other. The headend permits only one subscriber at a time to transmit. EPON deploys only one channel for downstream traffic and another channel for upstream that are both shared by all ONUs as shown in Figure 12.9 and Figure 12.10, respectively, and it is standardized by the IEEE 802.3ah working group.

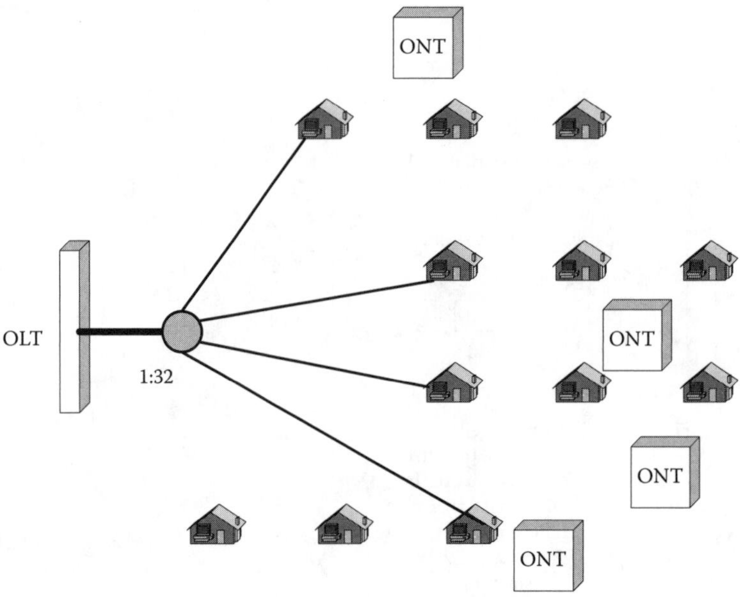

FIGURE 12.6 Typical BPON architecture.

The medium access control (MAC) of EPON supports time division multiplexing (TDM), where each ONU is allocated a fixed or dynamic time slot depending on the dynamic bandwidth allocation (DBA) scheme to transmit data to the optical line termination (OLT). The packets received from the different subscribers are buffered by each of the ONUs until they are transmitted in the assigned window.

Ethernet data streams are made up in most cases from a plethora of individual data frames; therefore, the overall transmission overhead for such data transaction can be assessed only in statistical terms. Therefore, it is necessary to consider the

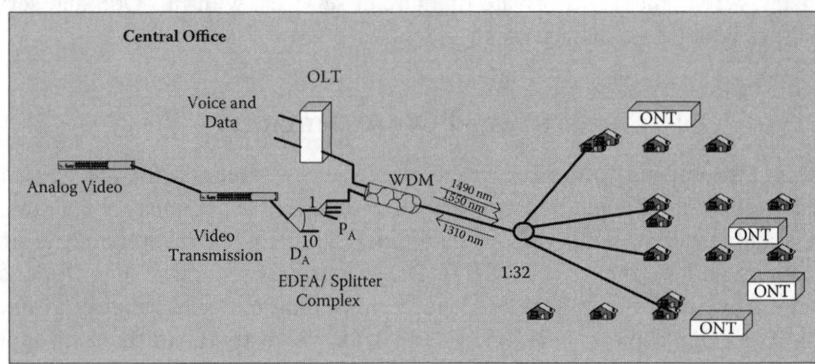

FIGURE 12.7 Typical BPON with video architecture.

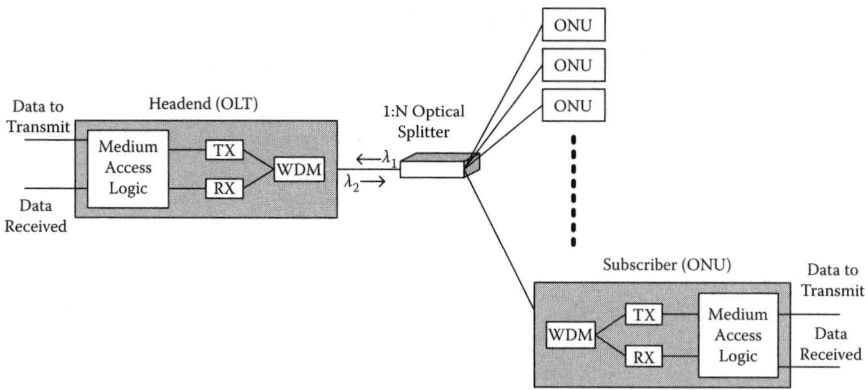

FIGURE 12.8 EPON configuration.

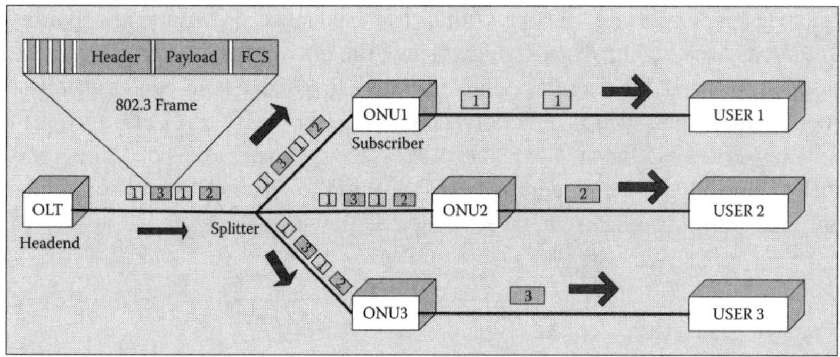

FIGURE 12.9 EPON downstream.

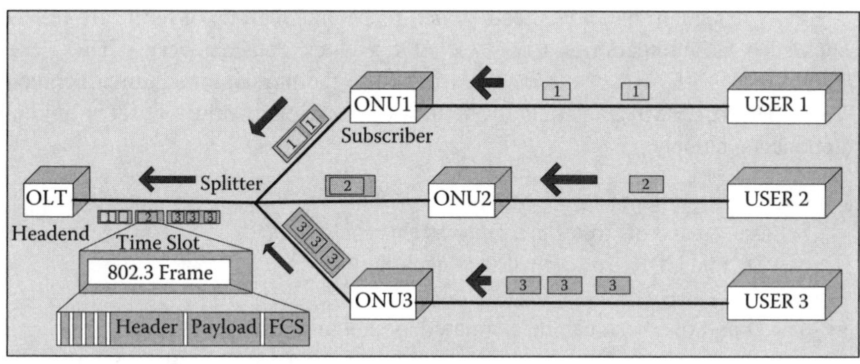

FIGURE 12.10 EPON upstream.

packet size distribution for the given Ethernet link. The average transmission overhead value can be estimated using Equation (12.1)

$$O_{avg} = 1 - \frac{\sum_{n=21}^{1500} n \times f(n)}{\sum_{n=21}^{1500} E(n) \times f(n)} \tag{12.1}$$

$$E(n) = (n;46)_{\max} + 38 \tag{12.2}$$

where n is the payload size (21 – 1500 bytes), *f(n)* is the probability distribution function (PDF) for the given network link, and *E(n)* is the total size of the encapsulated payload of size n, including Ethernet transmission overhead, and is given by Equation (12.2).

Equation (12.2) defines the Ethernet framing function where 38 bytes of the overload include frame preamble, header, frame check sequence (FCS), and inter packet gaps (IPG). In the case of control message overhead, the flow control message overhead represents the fraction of the overall available bandwidth lost due to in-band transmission of flow control messages, such as multipoint control protocol (MPCP) GATE and REPORT. This makes it possible for the flow control message overhead to depend directly on the number of active logical link identifiers (LLIDs) in the system and on the polling cycle length. The control overhead can be estimated as shown in Equation (12.3):

$$O_{control} = \frac{MPCPDU_{size} \times N_{ONU}}{cycle_{time}[s] \times data_{rate}[bps]} \tag{12.3}$$

where $MPCPDU_{size}$ is the standard MPCPDU message size, equal to 64 bytes, increased by Ethernet preamble and IPG, amounting to 84 bytes; N_{ONU} is the number of active optical network units (ONUs); $cycle_{time}[s]$ is the polling cycle length (seconds); and $data_{rate}[bps]$ is the effective EPON data rate (1 Gbps).

EPON has guard band overhead, which represents a group of hardware-related issues such as automatic gain control (AGC), clock data recovery (CDR), laser ON/OFF times, etc. All of these are attributed to the physical medium attachment (PMA). The IEEE 802.3ah standard defines the maximum values of its individual components, namely:

- Code alignment zone (CAZ) is set to 32 ns
- Laser on and off times are set to 521 ns each
- AGC and CDR are set to 400 ns each

The guard band overhead can be estimated as shown in Equation (12.4):

$$O_{control} = \frac{(L + AGC + CDR + CAZ) \times N_{ONU}}{cycle_{time}[s]} \tag{12.4}$$

where L is Laser ON and OFF time (1024 ns); AGC is AGC group (400 ns); CDR is CDR group (400 ns); CAZ is CAZ group (32 ns).

In the case of the discovery window overhead, the estimated value can be calculated as shown in Equation (12.5). It represents the bandwidth loss resulting from operation of the discovery protocol.

$$O_{Discovery} = T_D \times f_D \tag{12.5}$$

where T_D is the discovery window length, established at 500 μs; F_D is the discovery window frequency, ranging between 10 and 0.1 Hz, defined at 1 Hz.

The frame delineation overhead is concerned with the size variability of the typical data packets generated by user applications. The frame delineation overhead can be estimated using Equation (12.6). It is also possible to estimate the frame delineation overhead (Os_{avg}) using Equation (12.7) based on the calculated average slot remainder sizes.

$$S_{avg} = \frac{\sum_{r=1}^{S^{MAX}-1} r \times (1 - F_{Ethernet}(r))}{\sum_{s=S^{MIN}}^{S^{MAX}} s \times f_{Ethernet}(s)} \tag{12.6}$$

where S^{MIN}, S^{MAX} are the minimum (64) and maximum (1518) Ethernet frame sizes (bytes); $F_{Ethernet}(s)$ is the probability distribution function for Ethernet frames based on measured PSDs; and $F_{Ethernet}(r)$ is the cumulative distribution function for Ethernet frames.

$$Os_{avg} = \frac{S_{avg} \times 8[ns]}{cycle_{length}[ns]} \tag{12.7}$$

12.2.4 GPON—Gigabit PON Architecture

GPON systems enable the service providers to provide 2.5 Gbps converged data and voice services for the last mile, in a cost-effective and bandwidth-efficient manner.

Figure 12.11 shows the GPON access network. GPON emerged as a new solution in PON access market, offering unprecedented high-bit-rate support (up to 2.488 Gbps) while enabling the transport of multiple services, specifically data and TDM, in native formats and with extremely high efficiency. GPON is capable of supporting the higher bandwidth requirements of business and residential services by providing a nominal line rate of 2.4 Gbps downstream and 1.2 Gbps upstream. However, it can support up to 64 users, making its average bandwidth per user a bit lower than EPON. The GPON (G.984 standard) uses GPON encapsulation method (GEM). The GEM has lower overhead and uses simpler timing.

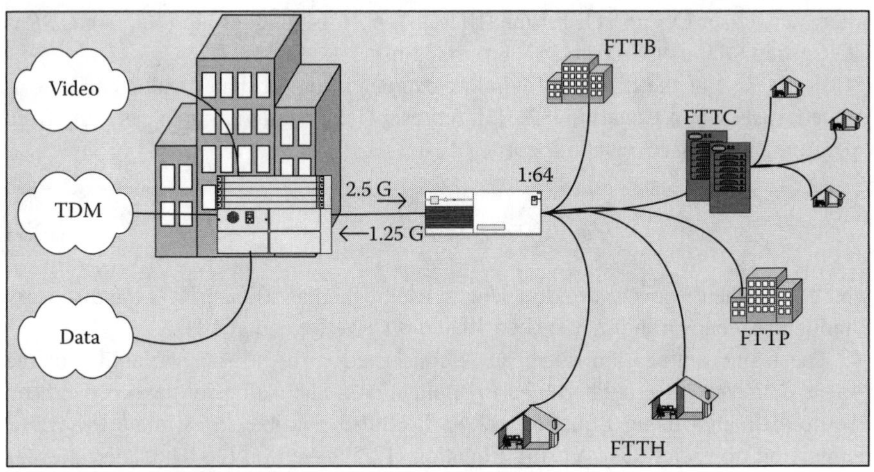

FIGURE 12.11 GPON access network.

12.2.5 WDM-PON Architecture

One of the future attractions of access network is WDM-PON. It requires the development of a low-cost WDM source in the optical network unit (ONU) for the practical implementation of WDM-PON. Figure 12.12 shows the schematic diagram of WDM-PON link. With this scheme, it is possible to have 1-Gbps digital signals for both up- and downstream and 20-Mbps subcarrier multiplexing (SCM) signal at 2.2 GHz for broadcasting in 10-km bidirectional link.

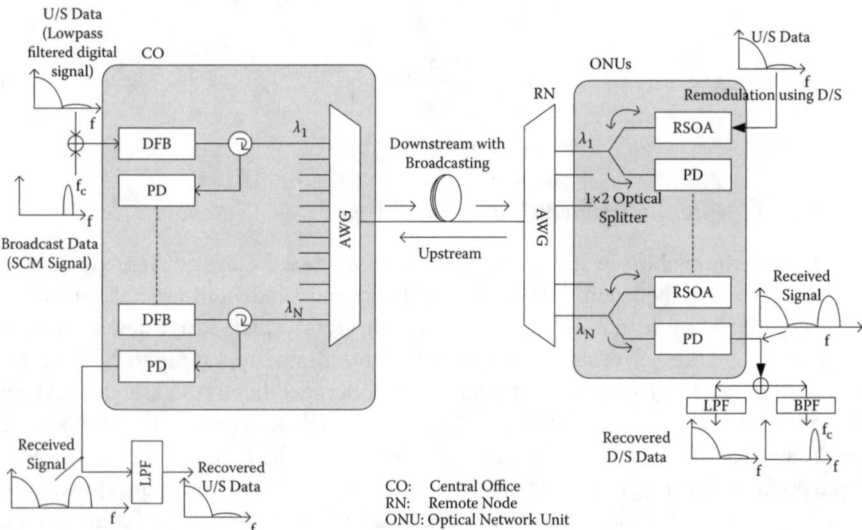

FIGURE 12.12 Schematic diagram of WDM-PON link.

In the central office (CO), the baseband digital signal for downstream data service and the SCM signal for broadcasting service are simultaneously modulated using a single distributed feedback (DFB) laser diode (LD). The OLT resides at the CO. It has an array of transmitters and receivers. Each transmitter-receiver pair is set at the wavelength band of the port of the multiplexing device, in this case, arrayed waveguide grating (AWG), to which the pair is connected. The AWG, which is 16 × 16 cyclic type with 100-GHz spacing, performs wavelength multiplexer/demultiplexer functions as a remote node (RN).

The photodetector (PD) in ONU detects a portion of the downstream source containing the broadcasting signal, and the digital and SCM signals are separated by the electrical filters. Then the residue of the downstream source becomes injected into the reflective semiconductor optical amplifier (RSOA) as a seeding source and remodulated by RSOA with the digital signal for upstream.

12.3 COMPARATIVE ANALYSIS OF PON TECHNOLOGIES AND ARCHITECTURES

There are two major functional differences in PONs. The first one is called the time division multiplexed PON (TDM-PON) and the second one is called WDM-PON. There is a long development history within the TDM-PON category such as APON, BPON, EPON, and GPON. In the TDM-PON techniques, the key concept is that it uses a single high-performance shared transceiver at the CO to communicate with n remote ONU transceivers as shown in Figure 12.13a. In this technique, the use of a $1 \times n$ power splitter to divide the optical power equally between the multiple ONUs is required.

Because each remote ONU uses the same upstream wavelength, they therefore must all take turns using dedicated and variable time slots where only a single ONU is allowed to transmit. The management and assignment of these individual transmission time slots are controlled by a relatively complex processor located at the OLT. The TDM-PON uses a single data wavelength to broadcast to all the users in the downstream direction. Using the address information located in the header bit streams, the ONUs identify their specific data packets.

While it is true that TDM-PON minimizes the number of required optical components, there are some challenges in performance. There exists in the optical power budget an approximate $1/n^2$ penalty. This happens firstly as result of a $1/n$ power loss through the optical power splitter, plus an additional $1/n$ penalty due to the receiver noise bandwidth that must be n times the average data rate to each ONU. Second, potential QoS issues may arise since n different users share the same data stream and a relatively complex algorithm is required for granting time slots to each of the users.

It is not surprising then, that this interaction or coupling of the n users into a single PON data channel can also raise some challenging management problems. As an example, if too many users in a PON decide to sign up for premium services relating to either data rate or QoS, challenging problems may occur. In addition to algorithm complexity, the opto-electronic hardware also needs to become significantly more complex due to its required burst-mode nature. As an example, the OLT receiver must quickly adjust both its gain sensitivity and clock synchronization for each ONU transmission because each will have a different time delay and link loss.

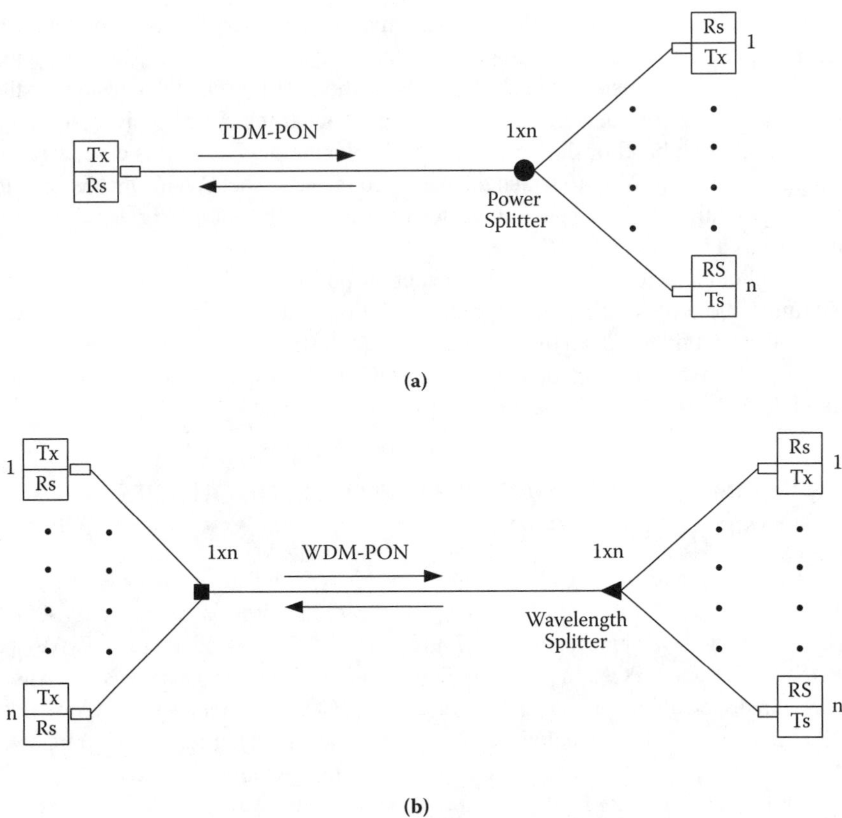

FIGURE 12.13 Functionality comparison between (a) a TDM-PON and (b) a WDM-PON access network.

In the TDM-PON discussed, there are obvious challenging problems that need solutions. WDM-PON does not have those challenging problems. This is because the wavelength splitter in WDM-PON replaces the power splitter in TDM-PON as shown in Figure 12.13b. Because of that, the splitting loss can be very small. In addition, because WDM-PON provides a point-to-point optical connection, the receiver noise penalty does not exist since the bandwidth of each receiver is matched to its data rate.

In addition, due to the direct point-to-point connectivity between end points, there are no QoS issues since each user is uncoupled from the others who share the PON. These features can be of high value if both business and residential customers share the same PON. Another relatively important advantage is the ability to completely characterize all the optical fiber paths in a WDM-PON using optical time domain reflectometer (WDM-OTDR). What makes this possible is the fact that at each wavelength, a single optical path exists between the CO and the remote ONU. In the case of a TDM-PON, the remote-node power splitter prevents an OTDR from separating and identifying the multiple Rayleigh backscatter signals from each of its *n* distribution fibers. Table 12.1 shows the different PON technologies and functionalities.

TABLE 12.1
Comparison of Different PON Technologies

	TDM-PON				
	APON	**BPON**	**EPON**	**GPON**	**WDM-PON**
Downstream	622 Mbps	622 Mbps	1.2 Gbps	2.5 Gbps	1.0 Gbps
Upstream	155 Mbps	155 Mbps	1.2 Gbps	1.2 Gbps	1.0 Gbps
Framing	ATM	ATM/Other Traffics	Ethernet	GEM/ATM	None specified
Maximum Bandwidth	622 Mbps	622 Mbps	1 Gbps	2.488 Gbps	1–10 Gbps per channel
Wavelength	1480–1500nm downstream, 1310nm upstream	1490–1550nm downstream, 1310nm upstream	1510 nm downstream, 1310nm upstream	1490nm downstream, 1310nm upstream	1554.6 nm downstream, 1480 nm upstream
Users/PON	32	32	16	64	100s
Efficiency	Can guarantee QoS for diverse network applications	ATM cell tax reduces efficiency	40% to 70% efficient (comparable to Ethernet networks)	80% to over 90% efficient	Highly efficient
Average Bandwidth per User	20 Mbps	20 Mbps	60 Mbps	40 Mbps	1–10 Gbps
Video	RF	RF	RF/IP	RF/IP	RF/IP
Applications	Voice, video, and data services, etc.	Voice, video, and data services, etc.	VoIP, standard and high-definition video (STV and HDTV), videoconferencing (interactive video), data traffic access network, etc.	Supports Internet protocol television (IPTV) and other video applications (videoconferencing, video chatting, online gaming, video cams, peer-to-peer applications, etc.	Video on demand, multimedia broadcasting, access and metro transport networks, etc.
Estimated Cost	Low	Low	Lowest	Medium	High
Upgradeability	Difficult	Difficult	Difficult	Difficult	Easy
Standard	ITU-T 983.1	ITU G.983.3	IEEE 802.3ah	ITU G.984	None

12.4 ADVANTAGES OF TDM PON

The advantages of TDM PON are as follows:

- It allows for a single upstream wavelength and a single transceiver in the OLT, resulting in a cost-effective solution.
- Since it uses the statistical multiplexing techniques, it uses its bandwidth effectively.
- Since it is possible for all ONUs to operate on the same wavelength, the components can be identical.
- Because it has intrinsic attributes, it is adequate for broadcasting application services.
- It both competes with and complements other subscriber access technologies like DSL or cable.
- Considering FTTP, because the access media is optical fiber, the bandwidth potential is many times greater than that of copper-based media.
- PONs share the costs of fiber and much of the equipment located with the service provider among several customers, while also eliminating expensive powered equipment between the service provider and these customers.
- The optical path is "transparent" to bit rate, modulation format (e.g., digital or analog), and protocol (e.g., SONET/SDH, IP, Ethernet). Such transparency results from nothing being installed between the service provider and the customer, which is specific to the bit rate, modulation format, and other features, allowing services to be mixed or economically upgraded as needed in the future.
- New services and/or new customers can be added by changing service-specific equipment only at the ends of the network, and only for those customers affected. Such flexibility is not the case in most of today's other access network architectures.

12.5 DISADVANTAGES OF TDM PON

Despite their advantages, PONs face significant obstacles on the road to success. The disadvantages of TDM PON are as follows:

- The fact that PONs share bandwidth among multiple subscribers lowers service costs and helps carriers efficiently amortize the equipment and operation expenses. However, any amount of upstream bandwidth transmitted over a PON will be divided up among the number of users at the customer site. Therefore, on a 155-Mbps PON link with four splits, each subscriber will receive 38.75 Mbps. Addition of splitters to links that have already been split leads to lowering of the final available bandwidth.
- In addition, the fact that PONs do not regenerate or convert optical signals mid-network makes them cheaper, but it also limits their reach. Without regeneration, light signals lose power quickly, consequently losing transmission capability.

- They use a downstream communication method that shares total bandwidth among users thereby resulting in a weak security.
- There is a weak fault tolerance since all users of PON are affected by a single ONT failure.
- There is limited transmission distance because of the sharing of a power splitter.
- The media sharing method between remote nodes results in complicated protocol and large overhead.
- The solution offered by PON may not be the ultimate solution because many customers are aggregated on a fiber for a low cost, which results in bandwidth reduction.

Due to these disadvantages and the availability of other broadband access alternatives, it is not expected that there will be a dramatic rise in the market for PONs soon.

12.6 ADVANTAGES OF WDM PON

The advantages of WDM PON are as follows:

- There are virtual point-to-point connections.
- The QoS and large bandwidth are guaranteed.
- There is line-rate independency and protocol transparency.
- It has a long transmission distance compared to TDM PON.
- It has aggregation of several wavelengths on a fiber.
- Using passive optical components of AWG results to high reliability.
- There is a high security and privacy in WDM PON.
- Owing to network configuration through optical fiber, there is available facility of operation and maintenance.
- Due to the filtering function of AWG, it has high fault tolerance and localization.

12.7 PON SECURITY ISSUES

> Definition 3: A secure PON architecture can be defined as the network that provides techniques to prevent or minimize the impact of active or passive security threats.

As defined, threats can be active or passive. In the case of active threats, the state of a system or network is changed. Denial of service and masquerading as an authorized entity are examples of active threats. Passive threats involve the unauthorized disclosure of information without changing the state of the system or network. Eavesdropping is one example of passive threat with the goal of capturing clear or encrypted messages online. Therefore, there is a need for security enhancements.

12.7.1 Possible Security Enhancements in PON

The possible security enhancement is based on using a PON architecture similar to that of Figure 12.1 or that of the PON domain encryption as shown in Figure 12.14. The downstream traffic in PONs is provided protection through wavelength hopping

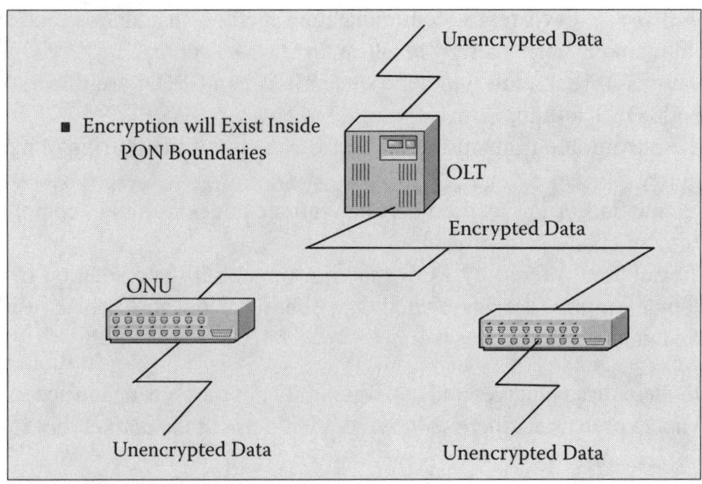

FIGURE 12.14 PON domain encryption.

(WH) and code sequencing (CS) techniques that diffuse motion picture expertise group (MPEG) packets over several wavelengths. MPEG is a family of standards used for coding audiovisual information such as music, video, and movies. Figure 12.1 illustrates the use of wavelength hopping as a way to provide protection for downstream traffic in PON. This idea provides a technique for protection against theft of service.

The OLT and ONU operations shown in Figure 12.1 and Figure 12.14 require tunable laser transmitters and filters across the associated processing of algorithms to enable the wavelength hopping scheme (WHS). The WH sequence generation for a specific ONU is based on mapping two matrices: wavelength grid matrix W_{mn}^{G} and code matrix C_{mn}^{ks} where G represents the grid number being used by the network, k is the ONU number, s represents the code matrix number assigned to a single ONU, which can range from 1 code matrix to many matrices depending on the security level of the ONU, and m and n are the number of rows and columns of the matrices respectively.

The total number of wavelengths (channels) available to populate W_{mn}^{G} matrix is dependent on channel frequency spacing (Δ) used from a center frequency of 193.1 T Hz as shown in Equation (12.8).

$$\text{Channel Frequency in T Hz} = 193.1 + F\,x\,\Delta \tag{12.8}$$

where F can be a positive or negative number including 0 and channel frequency spacing (Δ) can be 12.5 GHz, 25 GHz, 50 GHz, 100 GHz, or 200 GHz. The wavelength grid matrix W_{mn}^{G} is arranged as shown in Equation (12.9), where the elements of the matrix are channels generated by the relation in Equation (12.8)

$$\text{Wavelength Grid} = W_{mn}^{G} = \begin{pmatrix} \lambda_{11} & \cdots & \lambda_{1n} \\ . & . & . \\ \lambda_{m1} & . & \lambda_{mn} \end{pmatrix} \tag{12.9}$$

By permuting the elements (channels) of the matrix in Equation (12.9), we can have many different formats. Equation (12.10) illustrates the permutation possibilities for the wavelength channels in Equation (12.9). In this case, every matrix generated will have unique wavelength channels unique in its m and n location, and never repeated within the same matrix.

$$\text{Maximum wavelength grids } G_{max} = (n!)^m \tag{12.10}$$

Example 12.1

Suppose you have a channel spacing of $(\Delta) = 25$ GHz, $n = 13$, and 169 wavelengths can be arranged in the matrix of Equation (12.9). What will be the maximum matrices (G_{max}) provided?

Solution

Using Equation (12.10), G_{max} can be calculated as:

Maximum wavelength grids $G_{max} = (13!)^{13} = 2.1166 \times 10^{127}$ wavelength matrices.

The above example demonstrates a very good source to use as keys for network security. In this case, it is possible, for example, to have an hourly change of matrix as a key. The available keys can last for 2.416×10^{123} years before any key is reused. This should be able to provide a major security enhancement in PON.

The code matrix C_{mn}^{ks} can be determined as shown in Equation (12.11).

$$\text{Code Matrix} = C_{mn}^{ks} = \begin{pmatrix} a_{11} & a_{12} & \cdots & a_{n1} \\ a_{21} & a_{22} & \cdots & a_{2n} \\ & & \cdots & \\ a_{m1} & & \cdots & a_{mn} \end{pmatrix} \tag{12.11}$$

The code matrix C_{mn}^{ks} in Equation (12.11) is used in access networks with many users on a shared channel where multiple access interference (MAI) needs to be avoided or controlled. It is a two-dimensional time spreading (TS) and WH scheme where each row in Equation (12.11) has a single pulse 1. This scheme seem to have promising code types for generating code spaces that are large enough to prevent successful brute force code search attacks.

In determining the element a_{mn} values (0 or 1) of the code matrix of Equation (12.11), symmetric and asymmetric cases must be considered. When TS/WH is symmetric, $m = n$, while in the case of asymmetric TS/WH, $m \neq n$. The pulse placement operator shown in Equation (12.12) is used in deciding the values of a_{mn} (0 or 1) where a prime number $p = m = n$ in the symmetric case.

$$a_{mn} = (s.m)\ Mod\ (p) \quad s = 1, 2, 3, P, m = 1, 2, .P \tag{12.12}$$

where a set (s) number is the matrix number from the available S_{max} code matrices. The value of s varies from 1 to P and m is the row number where 1 needs to be placed.

For the asymmetric case, two different prime numbers P_s for spreading and P_h for hopping can be used. Equation (12.13) shows how the 1s can be placed in the asymmetric case. This helps to determine the place of pulse within a row of matrix that has ($2P_s - 1$) columns and P_s rows in the code matrix C_{mn}^{ks}.

$$C_{mn}^{ks} = \left(s\frac{m(m+1)}{2} \right) Mod(P), \quad s = 1,2,3, p-1; \quad m = 1,2, p \tag{12.13}$$

where s is the matrix number ($s \in S_{max}$).

It should be noted that correlation, which represents the degree of mutual interference between two codewords, and autocorrelation, which facilitates the detection of the desired signal and determines how well a codeword is detected at the intended receiver (detector) in the presence of mutual interference, are two basic requirements of any codes used in these security enhancements. There are other basic security measures that can be implemented, such as authentication and privacy issues. These are limited to protecting a shared medium. In the authentication case, it is based on 802.1x standard where the OLT is used as the authenticator while the ONU is used as the supplicant.

As shown in Figure 12.14, encryption exists inside the PON boundaries. The downlink data should be encrypted. The two different approaches that can be considered are adding MAC-level encryption and using higher level mechanisms like IPsec.[1]

APON uses a 24-bit key "churning" mechanism. Churning is a memoryless transformation of one byte to a different byte. The key strength is $2^{\text{key length}}$. However, last block problems may exist where the encryption block boundary may be dissimilar to the packet boundary. The last n bits ($n < 128$) will be XORed with the result of additional application environment standard (AES) over the next-to-last block. AES supports 128-, 192-, and 256-bit key length.

In the area of key management, there has to be periodic rekeying. In APON (churning), it is done every 1 s. In data over cable service interface specification–data encryption standard (DOCSIS–DES), it is done every 12 hours. In EPON (128-bit AES), it is done every 3^{17} years. There are two challenges: key distribution and key synchronization. In key distribution, OLT has a random number generator. OLT distributes the keys AES encrypted with a special key and there has to be an acknowledgement of the key distribution messages. In the key synchronization case, rekeying should not take place in the middle of a packet. Using precise time stamp is ineffective.

12.8 PASSIVE OPTICAL NETWORK STANDARDS

There are three basic standard versions of PONs. They are BPON, GPON, and EPON. The initial PON specifications defined by the International Telecommunications Union Full Service Access Network (ITU FSAN) committee used ATM as

---

[1] IPsec is a security architecture developed in the mid 1990s to resolve some of the issues of conducting secure transactions on the Internet, particularly business-to-business and electronic commerce transactions. The architecture encompasses protocols, associations, and algorithms for security, authentication, and encryption.

their layer 2 signaling protocol. As such, they became known as ATM-based PONs or APONs. FSAN decided that second-generation products should broaden the name to broadband PON or BPON.

In 1995, the North American operators initiated the BPON standards through the Full Service Access Network (FSAN) group. The results of their efforts were standardized later by ITU in the G.983 recommendations. The BPON standards specify the protocols from physical layer to QAM. The work of the IEEE P802.3ah Ethernet in the First Mile Task Force (FMTF) is now complete, with the approval of IEEE Std 802.3ah-2004 at the June 2004 IEEE Standards Board meeting.

The IEEE 802.3ah task force developed the EPON standard as part of the Ethernet in the First Mile (EFM) project. IEEE 802.1D and 802.1Q standards have focused on the mechanism to differentiate classes of services to a level comparable with ATM allowing Ethernet frames to carry a priority information field. In January 2003, the GPON standards were ratified by ITU-T. The standards are known as ITU-T Recommendations G.984.1, G.984.2, and G.984.3.

The following are some of the key PON standard recommendations:

- ITU-T Recommendation G.983.1: Broadband Optical Access Systems Based on Passive Optical Networks (PON), January, 2005.
- ITU-T Recommendation G.983.3: A Broadband Optical Access System with Increased Service Capability by Wavelength Allocation, March 2001.
- ITU-T Recommendation G.984.1: Gigabit-capable Passive Optical Networks (GPON): General Characteristics, March, 2003.
- ITU-T Recommendation G.984.2: Gigabit-capable Passive Optical Networks (GPON): Physical Media Dependent (PMD) Layer Specification, March, 2003.
- ITU-T Recommendation G.984.3: Gigabit-capable Passive Optical Networks (GPON): Transmission Convergence Layer Specification, February, 2004.
- ITU-T Recommendation G.984.4: Defines the ONU Management and Control Interface (OMCI) on the systems.
- ITU-T Recommendation G.114: One Way Transmission Time.
- IEEE, 802.3ah, Clause 64: Multi-Point MAC Control, IEEE Standard 2004.
- IEEE Standard 802.3ah: Telecommunications and Information Exchange between Systems–Local and Metropolitan Area Networks, 2004.
- IEEE 802.3 EFM: Ethernet PON and Security Considerations.

SUMMARY

1. One of the emerging new technologies in broadband communication systems is passive optical networks (PONs).
2. The difference between PON and the other optical network topologies is that the former is a point to multipoint (P2MP) topology while the latter is not.
3. Optical fiber carries much more information than conventional copper wire and is in general not subject to electromagnetic interference.
4. In a stand-alone system, a PON could deliver up to 622 Mbps downstream to the user and up to 155 Mbps upstream.

5. The main PON architectures are ATM PON (APON), broadband PON (BPON), Ethernet-based PON (EPON), gigabit PON (GPON), and wavelength-division-multiplexed passive optical network (WDM-PON).
6. APON is based on the ATM framework with a maximum reach of 20 km and is designed to be a cost-sensitive access solution.
7. Broadband passive optical networking (BPON) is a point-to-multipoint architecture for delivering last-mile connectivity without any active components in the distribution network.
8. EPONs deploy only one channel for downstream traffic and another channel for upstream that are both shared by all ONUs; and, they are standardized by IEEE 802.3ah working group.
9. GPON systems enable the service providers to provide 2.5-Gbps converged data and voice services for the last mile in a cost-effective and bandwidth-efficient manner.
10. The development of a low-cost WDM source in the optical network unit (ONU) is required for the practical implementation of WDM-PON.
11. There are two major functional differences in PONs. The first one is called the time division multiplexed PON (TDM-PON) and the second one is called WDM-PON.
12. PONs share the costs of fiber and much of the equipment located with the service provider among several customers, while also eliminating expensive powered equipment between the service provider and these customers.
13. The addition of splitters to links that have already been split leads to lowering of the final available bandwidth in PONs.
14. BPON, GPON, and EPON are the three basic standards of PON.

REVIEW QUESTIONS

12.1 The word "passive" implies that optical transmission has no power requirements or active electronic parts once the signal is going through the network.
   a. True
   b. False

12.2 FTTH stands for:
   a. Fiber-to-the-hall
   b. Fiber-to-the-home
   c. Fiber-to-the-house
   d. Fiber-to-the-hospital

12.3 Optical fiber does not carry much more information than conventional copper wire and is in general subject to electromagnetic interference.
   a. True
   b. False

12.4 In a PON, transmit and receive signals operate on different wavelengths allowing bidirectional operation over a single fiber.
   a. True
   b. False

12.5 In a stand-alone system, a PON could deliver up to ______ downstream to the user and up to ______ upstream.
   a. 622 Kbps downstream and 155 Kbps upstream
   b. 155 Mbps downstream and 622 Mbps upstream
   c. 622 Mbps downstream and 155 Mbps upstream
   d. 622 Mbps downstream and 155 Mbps upstream

12.6 The main TDM PON architectures are:
   a. ATM PON (APON)
   b. Broadband PON (BPON)
   c. Ethernet-based PON (EPON)
   d. Gigabit PON (GPON)
   e. All of the above

12.7 The wavelength-division-multiplexed passive optical network WDM-PON is not one of the architectures of the PONs.
   a. True
   b. False

12.8 The GPON G.984 standard uses GPON encapsulation method (GEM).
   a. True
   b. False

12.9 PONs share bandwidth among multiple subscribers, lower service costs, and help carriers efficiently amortize the equipment and operations expenses.
   a. True
   b. False

12.10 The IEEE 802.3ah task force developed the EPON standard as part of the Ethernet in the First Mile (EFM) project.
   a. True
   b. False

Answer: 12.1a, 12.2b, 12.3b, 12.4a, 12.5c, 12.6e, 12.7b, 12.8a, 12.9a, 12.10a

PROBLEMS

12.1 What does FTTx stand for in PON? Name and describe the various termination points associated with PON.

12.2 Name the different types of PON architectures and describe each of them.

12.3 Why are some of the architectures of PON classified as TDM-based PON?

12.4 What are the differences between TDM PON and WDM PON?

12.5 Compare and contrast the different architectures of PON.

12.6 What are the enabling technologies for the realization of PON?

12.7 Suppose you have a channel spacing of $(\Delta) = 50$ GHz, $n = 13$, and 169 wavelengths. What will be the maximum matrices provided?

12.8 If you are confronted with high demand for bandwidth, what choices do you have as a broadband engineer?

12.9 What advantages does optical fiber have over copper wire as a transmission medium? Describe an application in which one would prefer using optical fiber.

12.10 Discuss the core standards that make up PON.

12.11 Discuss the advantages of PON and give examples.

12.12 Discuss some of the disadvantages of TDM PON.

12.13 What are some of the security vulnerabilities of PON?

12.14 How can some of the security issues facing PON be mitigated? Discuss some of the enhancement issues involved.

12.15 Calculate the maximum wavelength matrices if the channel spacing is 25 GHz. In addition, n = 20 and m = 15.

12.16 What is the significance of the wavelength grid matrix W_{mn}^{G} and the code matrix C_{mn}^{ks} in PON security enhancement?

12.17 What are the limitations of some of the security measures, such as encryption, currently used in PON?

12.18 What are the functions of an ONU, OLT, and optical splitter? Where is each component normally located?

12.19 What characterizes a PON system and what are the motivations for using PON as a last-mile solution?

12.20 Why are the calculations of cross correlation and autocorrelation function important in the development of PON security enhancements?

Part 4

Wireless Networks

13 Wireless Data Services

A successful person is one who can lay a firm foundation with the bricks that others throw at him.

—David Brink

Wireless communication is undergoing an explosive growth. Voice traffic volume may be growing steadily, but wireless data services are quickly coming to dominate. We live in the Information Age, driven by our growing dependence on computers, cellular phones, pagers, fax machines, email, and the Internet. This demand for real-time information exchange is being made on an increasingly mobile workforce. Wireless communication networks include everything from cellular, personal communications system (PCS), and group system for mobile communications (GSM) networks to wireless LANs, wireless ATM, and satellite-based networks. Applications for wireless communications have found their way into the networking arena and the transfer of volumes of data. Although wireless data communication is one of the most promising communications technologies, the industry is still in its infancy.

Wireless communications is one of the fastest growing fields in engineering. It is communicating "without wires." Wireless communications are targeted to meeting certain needs, which include

- bypass physical barriers (roads, railroads, buildings, rivers, etc.)
- remote data entry
- mobile applications (cars, airplanes, etc.)
- connectivity to hard-to-wire places
- worldwide connectivity for voice, video, and data communications

In this chapter, we will consider various wireless communication systems. Such systems include:

- wireless local area networks (WLANs), which enable communication between stations without cables by means of radio frequency or infrared
- wireless asynchronous transfer mode (WATM), which facilitates the use of ATM technology for a broad range of wireless network access and internetworking scenarios
- wireless personal area network (WPAN), which refers to using a near-field electric field to send data across various devices using the human body as a medium
- cellular communications, which allows frequency reuse by dividing regions into small cells, each cell with a stationary radio antenna

Wireless personal communications services (PCS) and satellite communications, which are also wireless networks, will be treated separately in Chapters 14 and 15, respectively.

13.1 WIRELESS LAN

Wireless local area network (WLAN) is a new form of communication system.

> **Wireless local area network** (WLAN) is basically a local area network, confined to a geographically small area such as a single building, office, store, or campus, that provides high data connectivity to mobile stations.

Using electromagnetic airwaves (radio frequency or infrared), WLANs transmit and receive data over the air. A WLAN suggests less expensive, fast, and simple network installation and reconfiguration.

The proliferation of portable computers coupled with the mobile user's need for communication is the major driving force behind WLAN technology. WLAN creates a mobile environment for the PC and LAN user. It may lower LAN maintenance and expansion costs since there are no wires that require reconfiguration. Thus, WLANs offer the following advantages over the conventional wired LANs:

- Installation flexibility: allows the network to go where wire cannot go.
- Mobility: can provide LAN users with access anywhere.
- Scalability: can be configured in a variety of topologies to meet specific needs.

However, WLAN does not perform as well as wired LAN because of the bandwidth limitations and may be susceptible to electromagnetic interference and distance. While the initial investment in WLAN hardware can be higher than the cost of wired LAN hardware, overall installation expenses and life-cycle costs can be significantly lower.

13.1.1 Physical Layer and Topology

WLAN does not compete with wired LAN. Rather, WLANs are used to extend wired LANs for convenience and mobility. Wireless links essentially fill in for wired links using electromagnetic radiation at radio or light frequencies between transceivers. A typical WLAN consists of an access point and the WLAN adapter installed on the portable notebook. The access point is a transmitter/receiver (transceiver) device; it is essentially the wireless equivalent of a regular LAN hub. An access point is typically connected with the wired backbone network at a fixed location through a standard Ethernet cable and communicates with wireless devices by means of an antenna. WLANs operate within the prescribed 900 MHz, 2.4 GHz, and 5.8 GHz frequency bands. Most LANs use 2.4 GHz frequency bands because it is most widely accepted.

A wireless link can provide services in several ways including the following three:

- Replace a point-to-point connection between two nodes or segments on a LAN. A point-to-point link is a connection between two devices for transferring data. A wireless link can be used to bridge two LAN segments. Like a point-to-point link, the link connects two wireless bridges attached to the two LANs. Such an arrangement is useful for linking LANs in two buildings where a highway or river makes direct connection difficult.
- Provide a connection between a wired LAN and one or more WLAN nodes. In this case, a device is attached to the wired LAN to act as a point of contact (called access point) between the wired LAN and the wireless nodes as shown in Figure 13.1. The device can be a repeater, bridge, or router.
- Act as a stand-alone WLAN for a group of wireless nodes. This can be achieved using topologies similar to wired LAN, namely, a star topology can be formed with central hub controlling the wireless nodes, a ring topology with each wireless node receiving or passing information sent to it or a bus topology with each wireless capable of hearing everything said by all the other nodes.

13.1.2 Technologies

When designing WLANs, manufacturers have to choose from two main technologies that are used for wireless communications today: radio frequency (RF) and infrared (IR). Each technology has its own merits and demerits.

Radio frequency is used for applications where communications are over long distances and are not line-of-sight. In order to operate in the license-free portion of the frequency spectrum known as the ISM band (industrial, scientific, and medical), the RF system must use a modulation technique called "spread spectrum" (SS). Spread

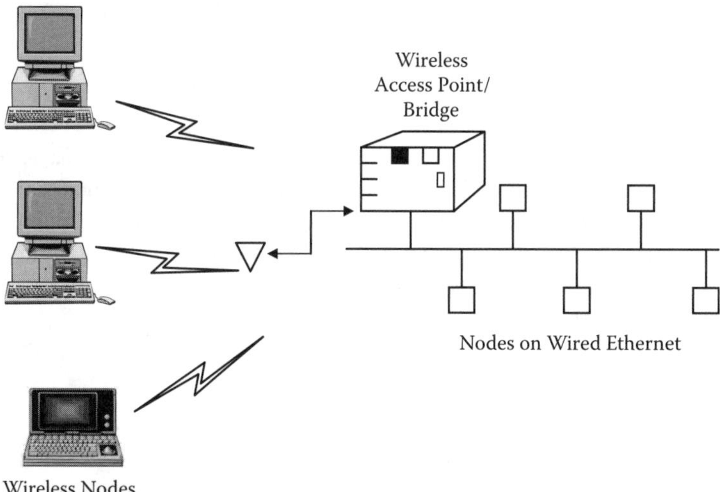

FIGURE 13.1 Connection of a wired LAN to wireless nodes.

spectrum is wideband radio frequency technology developed by the military during World War II for use in reliable, secure, mission-critical communications systems. An SS system is one in which the transmitted signal is spread over a frequency much wider than the minimum bandwidth required to send the signal. Using spread spectrum, a radio is supposed to distribute the signal across the entire spectrum. This way, no single user can dominate the band and collectively all users look like noise. The fact that such signals appear like noise in the band makes them difficult to find and jam, thereby increasing security against unauthorized listeners. There are two types of spread spectrum technology: frequency hopping and direct sequence.

Frequency hopping spread spectrum (FHSS) offers a current maximum data rate of 3 Mbps. It uses a narrowband carrier that changes frequency in a pattern known to both transmitter and receiver. It is based on the use of a signal at a given frequency that is constant for a small amount of time and then moves to a new frequency. The sequence of different channels for the hopping pattern is determined in pseudorandom fashion. This means that a very long sequence code is used before it is repeated, over 65,000 hops, making it appear random. Thus it is very difficult to predict the next frequency at which such a system will stop and transmit/receive data as the system appears to be a noise source to an unauthorized listener. This makes FHSS systems very secure against interference and interception. FHSS is characterized by low cost, low power consumption, and less range than DSSS but greater range than infrared. Most WLAN systems use FHSS.

Direct sequence spread spectrum (DSSS) takes a signal at a given frequency and spreads it across a band of frequencies where the center frequency is the original signal. The spreading algorithm, which is the key to the relationship of the spread range of frequencies, changes with time in a pseudorandom sequence. When the ratio between the original signal bandwidth and the spread signal bandwidth is very large, the system offers great immunity to interference. For example, if a 10-kbps signal is spread across 1 GHz of spectrum, the spreading ratio is 100,000 times or 50 dB. However, in the ISM band used in WLAN, the available bandwidth critically limits the ratio of spreading and so the advantages of the DSSS scheme against interference is greatly limited. It has been shown that for the WLAN system using DSSS, the spreading ratio is at best 10 times. DSSS is characterized by high cost, high power consumption, and more range than FHSS and infrared physical layers.

The second technology used in WLAN is infrared (IR), where the communication is carried by light in the invisible part of the spectrum. Infrared has a wavelength of 800 and 900 nm. It is primarily used for very short distance communications (less than 1 m), where there is a line-of-sight connection. Since IR light does not penetrate solid materials (it is even attenuated greatly by window glass), it is not really useful in comparison to RF in WLAN systems. However, IR is used in applications where the power is extremely limited such as a pager.

13.1.3 Standards

Although a number of proprietary, non-standard wireless LANs exist, standards have now been developed. Two international organizations have contributed to the development of standards for WLANs: the Institute of Electronics and Electrical Engineers (IEEE) and the European Telecommunications Standards Institute (ETSI).

In 1997, the IEEE 802.11 committee (http://ieee802.org/11) issued a standard for wireless LANs. The standard addresses the physical and MAC layers of the OSI model and includes the following:

- a transmission rate of up to 2 Mbps
- two different media for transmission over wireless LAN: infrared (IR) and radio frequency (RF)
- the media access control (MAC) protocol as carrier sense multiple access with collision avoidance (CSMA/CA), i.e., devices can interoperate with wired LANs via a bridge
- MAC protocol provides two service types: asynchronous, which is mandatory, and synchronous (or contention-free), which is optional
- MAC layer protocol is tied to the IEEE 802.2 logical link control (LLC) layer making it easier to integrate with other LANs
- three different physical layers: an optical-based physical-layer implementation that uses IR light to transmit and two RF-based physical-layer choices: direct sequence spread spectrum (DSSS) and frequency hopping spread spectrum (FHSS) both operating at 2.4 GHz industrial, scientific, and medical (ISM) frequency bands. (The ISM bands 902–928 MHz, 2400–2483.5 MHz, and 5725–5850 MHz do not require a license to operate.) The IEEE 802.11 specifications for DSSS wireless LAN are shown in Figure 13.2.
- added features to the MAC that can maximize battery life in portable clients via power-management schemes
- data security through which the wireless LANs can achieve wired-equivalent privacy

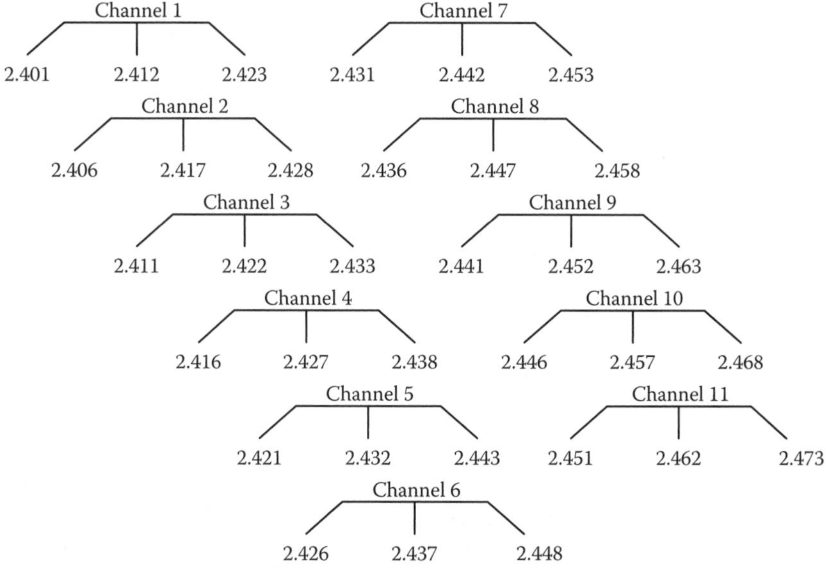

FIGURE 13.2 Eleven 22-MHz-wide channels for DSSS wireless LANs.

The standard basically defines the media and configuration issues, transmission procedures, throughput requirements, and range characteristics for WLAN technology. It avoids rigid requirements and gives room for vendors in the following areas: multiple physical media, common MAC layer irrespective of the physical layer, common frame format, power limit, and multiple on-air data rates.

There are three major problems encountered by an RF LAN. First, frequency allocation is limited for LANs. But since LANs operate with low power, frequency reuse is possible. Second, interference from other wireless LANs controlled by different organizations and other wireless sources is a problem. This problem can be controlled by using spread spectrum techniques. Third, security is at stake because RF signals can penetrate through the wall and hostile operators can intercept RF LAN communications. Encryption can be used to lessen this problem. IR LAN uses both laser diodes and light-emitting diodes as emitters. It is useful in high electromagnetic interference (EMI) environments. It is also secure since IR signal cannot penetrate the wall.

CSMA/CA is slightly different from carrier sense multiple access with collision detection (CSMA/CD), which is the MAC protocol used in Ethernet wired LAN. In CSMA/CA, when a node has something to transmit, it waits for silence on the network. When no other nodes are heard, it transmits and waits to receive an acknowledgement from the recipient node. If it fails to receive an acknowledgement within a time period, it assumes that collision has occurred and follows a process similar to CSMA/CD. Each node then waits for silence and only transmits after a random amount of waiting. While CSMA/CA protocol is slower than CSMA/CD due to the need for waiting for acknowledgement, it works well for wireless LANs. Also, WLANs operate in a strong multipath fading channel where channel characteristics can change resulting in unreliable communication.

The ETSI devoted its attention to RF wireless LANs and is close to finalizing its standard, which is based on the 2.4 GHz range used for spread-spectrum LANs in several European countries. European standard WLAN, called HiperLAN, will allow speeds of 24 Mbps.

Besides IEEE and ETSI, there are organizations that are more interested in the implementation and interoperability of WLAN products. Such organizations include Wireless LAN Alliance (WLANA at www.wlana.com) and Wireless Ethernet Compatibility Alliance (WECA at www.wi-fi.org or www.wirelessethernet.com). WLANA was formed in 1996 with 12 members as a trade association for wireless LAN vendors. WECA is a nonprofit manufacturing consortium with over 60 companies as members; it was formed in 1999 to certify interoperability of IEEE 802.11 products.

13.1.4 Applications

Offering obvious advantage of no wire installation costs, wireless LANs can be deployed in a dynamic environment where there is frequency reconfiguration of computer networks. Also, without the cables, excavation, and long installation waiting times, it is simpler to connect difficult-to-reach customers.

Although several products for RF and IR LANs are already available in the marketplace, the introduction of their applications is just beginning. Typical mobile users are assumed to be laptop or notebook computers and portable base stations. Services provided by WLANs include data applications such as over TCP/IP and multimedia applications.

The most prominent users of WLANs are those whose projects promise quick payoffs for adding mobility. Industries such as security services, banks, retail, manufacturing, and health care are more notable for deploying wireless LANs that allow workers to roam about while gathering information.

Mobile terminals—personal digital assistants (PDAs), specialized handheld terminals, and barcode scanners—connected to WLANs are increasingly being used to enhance business operations. It has become commonplace for WLANs to be used in applications such as

- printer sharing: linking to a distant printer within a department
- electronic mail: sending and receiving emails from anywhere
- health care: access to patient records from practically anywhere and location-independent claims processing
- financial services such as stock or community exchanges: implementing hand-held communicators in the trading room to increase the speed, accuracy, and reliability of its price reporting system
- factory control: data acquisition, inventory control, scoreboards, and robotics

Other applications include trading, banking, restaurants, retail industry, warehousing, manufacturing, education, office environments, petroleum industry, agriculture, and food services. Today, WLAN technology is becoming fairly mature. WLANs are becoming more widely recognized as a general-purpose connectivity alternative for a broad range of customers.

Still, the WLAN market remains small because the technology is new and so components are expensive and the data rates are low. On the one hand, it costs less than $100 to buy the network card needed to connect a PC to a wired Ethernet LAN with a data rate of 10 Mbps. On the other hand, the card needed to interface the same PC to wireless radio LAN costs $500, while the wireless hubs (access points) that connect the portable units to the wired network cost as much as $3000 each for a data rate of 1–2 Mbps. However, research groups are working hard to shrink radios into a chip that can be mass produced cheaply. If they succeed, the demand for radio LANs may follow the same trend as cellular phones in recent years.

13.2 WIRELESS ATM

Asynchronous transfer mode (ATM) technology is the result of efforts to devise a transmission and networking technology to provide high-speed broadband integrated services: a single infrastructure for data, voice, and video. Until recently, the integration of wireless access and mobility with ATM had received little attention.

Wireless ATM and other high-speed wireless networking technologies are motivated by the increasing demand for portable computing/telecommunications devices. In the early 1990s, researchers began to extend ATM capabilities into the wireless arena. Although wired ATM is becoming the technology of choice for broadband service integration, the area of wireless ATM is an emerging technology where there are still many issues to be resolved.

The concept of wireless ATM (WATM) was first proposed in 1992.

> **Wireless ATM** (WATM) is now regarded as the potential framework for next-generation wireless broadband communications that will support integrated, quality-of-service (QoS) multimedia services.

A lot of research and development has been done and many experimental wireless ATM network prototypes have been developed. WATM technology is currently migrating from the research stage to standardization and early commercialization. However, before introducing WATM, it is expedient that we briefly provide some background on ATM.

13.2.1 Overview of ATM

Asynchronous transfer mode (ATM) is the switching and multiplexing ITU-T standard for broadband integrated services digital network (BISDN). It is a packet-switching and multiplexing technique designed to handle voice, data, and video in a single physical channel. ATM is perceived as the transport mechanism for high-speed networks because of its support for bandwidth-intensive applications, its ability to carry different media types, and its ability to guarantee quality of service.

In an ATM network, the data is segmented into small, fixed-length cells. The small, fixed cell length makes it easier to bound the variation in delay across the system and also makes ATM networks easier to build than traditional packet-switching schemes. Each cell is 53 bytes with 5 bytes serving as the header, which comprises identification, control priority, and routing information, and the remaining 48 bytes contain the actual data. (Each cell header, called a label, contains its destination address.) ATM cells are routed based on header information that identifies the cell as belonging to a specific ATM virtual connection. ATM supports bidirectional transfer of cells, while preserving the order of transmission.

ATM networks are connection-oriented packet-switching networks designed for fiber optical links, which are characterized by a large bandwidth and a very reliable, almost error-free transmission. The ATM protocol requires that the QoS parameters are specified in the connection setup phase. These parameters are

- peak-to-peak cell delay variation (CDV)
- maximum cell transfer delay (CTD)
- cell loss ratio (CLR)
- cell error ratio (CER)

As a connection-oriented technique, a virtual circuit must be set up and established across the ATM network before any data can be transmitted from source to destination. An ATM connection is identified by two labels known as the "virtual channel identifier" (VCI) and "virtual path identifier" (VPI). The physical path can be a coaxial cable or fiber-based SONET. A physical path contains a certain number of virtual paths. Each virtual path has many virtual channels.

An ATM network consists of ATM switches interconnected by point-to-point ATM links. The ATM switch constitutes an important component of the ATM network. A switch may handle several hundred thousand cells per second at each switch port. In its simplest form, a switch has a number of links to receive and transmit cells. The primary function of an ATM switch is to route cells from an input port onto an appropriate output port at extremely high bit rate. By interconnecting ATM switches, an ATM network can be developed to span a building, campus, city, nation, or the globe.

There are two types of ATM connections:

- Permanent virtual connections (PVC): A PVC is a connection set up by some external mechanism, in which a set of switches between an ATM source and destination ATM systems are programmed with the appropriate VPI and VCI values. PVCs always require some manual configuration.
- Switched virtual connections (SVC): An SVC is a connection that is set up automatically through signaling protocol. All higher-layer protocols operating over ATM primarily use SVCs.

For each connection, there are four service classes identified as

- Constant bit rate (CBR): Cells are generated by the source at a constant rate, e.g., pulse code modulated (PCM) speech.
- Variable bit rate (VBR): Cells are generated at a non-constant rate, e.g., video.
- Available bit rate (AVR): Connection-oriented service utilizing the bandwidth remaining after CBR and VBR allocations.
- Unspecified bit rate (UBR): Packet-oriented, connectionless data without delay.

ATM networks are designed to support a wide range of applications with various service requirements and traffic requirements. A typical ATM network is shown in Figure 13.3, where the terminal equipment (TE) might be high-speed LANs.

13.2.2 Wireless ATM Architecture

The idea of WATM networks raises a number of challenges. First, integrated multimedia services demand high data rates (from 2 Mbps up), which are common in wired networks but still constitute a challenge for a wireless link. Second, ATM was designed with the assumption that the physical medium has a very low bit error rate (e.g., 10–9), whereas wireless communications suffer from very high bit error rate (e.g., 10–3) due to time variant multipath propagation, blocking, and interference. Additional protocol layers and special coding techniques are required to share the radio channel bandwidth and minimize the bit error rate. Third, the incorporation of mobility in the connection-oriented BISDN requires the dynamic reestablishment of the ATM virtual circuits (VCs) with short time interval for the mobile terminal handover from one macrocell to another. In addition, the reestablishment

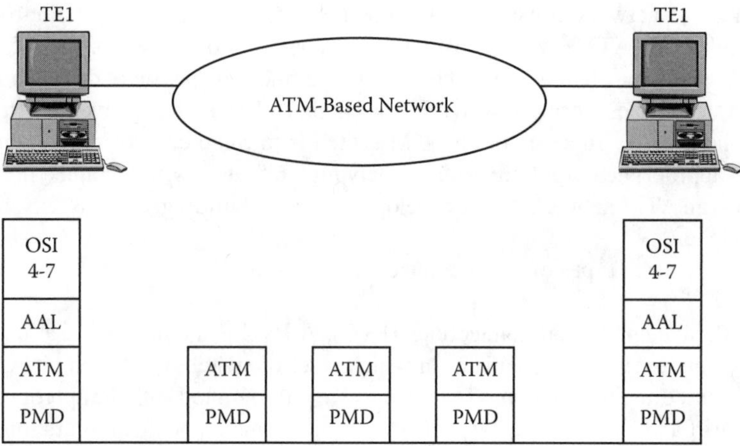

FIGURE 13.3 A typical ATM network.

of the VCs must ensure in-sequence and loss-free delivery of the ATMs in order to guarantee the QoS requirement on the connections. Moreover, the wireless channel is an expensive resource in terms of bandwidth, whereas ATM was designed for a bandwidth-rich environment. ATM effectively trades off bandwidth for simplicity in switching. Every ATM cell carries a header with an overhead of about 10 percent. Even this much overhead is considered too high for shared radio channels. If these challenges are resolved, there are significant advantages to wireless access.

The advantage of wireless ATM networks is definitely mobility resulting for the tetherless feature that cannot be supported by wired or fixed ATM networks. To enable nomadic access to ATM networks, additional functions are required in ATM networks for mobility support.

Wireless ATM networks are basically the wireless extension of fixed ATM networks. The 53-byte ATM cell is too big for a wireless ATM network. Therefore, WATM networks may use a 16- or 24-byte payload. Thus, in a wireless ATM network, information is transmitted in the form of a large number of small transmission cells called pico cells. Each pico cell is served by a base station, while all the base stations in the network are connected via the wired ATM network. The ATM header is compressed or expanded to a standard ATM cell at the base station. Base stations are simple cell relays that translate the header formats from the wireless ATM network to the wired ATM network. ATM cells are transmitted via radio frames between a central station (B-CS) and user radio modules (B-RM) as shown in Figure 13.4. All base stations operate on the same frequency so that there is no hard boundary between pico cells.

Reducing the size of the pico cells helps in mitigating some of the major problems related to wireless LANs. The main difficulties encountered are the delay spread due to multipath effects and the lack of a line-of-sight path, which results in high attenuation. Also, small cells have some drawbacks compared to large cells.

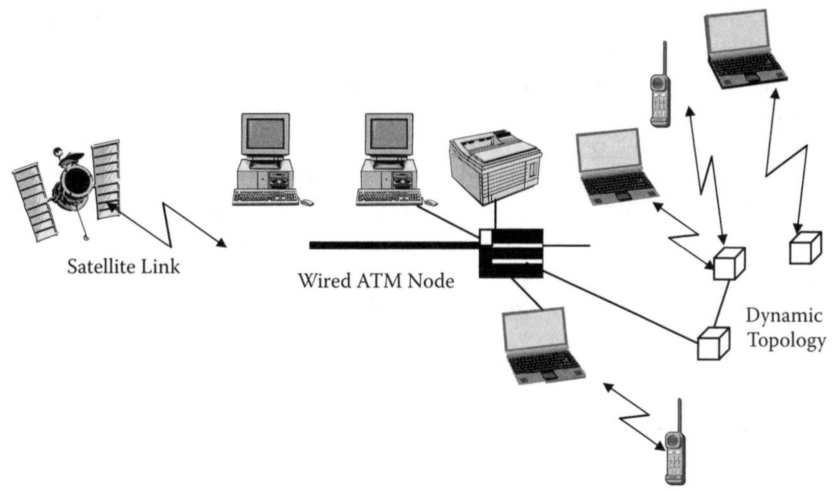

FIGURE 13.4 A typical wireless ATM network.

From Figure 13.4, we notice that a wireless ATM typically consists of three major components: (1) ATM switches with standard UNI/NNI capabilities, (2) ATM base stations, and (3) wireless ATM terminal with a radio network interface card (NIC). There are two new hardware components: ATM base station and WATM NIC. The new software components are the mobile ATM protocol extension and WATM UNI driver.

In conventional mobile networks, transmission cells are "colored" using frequency-division multiplexing or code-division multiplexing to prevent interference between cells. Coloring is considered a waste of bandwidth because in order for it to be successful there must be areas between re-use of the color in which it is idle. These inactive areas are wasted rather than used for transmission.

Wireless ATM architecture is based on integration of radio access and mobility features. The idea is to fully integrate new wireless PHY, medium access control (MAC), data link control (DLC), wireless control, and mobility signaling functions into the ATM protocol stack.

The requirements for the physical (PHY) layer include

- Frequency band: It is expedient to use the appropriate available band, which is 5 GHz for unlicensed in-building wireless ATM access.
- Data rates: High data rates are less essential for mobile devices with limited power and processing budget. The ATM Forum has approved 25 Mbps as a standard PHY data rate for low-cost ATM solutions.
- Modulation efficiency: Radio bandwidth is a rare resource. It is therefore important, especially at 5 GHz, to use an efficient modulation technique.

As for the medium access control (MAC), a wireless ATM needs to have some method of accessing the shared channel. The requirements in a WATM system are very stringent as the MAC protocol needs to provide guarantees for both asynchronous and isochronous traffic types in the face of an unreliable channel. The MAC protocol must also facilitate efficient handover between base stations for mobile units and allocate the available bandwidth in a flexible and dynamic manner. Since one of the major advantages of ATM is its ability to give QoS guarantees to connections, the MAC protocol of a WATM network should support all the service categories including constant bit rate (CBR), available bit rate (ABR), and unspecified bit rate (UBR). Also, since many WATM devices are portable, battery powered with limited power budgets, it is important to consider power saving techniques in all radio access protocol layers.

For the data link control (DLC), the reliability of the radio channel needs to be increased using link level error control procedures. Techniques to recover from bit errors include the use of forward error control (FEC) and cyclic redundancy check (CRC) in combination with an automatic repeat request (ARQ). As mentioned earlier, ATM protocol was designed with the assumption of highly reliable links and point-to-point transmission, which contrasts with the characteristics of the wireless link. To overcome these incompatibilities, the data link control (DLC) layer is introduced between the ATM layer and the physical layer, as shown in Figure 13.5. Wireless ATM will need a custom data link layer protocol due to high error rate and different packet size for WATM. As mentioned earlier, WATM may use a 16- or 24-byte payload since 53 bytes may be too long. The data link protocol may contain service type definition, error control, segmentation and reassembly, and handoff support. The service type will indicate whether a cell is of type control, constant bit rate (CBR), variable bit rate (VBR), etc. Error control is needed due to interference and poor physical level characteristics of the physical medium. Segmentation and reassembly is required because WATM may use 16- or 24-byte cells. Handoff occurs when the mobile unit moves from one area to another and should be transparent.

Because eavesdropping and spoofing are in general harder to detect in wireless systems than wired, security must be carefully considered. Encryption must be included at the physical layer; this will, of course, increase hardware cost. Also, a set of requirements must be defined for the handover process. The handover protocol enables a wireless terminal or wireless network to seamlessly move between access points (AP) while retaining the negotiated QoS of its connections. This can be achieved by the efficient rerouting and switching of the wireless terminal's VCs to the appropriate AP in the wired ATM network.

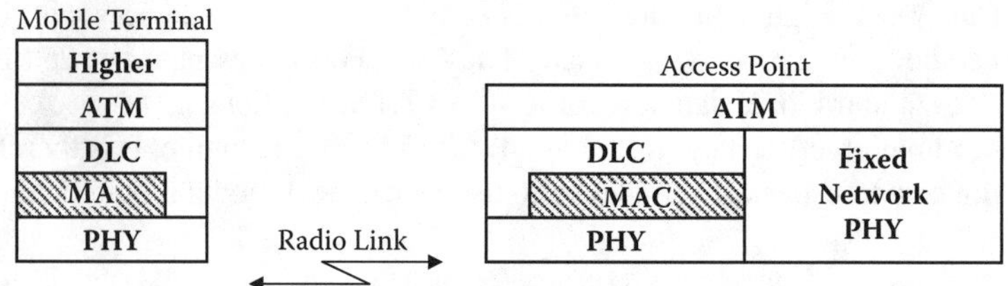

FIGURE 13.5 Wireless ATM protocol architecture.

Potential applications of wireless ATM networks include

- private ATM LANs
- portable enhanced television services
- wireless internet access and mobile network computers
- cordless telephony
- tactical networks

Wireless mobile ATM (WMATM) is naturally evolving from WATM. It is extending broadband multimedia services to mobile wireless terminals. Wireless ATM is not as mature as wireless LAN. No standards have been defined by either ITU-T or the ATM Forum. However, the ATM Forum's WATM Working Group (started in June 1996) is developing specifications that will facilitate the deployment of WATM.

13.3 WIRELESS PAN

Wearable computer devices such as cellular phones, personal digital assistants (PDAs), pagers, personal stereo, pocket video games, and notebook computers are getting smaller in size and less expensive. There is an ever-increasing use of these personal devices and the number of these devices is increasing as well. However, there is currently no means for these devices to share or exchange data among themselves. To facilitate communications among these personal devices, a new wireless communications technology, known as "wireless personal area network," is emerging.

> **Wireless personal area network** (WPAN) enables communication or information exchange between computing devices that may be portable or mobile and could be worn or carried by individuals.

WPAN operates in the license-free radio frequency band of 2.4 GHz (2400 to 2483.5 MHz), with ranges up to 10 m and data rates up to 1 Mbps. It ties together closely related objects, a function that is fundamentally different from the objective of a wireless LAN. Thus, WPAN must coexist with other radio technologies, like WLAN, that operate on the same frequency band.

WPAN may be regarded as a body network. It uses a near-field electric field to send data across various devices using the electrical conductivity of the human body as a medium. As shown in Figure 13.6, WPAN enables communication between two devices using the human body as a medium by using a transmitter and receiver, both battery powered. The transmitter, which needs not be in direct contact with the skin, is capacitively coupled through the body to the receiver. The system electrostatically induces picoamp currents (too little to be felt by the body or to have any biological effects) into the body, which is used to conduct the modulated currents. A typical prototype may be slightly larger and thicker than a credit card and consume 1.5 mW to power its coupling electrodes, which are not in direct contact with the skin but placed in shoes, wallets, pockets, etc.

WPAN was developed at MIT Media Laboratory and later supported by IBM Almaden Research Laboratory. It is being standardized by the IEEE 802.15 committee.

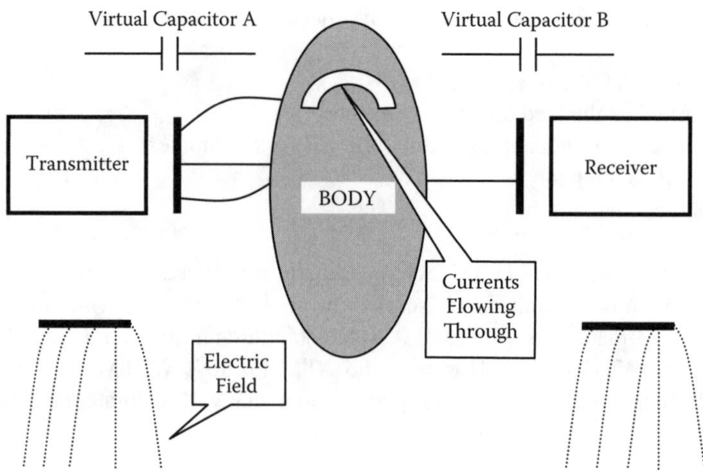

FIGURE 13.6 A wireless personal area network.

The committee will address the requirements for personal area networking of PCs, PDAs, peripherals, cell phones, pagers, and consumer electronic devices to communicate and interoperate with one another.

WPAN is different from wireless LAN. First, WPANs target primarily the vast consumer market and are used for ease of connectivity of personal wearable or handheld devices. Second, WPAN is optimized for low complexity, low power, and low cost. It does not require an access point like WLAN. Third, WPAN's close range throughput of 1 Mbps does not compare with the 11 Mbps the IEEE 802.11 WLAN offers. It has small coverage, typically about 10 m and connects only a limited number of devices.

WPAN and Bluetooth have been used interchangeably in many technical articles. Although the two technologies are similar, they should not be confused. Bluetooth is a far-field radio technology that is being promoted by the Bluetooth special interest group (SIP), which has more than 75 members. Bluetooth technology is for short-range, low-cost radio links between PCs, mobile phones, and other electronic devices. It provides a 10-m personal bubble that supports simultaneous transmission of both data and voice for multiple devices.

Some interesting applications of PAN have been proposed. These include:

- Transfer of information between all the electronic devices a gadget-hungry person may carry. Such devices include pagers, cellular phones, personal digital assistants (PDAs), identification badges, and smart cards.
- Exchange of business cards by just shaking hands. The electronic card is transferred automatically from one card device via the body to the other person's card device.
- Office and domestic automation at home or at the office. One is able to control lighting, heating, and even locks with a gentle touch. These devices

and appliances are programmed to communicate with each other with little or no human intervention.

- Telemedicine with intelligent sensors monitoring specific physiological signals such as EEG, ECG, and GSR and performing data acquisition. Such a collection of wearable medical sensors can communicate using PAN.

Other potential applications include sensor and automation needs, interactive toys, and location tracking for smart tags and badges.

A major problem with PAN is security. Since transferring data across devices has become simple, a lot of work on encryption needs to be done.

PAN is in the prototyping and standardization stage today and most of the work has been done in the lower layers. More work needs to be done at the higher layers to resolve issues like security, authentication, and reliability. Like any other technology, the success of PAN does not depend on what it promises to achieve but on what it is actually able to deliver.

13.4 CELLULAR COMMUNICATIONS

The conventional approach to mobile radio involved setting up a high-power transmitter on top of the highest point in the coverage area. The mobile telephone must have a line-of-sight to the base station for proper coverage. Line-of-sight transmission is limited to as much as 40 to 50 miles on the horizon. Also, if a mobile travels too far from its base station, the quality of the communications link becomes unacceptable. These and other limitations of conventional mobile telephone systems are overcome by cellular technology.

Areas of coverage are divided into small hexagonal radio coverage units known as cells. A cell is the basic geographic unit of a cellular system.

> A **cellular communications system** employs a large number of low-power wireless transmitters to create the cells.

These cells overlap at the outer boundaries, as shown in Figure 13.7. Cells are base stations transmitting over small geographic areas that are represented as hexagons. Each cell size varies depending on the landscape and tele-density. Those stick towers one sees on hilltops with triangular structures at the top are cellular telephone sites. Each site typically covers an area of 15 miles across, depending on the local terrain. The cell sites are spaced over the area to provide a slightly overlapping blanket of coverage. Like the early mobile systems, the base station communicates with mobiles via a channel. The channel is made of two frequencies: one frequency (the forward link) for transmitting information to the base station and the other frequency (the reverse link) to receive from the base station.

13.4.1 Fundamental Features

Besides the idea of cells, the essential principles of cellular systems include cell splitting, frequency reuse, handover, capacity, spectral efficiency, mobility, and roaming.

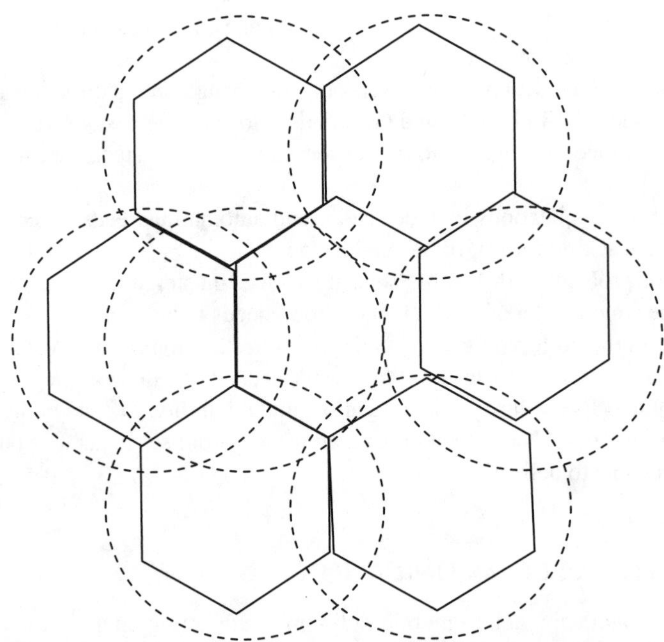

FIGURE 13.7 A typical wireless seven-cell pattern; cells overlap to provide greater coverage.

- Cell splitting: As a service area becomes full of users, the single area is split into smaller ones. This way, urban regions with heavy traffic can be split into as many areas as necessary to provide acceptable service, while large cells can be used to cover remote rural regions. Cell splitting increases the capacity of the system.
- Frequency reuse: This is the core concept that defines the cellular system. The cellular telephone industry is faced with a dilemma: services are growing rapidly and users are demanding more sophisticated call-handling features, but the amount of the electromagnetic (EM) spectrum allocation for cellular service is fixed. This dilemma is overcome by the ability to reuse the same frequency (channel) many times. Several frequency reuse patterns are in use in the cellular industry, each with its advantages and disadvantages. A typical example is shown in Figure 13.8, where all the available channels are divided into 21 frequency groups numbered 1 to 21. Each cell is assigned three frequency groups. For example, the same frequencies are reused in cell designated 1 and adjacent locations do not reuse the same frequencies. A cluster is a group of cells; frequency reuse does not apply to clusters.
- Handoff: This is another fundamental feature of cellular technology. When a call is in progress and the switch from one cell to another becomes necessary, a handoff takes place. Handoff is important because as a mobile user travels from one cell to another during a call, as adjacent cells do not use the same radio channels, a call must be either dropped or transferred

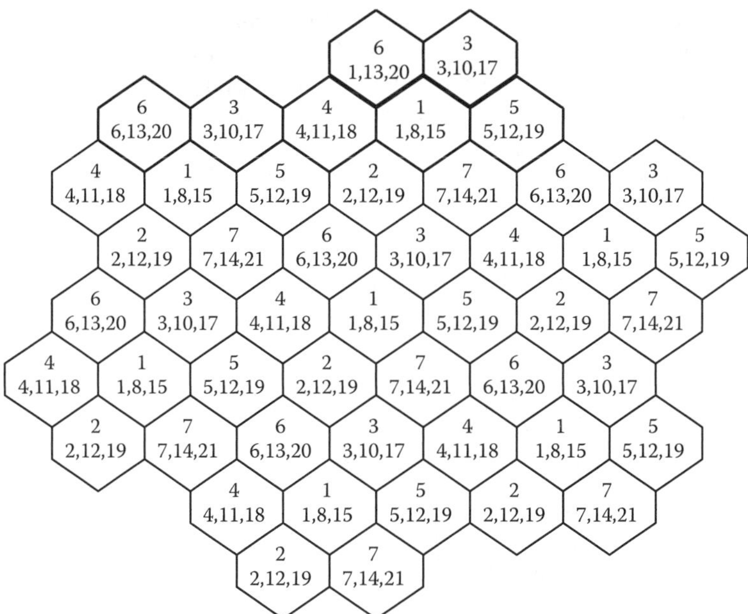

FIGURE 13.8 Frequency reuse in a seven-cell pattern cellular system.

from one channel to another. Dropping the call is not acceptable. Handoff was created to solve the problem. A number of algorithms are used to generate and process a handoff request and eventual handoff order. Handing off from cell to cell is the process of transferring the mobile unit that has a call on a voice channel to another voice channel, all done without interfering with the call. The need for handoff is determined by the quality of the signal, whether it is weak or strong. A handoff threshold is predefined. When the received signal level is weak and reaches the threshold, the system provides a stronger channel from an adjacent cell. This handoff process continues as the mobile moves from one cell to another as long as the mobile is in the coverage area.

- Mobility and roaming: Mobility implies that a mobile user while in motion will be able to maintain the same call without service interruption. This is made possible by the built-in handoff mechanism that assigns a new frequency when the mobile moves to another cell. Due to several cellular operators within the same region using different equipment while a subscriber is only registered with one operator, some form of agreement is necessary to provide services to subscribers. Roaming is the process whereby a mobile moves out of its own territory and establishes a call from another territory.
- Capacity: This is the number of subscribers that can use the cellular system. For an FDMA system, the capacity is determined by the loading (number of calls and the average time per call) and system layout (size of cells and

amount of frequency reuse utilized). Capacity expansion is required because cellular systems must serve more subscribers. It takes place through frequency reuse, cell splitting, planning, and redesigning of the system.

- Spectral efficiency: This a performance measure of the efficient use of the frequency spectrum. It is the most desirable feature of a mobile communication system. It produces a measure of how efficiently space, frequency, and time are used. Expressed in channels/MHz/km^2, channel efficiency is given by

$$\eta = \frac{\text{Total no. of channels available in the system}}{\text{Bandwidth} \times \text{Total coverage area}}$$

$$\eta = \frac{\frac{B_w}{B_c} \times \frac{N_c}{N}}{B_w \times N_c \times A_c} = \frac{1}{B_c N A_c} \tag{13.1}$$

where B_w is the bandwidth of the system in MHz, B_c is the channel spacing in MHz, N_c is the number of cells in a cluster, N is the frequency reuse factor of the system, and A_c is area covered by a cell in km^2.

13.4.2 Cellular Network

A typical cellular network is shown in Figure 13.9. It consists of the following three major hardware components:

- Cell site (base stations): The cell site acts as the user-to-MTSO interface, as shown in Figure 13.9. It consists of a transmitter and two receivers per channel, an antenna, a controller, and data links to the cellular office. Up to 12 channels can operate within a cell depending on the coverage area.
- Mobile telephone switching office (MTSO): This is the physical provider of connections between the base stations and the local exchange carrier. MTSO is also known as mobile switching center (MSC) or digital multiplex switch-mobile telephone exchange (DMS-MTX) depending on the manufacturer. It manages and controls cell site equipment and connections. It supports multiple-access technologies such as AMPS, TDMA, CDMA, and CDPD. As a mobile moves from one cell to another, it must continually send messages to the MTSO to verify its location.
- Cellular (mobile) handset: This provides the interface between the user and the cellular system. It is essentially a transceiver with an antenna and is capable of tuning to all channels (666 frequencies) within a service area. It also has a handset and a number assignment module (NAM), which is a unique address given to each cellular phone.

13.4.3 Cellular Standards

Due to the rapid development of cellular technology, different standards have resulted. These include:

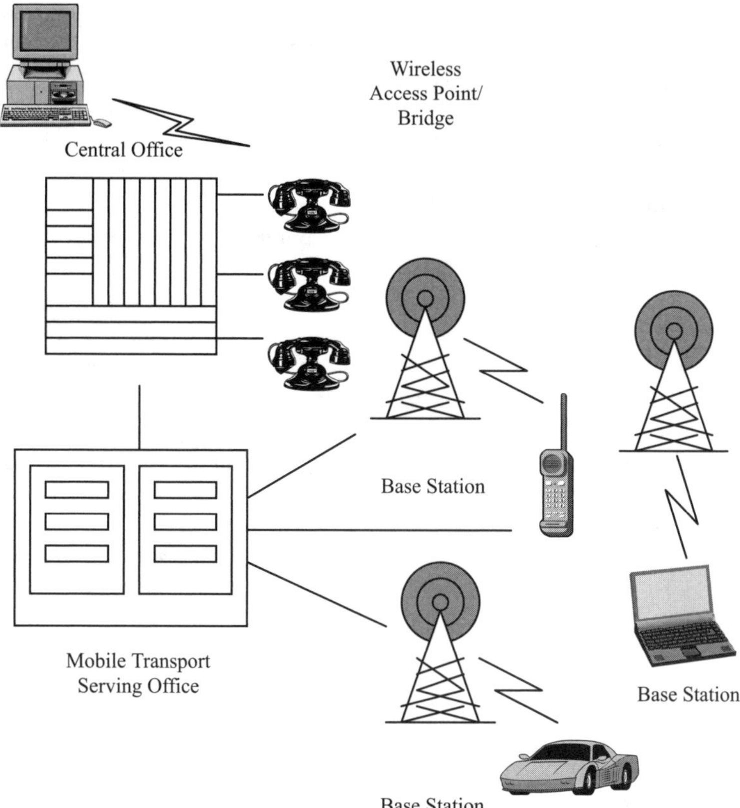

FIGURE 13.9 A typical cellular network.

- AMPS (advanced mobile phone system): This is the standard introduced in 1979. Although it was developed and used in North America, it has also been used in over 72 countries. It operates in the 800-MHz frequency band. It is based on frequency-division multiple access (FDMA). The mobile transmit channels are in the 825–845 MHz range, while the mobile receive channels are in the 870–890 MHz range. There is also the digital AMPS, which is also known as time-division multiple access (TDMA) (or IS-54). FDMA systems allow for a single mobile telephone to call on a radio channel; each voice channel can communicate with only one mobile telephone at a time. The TDMA systems allow several mobile telephones to communicate at the same time on a single radio carrier frequency. This is achieved by dividing their signal into time slots.
- CDMA (code-division multiple access): This is an alternative North American cellular standard (IS-95) introduced in 1994. It is based on the spread-spectrum technique, which allows many users to access the same band by assigning a unique orthogonal code to each user.

- GSM (global system for mobile communications): This is a digital cellular standard developed in Europe and designed to operate in the 900-MHz band. It is a globally accepted standard for digital cellular communication. It utilizes a 200-kHz channel divided into 8 time slots with FDM (frequency division multiplexing). The technology allows international roaming and provides integrated cellular systems across different national borders. GSM is the most successful digital cellular system in the world. It is estimated that many countries outside Europe will join the GSM partnership.
- PDC (personal digital cellular): This is a digital cellular standard developed in Japan. It was designed to operate on 800-MHz and 1.5-GHz bands.
- NMT (Nordic mobile telephone): This is the cellular standard developed in the Nordic nations of Denmark, Finland, Norway, and Sweden in 1981 but now deployed over 40 countries in Europe, Asia, and Australia. It was designed to operate in the 450- and 900-MHz frequency bands. NMT 450 and NMT 900 systems can coexist, allowing them to use the same switching center.

In many European countries, the use of GSM has allowed cross-country roaming. However, global roaming has not been realized because there are too many of these incompatible standards.

13.5 CELLULAR DIGITAL PACKET DATA

Cellular digital packet data (CDPD) is the latest in wireless data communication. CDPD systems offer one of the most advanced means of wireless data transmission technology.

> **Cellular digital packet data** (CDPD) is a cellular standard aimed at providing Internet protocol (IP) data service over the existing cellular voice networks and circuit-switched telephone networks.

The technology solves the problem of business individuals on the move who must communicate data between their work base and remote locations.

The idea of CDPD was formed in 1992 by a development consortium with key industry leaders including IBM, six of the seven Regional Bell Operating Companies, and McCaw Cellular. The goal was to create a uniform standard for sending data over existing cellular telephone channel. The Wireless Data Forum (www.wirelessdata.org), formerly known as CDPD Forum, has emerged as a trade association for wireless data service providers and it currently has over 90 members. CDPD has been defined by the CDPD Forum CDPD Specification R1.1 and operates over AMPS.

By building CDPD as an overlay to the existing cellular infrastructure and using the same frequencies as cellular voice, carriers are able to minimize the capital expenditures. It costs approximately $1 million to implement a new cellular cell site and only about $50,000 to build the CDPD overlay to an existing site.

CDPD is designed to exploit the capabilities of the advanced cellular mobile services (AMPS) infrastructure throughout North America. One weakness of cellular telephone channels is that there are moments when the channels are idle (roughly

30% of the air time is unused). CDPD exploits this by detecting and utilizing the otherwise wasted moments by sending packets during the idle time. As a result, data is transmitted without affecting voice system capability. CDPD transmits digital packet data at 113.2 kbps using idle times between cellular voice calls on the cellular telephone network.

CDPD has the following features:

- It is an advanced form of radio communication operating in the 800- and 900-MHz bands.
- It shares the use of the AMPS radio equipment on the cell site.
- It supports multiple, connectionless sessions.
- Its airlink transmission has a 113.2-kbps raw data rate.
- It utilizes the Internet protocol (IP) and the OSI connectionless network protocol (CLNP).
- It is fairly painless for users to adopt. To gain access to CDPD infrastructure, one only requires a special CDPD modem.
- It supports both the TCP/IP protocols as well as the international set of equivalent standards.
- It was designed with security in mind unlike other wireless services. It provides for encryption of the user's data as well as conceals the user's identity over the air link.

CDPD provides the following services:

- Data rate of 113.2 kbps.
- Connectionless as the basic service; a user may build a connection-oriented service on top of that if desired.
- All three modes of point-to-point, multicast, and broadcast are available.
- Security, which involves authentication of users and data encryption.

13.5.1 Network Architecture

CDPD is a packet-switched data transfer technology that employs radio frequency (RF) and spectrum in existing analog mobile phone system such as AMPS. The CDPD overlay network is made of some major components that operate together to provision the overall service. The key components that define CDPD infrastructure are illustrated in Figure 13.10. They are

- Mobile end system (MES): This is the subscriber's device for gaining access to the wireless communication services offered by a CDPD service. It is any mobile computing device that is equipped with a CDPD modem. Examples of an MES are laptop computers, palmtop computers, and personal digital assistants (PDAs), or any portable computing devices. The MES transmits data over the air link to the mobile data base station (MDBS) located in the cell site. It is the MES that is responsible for monitoring the received signal strength of the cellular channel and deciding to initiate a transfer or handoff from one cell to another cell.

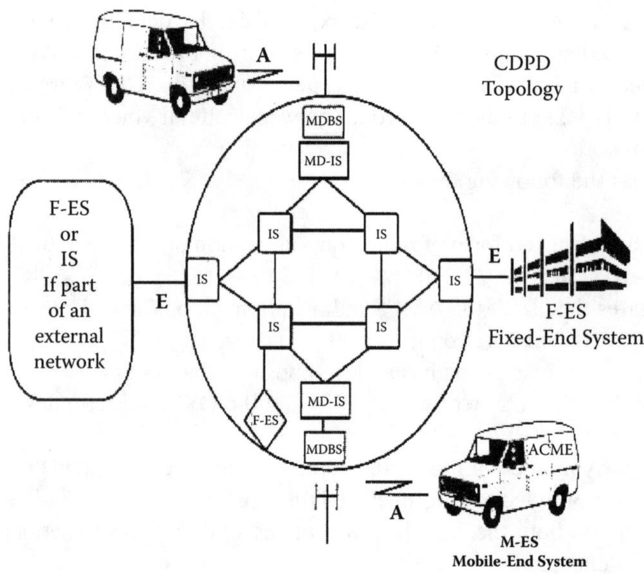

FIGURE 13.10 Major components of a CDPD network.

- Fixed end system (FES): This is a stationary computing device (e.g., a host computer, a UNIX workstation, etc.) connected to landline networks. The FES is the final destination of the message sent from an MES.
- Intermediate system (IS): This is made up of routers that are CDPD compatible. It is responsible for routing data packets into and out of the CDPD service provider network. It may also perform gateway and protocol conversion functions to aid network interconnection.
- Mobile data base station (MDBS): CDPD uses a packet-switched system that splits data into small packets and sends them across the voice channel. This involves detecting idle time on the voice channel and sending the packets on the appropriate unoccupied voice frequencies. This detection of unoccupied frequencies and sending of packets is done by the MDBS. Thus, the MDBS is responsible for relaying data between the mobile units and the telephone network. In other words, it relays data packets from the MES to the MDIS and vice versa. It is responsible for RF channel management. It is housed at the AMPS cell site. It uses the same antenna as the existing cellular network to receive a radio signal from the mobile unit and turns it into digital data. The MDBS provides bidirectional communication for each mobile unit using the reverse and forward channels and it can communicate with up to 16 mobile units in a sector. Forward channels are contentionless while the reverse channels are shared. The mobile units access the reverse channel using the slotted digital sense multiple access with collision detection (DSMA/CD) protocol similar to the CSMA/CD protocol used in Ethernet. With DSMA/

CD, packets are transmitted when the channel is free. Otherwise, the packets are rescheduled for transmission at some random time later as if collision has occurred.

- Mobile data intermediate system (MDIS): MDBSs that service a particular cell can be grouped together and connected to the backbone router, also known as the mobile data intermediate system (MDIS). The MDIS units form the backbone of the CDPD network. All mobility management functions are taken care of by MDIS. In other words, the MDIS is responsible for keeping track of the MES's location and routing data packets to and from the CDPD network and the MES appropriately. Authorization, following authentication, is also an MDIS function. If an MDIS receives a data packet addressed to a mobile unit within its domain, it sends the packet to the appropriate MDBS. If the data packet is not within its domain, it forwards it to the appropriate MDIS.

Very little new equipment is needed for CDPD service since existing cellular networks are utilized. Only the MDBSs are to be added to each cell. One can purchase CDPD cellular communication systems for Windows or MS-DOS computers. The hardware can be a hand-held AMPS telephone or a small modem that can be attached to a notebook computer. One would need to put up the antenna on the modem.

In order to effectively integrate voice and data traffic on the same cellular network without degrading the service provided for the voice customer, the CDPD network employs a technique known as channel hopping. When a mobile unit wants to transmit, it checks for an available cellular channel. Once a channel is found, the data link is established and the mobile unit can use the assigned channel to transmit as long as the channel is not needed for voice communication. Because voice is king, data packets are sent after giving priority to voice traffic. Therefore, if a cellular voice customer needs the channel, it will take priority over the data transmission. In that case, the mobile unit is advised by the MDBS to "hop" to another available channel. If there are no other available channels, then extra frequencies purposely set aside for CDPD can be used. This is a rare situation because each cell typically has 57 channels and each channel has an average idle time of 25–30%. The process of establishing and releasing channel links is called channel hopping and it is completely transparent to the mobile data unit. It ensures that data transmission does not interfere with voice transmission. It usually occurs within the call setup phase of the voice call. The major disadvantage of channel hopping is the potential interference to the cellular system.

CDPD has been referred to as an "open" technology because it is based on the open systems interconnection (OSI) reference model. The CDPD network is comprised of many layers: layer 1 is the physical layer; layer 2 is the data link layer; and layer 3 is the network layer; etc. For example, the physical layer corresponds to a functional entity that accepts a sequence of bits from the medium access control (MAC) layer and transforms them into a modulated waveform for transmission onto a physical 30-kHz RF channel. The network can use either the ISO connectionless network protocol (CLNP) or the transmission control protocol/Internet protocol (TCP/IP).

13.5.2 Applications

CDPD has the following advantages. First, it points to a truly mobile computing environment. Second, by using idle time in voice signals, the cost of data transmission is kept low. Also, cost is based on the volume of data transferred, not on the connection time. Third, because CDPD is an open, nonproprietary standard, it promotes broad availability of hardware and software. It can be used with the existing cellular networks around the world.

CDPD has three major disadvantages. First, CDPD signal is transmitted at roughly 113.2 kbps so that effective throughput of the system is only about 13.6 kbps due to the addressing data required for each data packet and the amount of data required to keep the system reliability at a respectable level. The CDPD encodes each block using a systematic Reed-Solomon forward-error-correcting code. Second, redundant MDBS may be needed in each cell in order to improve network reliability. Third, since CDPD is overlaid on the cellular systems, its design is subject to the constraint that no changes should need to be made to the existing cellular systems.

As an integrated technology, CDPD can be applied in a variety of ways. But CDPD is applicable for the most part for short bursty-type data applications, not for large file transfers. Generally used as a tool for businesses, CDPD holds promises for improving law enforcement communications and operations. The major applications include the following:

- Transaction applications: These might include credit card verification and charges; dispatch; insurance claims; lottery dispensing machines; ATM machines; remote loan application software; fleet management; package pickup, delivery, and tracking; electronic mail notification and delivery; telemetry; field sales and services; and information retrieval services. Encryption would ensure that these transactions are safe.
- Batch applications: These might include file transfer and statistical information transfer.
- Broadcast applications: These might include general information services (local, state, or international news), weather forecast, traffic advisories, advertising, and private bulletin boards.
- Transportation: Messages can be sent to taxis or company trucks.
- Internet access: CDPD provides access to the Internet since both CDPD and Internet are based on IP.
- Law enforcement applications: CDPD has the potential to help police departments operate in a far more efficient manner. An officer on shift may use a CDPD-equipped car on his regular patrol to enhance officer safety and efficiency. The officer, linked to headquarters using CDPD, can perform record checks, transmit reports, and send messages to other officers without worrying about compromising privacy or security. The CDPD technology is moving police organizations toward being "paperless," which could significantly increase the operating efficiency of a department covering a large jurisdiction with few officers.

For these applications, CDPD advantages include automated work order routing, transaction documentation, fraud reduction, and improved customer service. As newer technologies evolve, CDPD could be leveraged against those technologies to support wireless data communications. For now, CDPD can coexist with PCS and CDMA-based infrastructure.

13.6 WIMAX

In this section, we consider another wireless standard. WiMAX is the acronym for World Interoperability for Microwave Access. It aims to provide business and consumer wireless broadband service on the scale of the metropolitan area (MAN). Its coverage is around 30 to 50 km, which makes it a metropolitan area access technology. It can provide as high as 100-Mbps data rates in a 20-MHz bandwidth. WiMAX operates in the 2 GHz to 11 GHz frequency range. Figure 13.11 shows how WiMAX's attributes are applied in a variety of ways.

> **WiMAX** is a global wireless access technology that addresses interoperability across products based on the IEEE 806.16e standard, which stipulates the radio-frequency technology for wireless metropolitan area networks.

WiMAX is the commercialization of the IEEE 802.16 (an evolving standard initiated at the National Institute of Standards and Technology) and HiperMAN

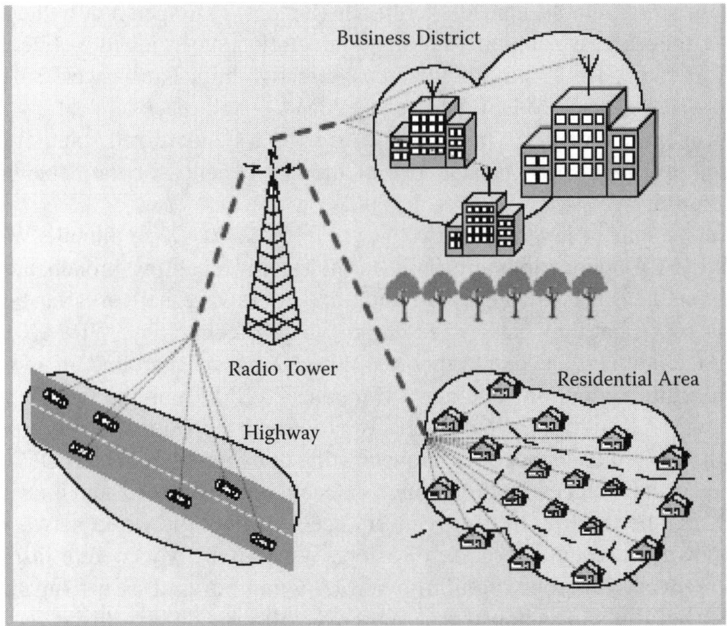

FIGURE 13.11 WiMAX applications.

standards. It should not be confused with Wi-Fi, which is a local area network (LAN) standard developed by IEEE 802.11 working group and is designed to be used indoors at close range. The key difference is that WiMax has a maximum range of some miles instead of Wi-Fi's few hundred feet. Unlike other wireless standards, which address transmission over a single frequency range, WiMAX permits data transport over multiple broad frequency ranges. This maximizes the technology's ability to transmit over the frequencies that will avoid interfering with other wireless applications.

The WiMAX standard has attracted a great deal of attention and interest. The WiMAX Forum (www.wimaxforum.org) is a non-profit organization formed in 2001 by a group of more than 300 industries to promote compatibility and interoperability of broadband wireless products. Members include AT&T, Intel, Cisco, Motorola, Samsung, Fujitsu, Siemens Mobile, British Telecommunications, France Telecom, and Qwest Communications. The forum promotes global WiMax deployment and makes WiMAX the standardized technology of choice for broadband wireless. Although WiMAX will support traffic based on various technologies (such as Ethernet and IP), the forum will certify compliance only for IP-related products. Because not all aspects of the technology have been finalized, deployment is a couple of years away. WiMAX must compete with proprietary broadband wireless technologies already established. Intel's sustained marketing push on WiMAX can only mean WiMAX-enabled laptops and PDAs will be everywhere in a couple of years.

The primary application of WiMAX is high-speed Internet access. WiMAX also promises to be an alternative 3G mobile Internet access service. It will offer high-speed data services to all nomadic terminals (laptops, PDAs, etc.). It will develop a powerful radio access solution with many integration synergies in mobile or fixed network architecture. It can provide a cost-effective broadband access solution in areas beyond the reach of DSL and cable. WiMAX will reach its peak by making portable Internet a reality. The ability to operate a standardized solution in both a licensed and unlicensed band is one of the key benefits for the deployment of WiMAX around the world.

WiMAX can be applied in two ways: fixed WiMAX or mobile WiMAX. Fixed WiMAX applications are point-to-multipoint, enabling broadband access to homes and businesses. Point-to-multipoint microwave networks have existed for many years, but they have relied on proprietary technology; WiMAX offers a standard technology. The fixed version of WiMAX has been around for a few years but with limited adoption. The newer version of WiMAX is the mobile version. Mobile WiMAX allows the full mobility of cellular networks at true broadband speeds. It is based on orthogonal frequency division mupliplexing (OFDM), which has inherent advantages in throughput, latency, and advanced antenna support. WiMAX can be used in disaster recovery scenes where the wired networks have broken down. In recent hurricane disasters, WiMAX networks were installed to help in recovery missions. Similarly, WiMAX can be used as backup links for broken wired links. It is hoped that WiMAX will provide broadband services in rural areas that are underserved by cable TV and DSL providers. While a number of factors will play in the success of WiMAX, most experts consider spectrum availability and competition as essential.

13.7 WIRELESS STANDARDS

The 802.11 wireless LAN working group, which has 250 supporting companies and 650 active members, has produced a series of wireless standards. The following list of standards is a summary. The term 802.11x is also used to denote any of this set of standards.

802.11

This is the original 1997 2.4-GHz wireless Ethernet standard, which runs at 1 or 2 Mbps. There were two versions, frequency hopping and direct sequence.

802.11a

55 Mbps in the 5 GHz band. Same speed as 802.11g close up, but gets slightly slower as the distance increases. The standard is fixed, but implementations in Europe are still under discussion. Although not quite as good as 802.11g in theory, in practice it is likely to be as good or better. Products that adhere to this standard are considered "Wi-Fi certified." "Wi-Fi" stands for wireless fidelity.

802.11b

11 Mbps in the 2.4 GHz band, the first popular standard and still by far the most used. For a while, it was also known as Wi-Fi, but now 802.11g and 802.11a are also known by that name. Wireless hotspots, domestic wireless broadband gateways, and company WLANs were nearly 100 percent 802.11b in early 2003. The 802.11b standard in wireless technology was the first wave to hit big in wireless development, due in large part to Lucent and Apple's adoption.

802.11c

This is a network interoperability standard that deals with bridging in wireless bridges or access points. It deals with MAC level bridging—to include 802.11 frames, thus helping with quality of service and filtering issues. This standard is a supplement to IEEE 802.1d that adds requirements associated with bridging the MAC layers.

802.11d

This is also called the global harmonization standard. Different countries have different parts of the 2.4 and 5 GHz bands available for unlicensed wireless networking. 802.11d seeks to help create standards that will be approvable in as many different countries as possible. It is used in countries where systems using other standards in the IEEE 802.11 family are not allowed to operate.

802.11e

The standard is considered of critical importance for delay-sensitive applications, such as voice over wireless IP and streaming multimedia. As of late 2005 it had been approved as a standard that defines a set of QoS enhancements for LAN applications.

802.11f

Inter-access point roaming protocol. Will let you move through a wireless LAN with multiple access points from different manufacturers, and maintain your connection. This is not finished yet.

802.11g

55 Mbps in the 2.4 GHz band. Bluetooth and 802.11g share the 2.4 GHz band with 802.11b. It offers transmission over relatively short distances at up to 54 Mbps, compared with the 11 Mbps theoretical maximum with the earlier 802.11b standard. 802.11g is an easy choice for corporate sites and home users to adopt because it does not require an upgrade to client equipment. 802.11g is backward-compatible with 802.11b, and it offers speeds similar to those of 802.11a.

802.11h

H stands for HiperLAN—the European standard wireless LAN. It was originally designed to address European regulations but is now applicable in many other countries. Its main features are frequency and power management to make sure that 802.11a networks do not interfere with radar and satellite services.

802.11i

This standard focuses on security. It is designed to provide secured communication of wireless LAN as defined by all the IEEE 802.11 specifications. It includes key management and distribution, encryption, and authentication. It makes use of the advanced encryption standard (AES) block cipher.

802.11 IR

802.11 for infrared. Developed at the same time as 802.11 itself, and supporting 1 or 2 Mbps. Still there are no products on the market that support this standard.

802.11j

An equivalent of 802.11h for the Japanese regulatory environment.

802.11k

A recently started project to standardize the way 802.11a, b, and g networks report measurements of radio and network conditions to other parts of the network stack and new applications. It should be good for network management, fault finding, and other diagnostic efforts.

802.11x

This is used as a journalistic shorthand for any 802.11 standard.

All the 802.11 specifications use the Ethernet protocol and carrier sense multiple access with collision avoidance (CSMA/CA) for path sharing.

13.8 GENERATIONS OF WIRELESS NETWORKS

This section discusses the history and evolution of wireless networks from the first generation (1G) to fourth generation (4G). Table 13.1 presents a short historical summary of the technologies.

First generation: Analog systems are usually referred to first-generation (1G) systems. Most of the devices that came from this generation had military/defense as their origin and then moved to civilian services. Almost all of them are analog systems where voice was considered to be the main traffic. Some of the 1G standards are AMPS, NMT, Hicap, CDPD, Mobitex, and DataTac.

Second generation: All the standards belonging to this generation are commercial based and they are digital in form. Two main groups have evolved, one from Europe and another from America. The second-generation standards are GSM, D-AMPS, IS-54, IS-95, PDC, CSD, PHS, and CDMA2000 (EDGE). At present second-generation (2G) technology GSM is widely used worldwide. However, the problem with 2G technology is that the data rates are limited. This makes it inefficient for data transfer applications like video conferencing, music, or video downloads, etc. Various new technologies have been developed to increase the speed. The first is 2.5G (GPRS) technology that allows data transfer at a better rate than GSM.

Third generation: This deals with systems that integrate voice and data, which provides higher data rates than 2G systems. The systems in this standard are basically a linear enhancement of 2G systems. Currently, transition is happening from 2G to 3G systems. Some of the 3G standards are W-CDMA, UMTS (3GSM), FOMA, 3.5G - HSDPA, and 3.75G - HSUPA. The maximum theoretical data transfer with this 3G technology is 2 Mbps.

TABLE 13.1
A Summary of Network Generations

Technology	1G	2G	3G	4G
Design Begin	1970	1980	1990	2000
Implementation	1984	1991	2002	2010?
Service	Analog voice	Digital voice	Packetized data	IP-oriented, multimedia
Standards	AMPS, TACa, NMT, etc.	TDMA, CDMA, PDC EDGE	WCDMA, CMDA2000	Single standard
Data bandwidth	1.9 kbps	14.4 kbps	2 Mbps	200 Mbps
Multiplexing	FDMA	TDMA, COMA	TDMA, CDMA	CDMA?
Core network	PSTN	PSTN	Packet network	Internet

Source: J. Ibrahim, "4G Features," *Bechtel Telec. Technical Journal*, vol. 1, no. 1, Dec. 2002, pp. 11–14.

AMPS = advanced mobile phone service; CDMA = code division multiple access; EDGE = enhanced data for global evolution; FDMA = frequency division multiple access; GSM = global system for mobile; NMT = Nordic mobile telephone; PDC = personal digital cellular; PSTN = public switched telephone network; TACS = total access communication system; TDMA = time division multiple access; WCDMA = wideband CDMA

Fourth generation: This new generation is expected to complement and replace the 3G systems. The features of a 4G system might be summarized with one word: integration. The system will also serve as an open platform where the new innovations can go with it. Some of the standards that pave the way for 4G systems are WiMax, WiBro, and the proposed 3GPP long-term evolution work-in-progress technologies such as HSOPA. The 4G technology, which is in its infancy, is supposed to allow data transfer up to 100 Mbps outdoors and 1 Gbps indoors.

At present the download speed for mobile Internet connections in Japan is between 9.6 kbps for 2G cellular, up to 128 kbps for PHS, typically 200 kbps for DoCoMo and Vodafone 3G cellular, and 2.4 Mbps for KDDI/AU CDMA2000-1x-WIN service. However, in actual use the data rates are usually slower, especially in crowded areas, or when the network is "congested." 4G mobile data transmission rates are planned to be up to 20 Mbps.

Fourth generation (4G) is as a network that operates on Internet technology, combines it with other applications and technologies such as Wi-Fi and WiMAX, and runs at speeds ranging from 100 Mbps to 1 Gbps.

The 4G working groups have defined the following as the objectives of the 4G wireless communication standard:

- high network capacity
- nominal data rate at high speeds (100 Mbps at stationary conditions and 20 Mbps at 100 miles/hr)
- smooth handoff across heterogeneous networks
- seamless connectivity and global roaming across multiple networks
- high quality of service for next-generation multimedia support (real-time audio, high-speed data, HDTV video content, mobile TV, etc.)
- interoperable with the existing wireless standards

In addition, 4G systems will be fully IP-based wireless Internet. The 4G systems should utilize the network resources to meet the minimal requirements of all the 4G-enabled users. 4G networks are becoming efficient and cost-effective solutions for those wanting an IP-based high-speed data component in the mobile network.

SUMMARY

1. Wireless LAN allows laptop personal computers and LAN users to link up through radio waves or infrared links, eliminating the need for restrictive cabling.
2. Wireless ATM networks are the extension of ATM technology to wireless communications.
3. Personal area network (PAN) is a near-field intrabody communication. It is a wireless connection between PCs, peripherals, and portables that will let the devices share information without having to make a physical connection.
4. Cellular systems operate on the principles of cell, frequency reuse, and handoff.

5. WiMAX is the commercialization of the IEEE 806.16e standard, which stipulates the radio-frequency technology for wireless metropolitan area networks.
6. The 802.11 working group has developed a series of wireless LAN standards.
7. Fourth-generation (4G) wireless networks will not be a revolution in technology, but more an evolution where new services and applications will emerge, and new business models will be developed.

REVIEW QUESTIONS

13.1 Wireless networks are designed to meet the following needs except:
  a. Bypass physical barriers
  b. Mobility
  c. Connectivity to hard-to-wire places
  d. Low cost

13.2 The wireless LAN uses this as the transmission means:
  a. Microwave
  b. Radio waves
  c. Infrared waves
  d. Any electromagnetic waves

13.3 Which of the following is an advantage or benefit of infrared transmission?
  a. More secure
  b. Immune to electromagnetic interference
  c. Has high bandwidth
  d. Cannot propagate through opaque objects

13.4 The access mechanism for wireless LAN as defined by IEEE 802.11 is:
  a. Token passing
  b. CSMA
  c. CSMA/CD
  d. CSMA/CA

13.5 Wireless transmission is less prone to error than wired transmission.
  a. True
  b. False

13.6 Infrared can be used for long-distance communication.
  a. True
  b. False

13.7 WPAN is not characterized by:
  a. Ease of connectivity of personal wearable devices
  b. Optimized for low power and low cost
  c. Requires access points
  d. Small coverage, say 10 m

13.8 Which of these standards allows for international roaming?
  a. AMPS
  b. CDMA
  c. GSM
  d. PDC

13.9 Which of these standards share the same band?
   a. 802.11a
   b. 802.11b
   c. 802.11d
   d. 802.11g

13.10 Which of these standards is for the Japanese regulatory environment?
   a. 802.11a
   b. 802.11b
   c. 802.11h
   d. 802.11j

Answer: 13.1d, 13.2b,c, 13.3d, 13.4d, 13.5b, 13.6b, 13.7c, 13.8c, 13.9b,d, 13.10d

PROBLEMS

13.1 Compare and contrast CSMA/CD and CSMA/CA.

13.2 Compare and contrast RF LAN and IR LAN.

13.3 What are the challenges facing wireless ATM networks?

13.4 Describe the requirements for the PHY, MAC, and DLC layers of a wireless ATM network.

13.5 How is WPAN different from WLAN?

13.6 Explain the concepts of cell splitting, frequency reuse, and handoff in a cellular communications system.

13.7 A cellular system has the following parameters: channel spacing is 50 kHz, the radius of each cell is 2 km, and has a frequency reuse factor of 4. Calculate its spectral efficiency.

13.8 Describe the two digital standards developed in North America for cellular systems.

13.9 What is GSM? Why is the dream of global roaming not yet realized?

13.10 Explain the three basic types of cellular technologies: FDMA, TDMA, and CDMA.

13.11 CDPD is a blend of digital data transmission and analog cellular technology. Explain.

13.12 Describe the following CDPD elements: MES, MDBS, and MDIS.

13.13 How does channel hopping work in CDPD?

13.14 What are the services provided by CDPD?

13.15 What are the advantages and disadvantages of CDPD networks?

13.16 Using Web resources, write a one-page summary of the Bluetooth network.

13.17 Using Web resources, describe IEEE 802.11u wireless standard.

13.18 Using Web resources, describe a wireless PBX.

13.19 Using Web resources, discuss wireless local loop (WLL).

13.20 How fast is wireless data networking?

13.21 Briefly describe WiMAX.

13.22 How does WiMAX avoid interfering with other wireless systems?

13.23 What are the major differences between 3G and 4G wireless systems?

14 Personal Communications Service

> Noble life demands a noble architecture for noble uses of noble men. Lack of culture means what it has always meant: ignoble civilization and therefore imminent downfall.
>
> —Frank Lloyd Wright

The personal communications service (PCS) that began in the mid-1990s is based on low earth orbiting (LEO) satellites. It is a high-frequency, low-power, standards-based, wireless mobile communication system. It mainly operates within the range of 1800 to 1900 MHz. The principal technology base for the PCS comes from wireless analog cellular, the emerging and almost well-established digital wireless cellular systems. It has also evolved from the global systems for mobile communications (GSM) networks. The implementation of the PCS is based on the interconnection of various functional components that are operating in accordance with developed standards so that interoperability and reliability of the systems can be achieved.

The past few decades have seen a rapidly evolving usage of the PCS technology worldwide and it is gaining momentum. PCS undoubtedly, is seen widely as one of the most significant growth areas within the world of telecommunications. This growth has the potential to continue for decades to come.

> **PCS** can be defined as a set of capabilities that allows some combination of service profile management, personal mobility, and terminal mobility.

The ability of a user to access and manipulate the user's service profile can be referred to as services profile management. In the case of personal mobility, it is the ability of the user to access the telecommunications services at any terminal on the basis of a unique personal number. The ability of a terminal to access telecommunication services from different locations and while in motion is known as terminal mobility.

> **PCS** can be defined as a communication system that offers the capability to communicate with everybody, at any time, from any place, with any suitable terminal, and using a personalized set of features.

Radio access techniques are used in realizing these basic characteristics, especially, the "anywhere" and "anytime" characteristics.

14.1 PERSONAL COMMUNICATIONS SYSTEMS

The GSM (global system for mobile communications) digital network has pervaded Europe and Asia. A comparable technology known as personal communication systems (PCS) is beginning to make inroads in the United States. PCS is an advanced phone service that combines the freedom and convenience of wireless communications with the reliability of the legacy telephone service. Both GSM and PCS promise clear transmissions, digital capabilities, and sophisticated encryption algorithms to prevent eavesdropping.

Personal communication services (PCS) is a new concept that will expand the horizon of wireless communications beyond the limitations of current cellular systems to provide users with the means to communicate with anyone, anywhere, anytime. It is called PCS by the FCC or personal communications networks (PCN) by the rest of the world. Its goal is to provide integrated communications (such as voice, data, and video) between nomadic subscribers irrespective of time, location, and mobility patterns. It promises near-universal access to mobile telephony, messaging, paging, and data transfer.

Personal communication enables the internetworking of mobile and fixed network capabilities to allow location/terminal-independent access to services with seamless operation. This is achieved by

- personal mobility: permitting users to access subscribed services, originate/receive calls at any location on any terminal
- terminal mobility: enabling mobile terminals at any location to be located and identified by the network
- service portability: allowing management of user service profiles

PCS/PCN networks and the existing cellular networks should be regarded as complementary rather than competitive. One may view PCS as an extension of the cellular to the 1900 MHz band using identical standards. Major factors that separate cellular networks from PCS networks are speech quality, complexity, flexibility of radio-link architecture, economics of serving high-user-density or low-user-density areas, and power consumption of the handsets. Table 14.1 summarizes the differences between the two technologies and services.

TABLE 14.1
Comparison of Cellular and PCS Technologies

Cellular	PCS
Fewer sites required to provide coverage	More sites required to provide coverage (e.g., a 20:1 ratio)
More expensive equipment	Less expensive cells
Higher costs for airtime	Airtime costs dropping rapidly
High antenna and more space needed for site	Smaller space for the microcell
Higher power output	Lower power output

PCS offers a number of advantages over traditional cellular communications:

- a truly personal service, combining lightweight phones with advanced features such as paging and voice mail that can be tailored to each individual customer
- less background noise and fewer dropped calls
- an affordable fully integrated voice and text messaging that works just about anywhere, anytime
- a more secure all-digital network that minimizes chances of eavesdropping or number cloning
- an advanced radio network that uses smaller cell sites
- a state-of-the-art billing and operational support system

14.2 BASIC FEATURES OF PCS

PCS refers to digital wireless communications and services operating at broadband (1900 MHz) or narrow-band (900 MHz) frequencies. Thus there are three categories of PCS: broadband, narrowband, and unlicensed. Broadband PCS addresses both cellular and cordless handset services, while narrowband PCS focuses on enhanced paging functions. Unlicensed service is allocated from 1890 to 1930 MHz and is designed to allow unlicensed short-distance operation.

The salient features that enable PCS to provide communications with anyone, anywhere, anytime include

- Roaming ability: The roaming service should be greatly expanded to provide universal accessibility.
- Diverse environment: Users must be able to use the PCS in all types of environment, e.g., urban, rural, commercial, residential, mountains, and recreational areas.
- Various cell size: With PCS, there will be a mix of broad types of cell sizes: the pico cell for low-power indoor applications, the micro cell for lower-power outdoor pedestrian application; macro cell for high-power vehicular applications; and super macro cell with satellites, as shown in Figure 14.1. For example, a miro cell of a PCS has a radius of 1 to 300 m.
- Portable handset: PCS provides a low-power radio, switched access connection to the public switched telephone network (PSTN). The user should be able to carry the handset outside without having to recharge its battery.
- FCC frequency allocation: The FCC frequency allocation for PCS usage is significant. FCC allocated 120 MHz for licensed operation and another 20 MHz for unlicensed operation, amounting to a total of 140 MHz for PCS, which is three times the spectrum currently allocated for cellular networks. The FCC's frequency allocation for PCS is shown in Table 14.2 and Table 14.3 for licensed and unlicensed operators. This is also illustrated in Figure 14.2, where MTA and BTA denote major trading area and basic trading area respectively. We notice that all MTAs are allocated a 15 MHz band, while all BTAs are allocated 5 MHz bands. Six different bands are assigned for MTAs and BTAs.

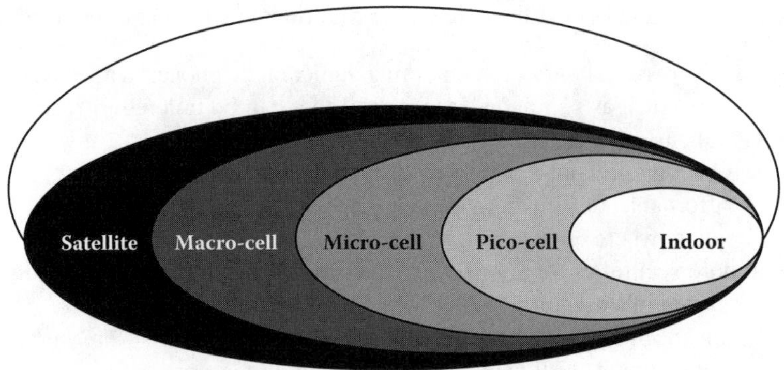

FIGURE 14.1 Various cell sizes.

TABLE 14.2
The PCS Frequency Bands for Licensed Operation

Block	Spectrum Low Side (MHz)	Spectrum High Side (MHz)	Bandwidth (MHz)
A	1850–1865	1930–1945	30
D	1865–1870	1945–1950	10
B	1870–1885	1950–1965	30
E	1885–1890	1965–1970	10
F	1890–1895	1970–1975	10
C	1895–1910	1975–1990	30
Total			120

TABLE 14.3
The PCS Frequency Bands for Unlicensed Operation

Block	Spectrum	Bandwidth (MHz)
Isochronous	1910–1920	10
Asynchronous	1920–1930	10
Total		20

MTA	BTA	MTA	BTA	BTA	BTA	Data	Voice	MTA	BTA	BTA	BTA	BTA
A 15 MHz	D 5 MHz	B 15 MHz	E 5 MHz	F 5 MHz	C 15 MHz	10 MHz	10 MHz	A 15 MHz	D 5 MHz	E 5 MHz	F 5 MHz	C 15 MHz

1850 (MHz) 1910 (MHz) 1930 (MHz) 1990 (MHz)

|<--------**Lower Band**, 60 MHz, Licensed-------->|<-Unlicensed->|<-------**Upper Band**, 60 MHz, Licensed-------->|

FIGURE 14.2 North American PCS frequency bands.

To use the PCS licensed frequency band, a company must obtain a license from the FCC. To use the unlicensed (or unregulated) PCS spectrum, a company must use equipment that will conform to the FCC unlicensed requirements, which include low power transmission to prevent interference with other users in the same frequency band.

14.3 PCS SOLUTIONS

PCS solutions provide a personalized set of communications services through an array of cellular communicators for various individuals. As users move from one cell location to another, antennas in the various areas of coverage pick up the signals from these users. These picked up signals are then forwarded to base stations that have links to wireless and/or wireline networks. The digital cellular technologies and the cellular communicators are utilized in the PCS deployment. These cellular communicators enable mobile Internet protocol (IP), Web browsing, and voice services applications.

PCS operates between a spectrum of 1.850 and 1.990 GHz frequencies in the United States. This includes three spectral bands, namely three 30 MHz, three 10 MHz, and three 20 MHz allocations, respectively, for applications that are exempted from licenses. The lower frequency block is called the cellular band and the higher RF block is called the PCS band within the United States. Third-generation PCS operations were initiated by the European Union in 2000 by the International Mobile Telecommunications (IMT) band.

Overlapping clusters of micro cells and pico cells are employed by PCS implementations. This is because the demand for PCS extended cellular services must be considered and accommodated. In urban areas that are densely populated, transmission rates of about 36 Kbps and higher are considered and enabled. Cellular digital packet data (CDPD) technology and GSM are used to support most of these PCS applications, services, and operations.

14.4 PCS ARCHITECTURE

A PCS network is a wireless network that provides communication services to PCS subscribers. The service area of the PCS network is populated with base stations. The base stations are connected to a fixed wireline network through mobile switch centers (MSCs). Like a cellular network, the radio coverage of a base station is called a cell. The base station locates a subscriber or mobile unit and delivers calls to and from the mobile unit by means of paging within the cell it serves.

PCS architecture resembles that of a cellular network with some differences. The structure of the local portion of a PCS network is shown in Figure 14.3. It basically consists of five major components:

- terminals installed in the mobile unit or carried by pedestrians
- cellular base stations to relay signals
- wireless switching offices that handle switching and routing calls
- connections to PSTN central office
- database of customers and other network-related information

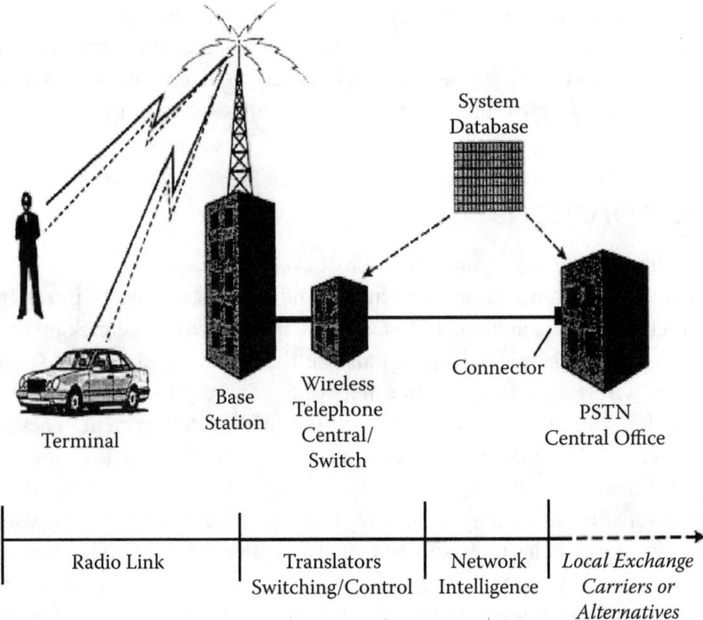

FIGURE 14.3 Structure of PCS network.

Since the goal of PCS is to provide anytime-anywhere communication, the end device must be portable and both real-time interactive communication (e.g., voice) and data services must be available.

The shape and type of any PCS architecture and/or environment can largely be determined by what type of application is considered for implementation, such as military or commercial. It is also important to note that there are some key overall attributes that service providers must provide to satisfy their customers. Some of these include

- Having screening capabilities that will allow the user to be able to control incoming calls and their accessibility.
- Personal numbers that will allow any user to be reached irrespective of that user's location.
- Portability of user's service profile. This allows the users to move around locally and on long distance.
- Operations that are seamless. This will make it possible for users to have uninterrupted service that may occur as the users move from one service provider area to another.

14.4.1 Taiwan PCS Architecture

Figure 14.4 shows a type of commercial PCS architecture that was under consideration in Taiwan around 1993. This architecture can be supported by integrating wireless

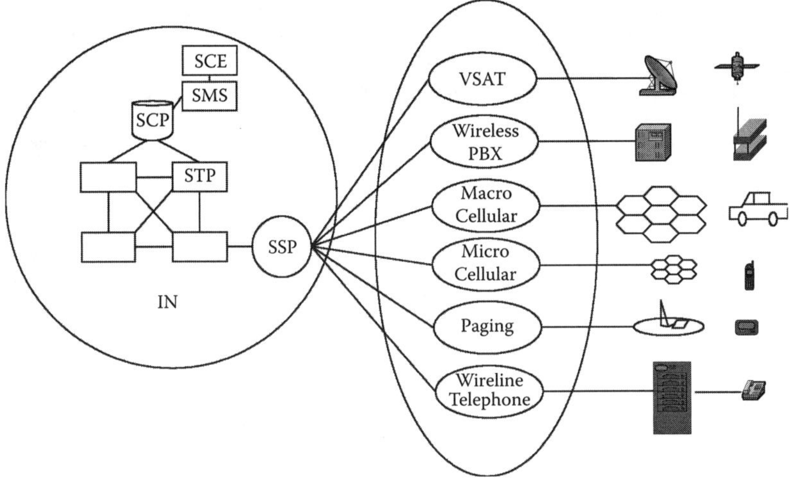

IN: Intelligent Network
VSAT: Very Small Aperture
SCE: Service Creation Environment
SCP: Service Control Point
SMS: Service Management System
STP: Service Transfer Point

FIGURE 14.4 The architecture of PCS.

and wireline systems on the basis of the intelligent network (IN). This IN is responsible for providing the network functions of personal and terminal mobility. As can be seen from Figure 14.4, this architecture has IN and consists of many different types of access systems such as digital cordless systems, satellite communication systems, wireline systems, macro- and micro-cellular systems, and paging systems.

In Figure 14.5, the system architecture of the IN/PCS prototype is shown. The service control point (SCP) provides credit telephones service (CTS) and advanced free phone (AFP). These system applications can be enhanced by the mobile communications abilities. This includes equipment identity register (EIR), visitor location register (VLR), and home location register (HLR). The common channel signaling system 7 (CCS7) connects all of the mentioned registers. Subscribers of the public switched telephone network (PSTN) are linked to the toll networks by the IN using service switched point (SSP). A mobile switching center communicates with the base station (BS).

14.4.2 Tactical PCS System Architecture

In the tactical PCS system architecture, some key requirements are needed. The entire system architecture is driven by these key requirements. They include

- being scaleable and compact to respond to the force structure
- providing directory number portability
- distributing switching and mobility management functions being employed
- depending on the operational scenario, providing both micro- and macro-cellular coverage

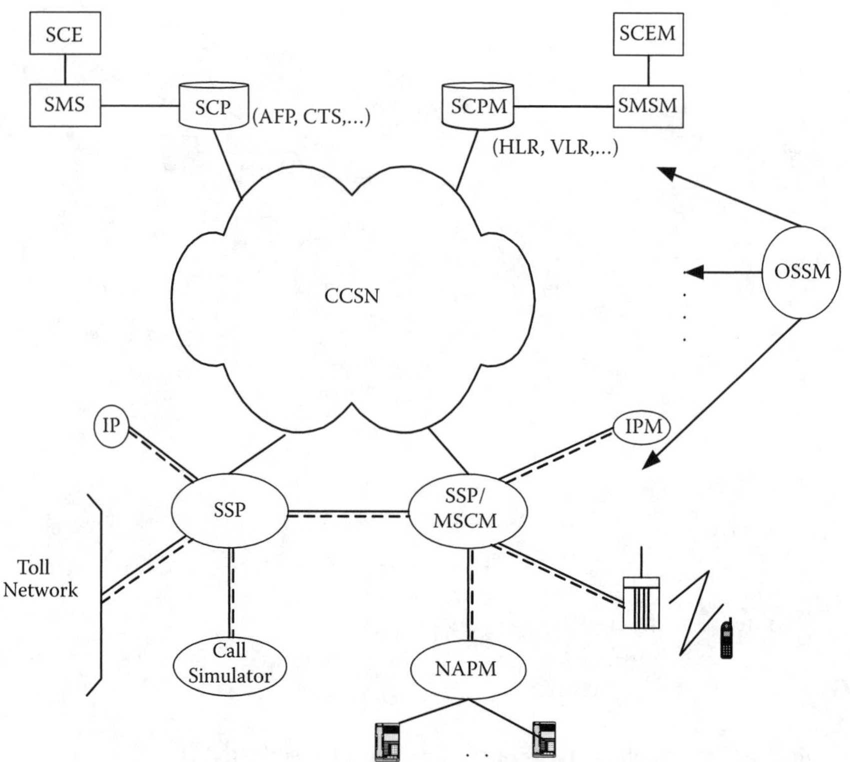

FIGURE 14.5 The system architecture of IN/PCS prototype.

Figure 14.6 and Figure 14.7 show the network connections and the PCS integration with the mobile subscriber equipment (MSE), respectively. It is important to integrate the PCS with the MSE system in order to maintain and seamlessly interconnect to the MSE network. The major considered network functions are

- handoff
- routing
- affiliation and registration
- precedence and preemption
- numbering

In the PCS satellite communication implementation, the LEO satellites play a significant part in the tactical system architecture as shown in Figure 14.8.

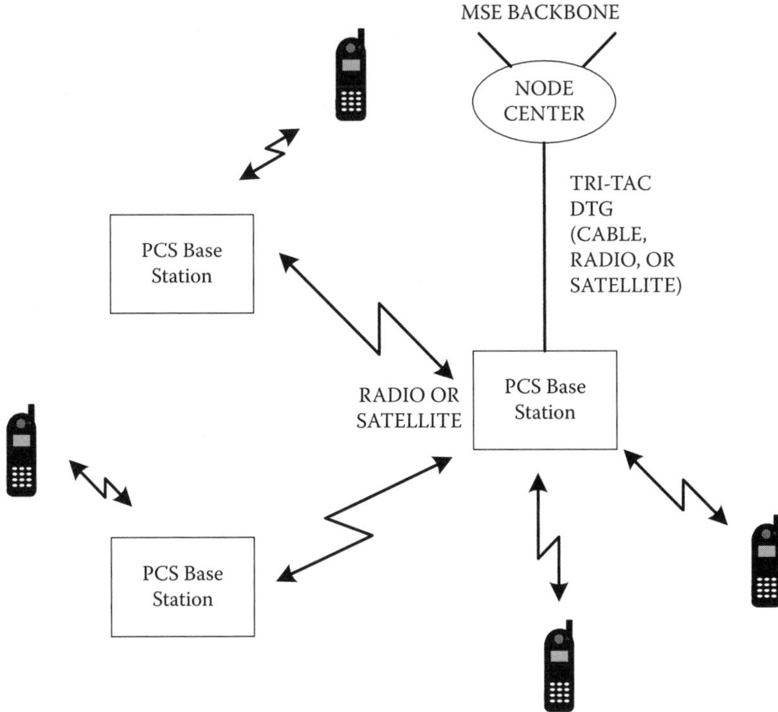

FIGURE 14.6 Network connections.

14.4.3 PCS Architecture for Cellular, Generic Cable, and Bellcore Systems

As shown in Figure 14.9, this type of architecture is being used by some of the cellular system operators. It uses type 1 and type 2 interface connections. The point of interface (POI) of a trunk that is between a wireless services provider (WSP) and a local exchange carrier (LEC) end office (EO) switching system is known as a type 1 interface. Type 2 is of two types, 2A and 2B, respectively. The interface at the POI of a trunk between a WSP and a LEC tandem switching system is called type 2A, while type 2B is the interface at the POI of a trunk between a WSP and LEC EO switching system.

Figure 14.10 and Figure 14.11 show the PCS architectures for generic cable and Bellcore systems. As shown in Figure 14.10, the PCS cable architecture can also use a type 1 interconnection that comes from an equipped base station. At the network interface to the switch, type 1 can be applied as shown in Figure 14.11. Connections to the numbers in the directory is served by the LEC EO and other carriers via the interconnection arrangement as shown in Figure 14.12.

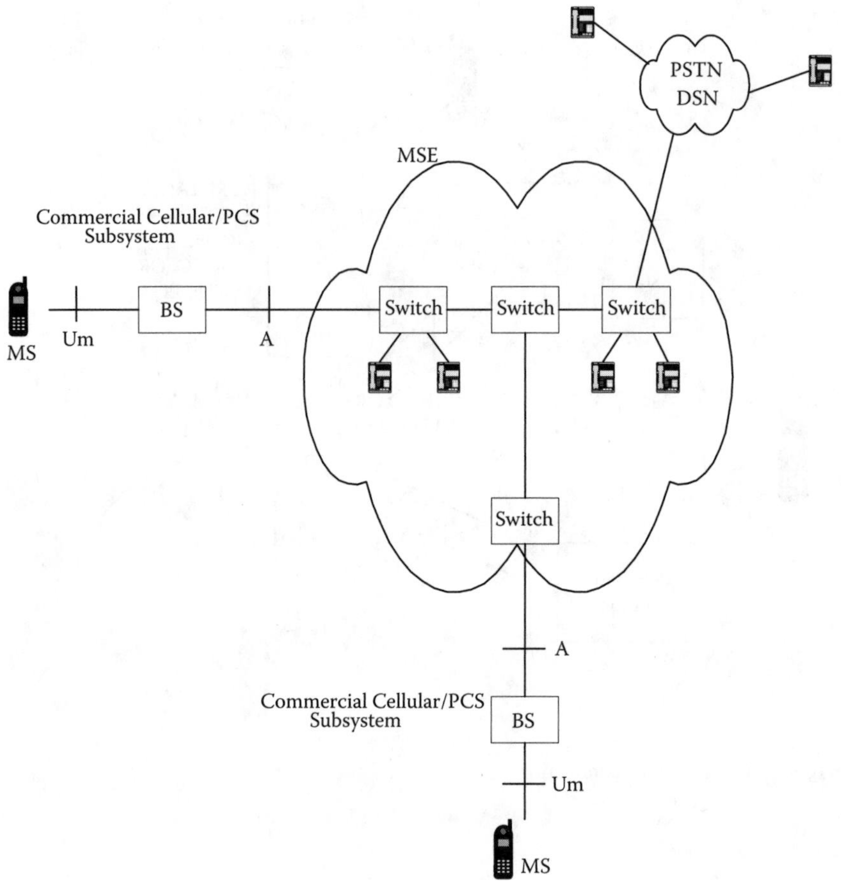

FIGURE 14.7 PCS integration with MSE.

14.4.4 PCS-Capable Switch Architecture

In the PCS-capable switch architecture shown in Figure 14.13, most of the functionalities are performed by the switch, which includes the registration, the visited location register (VLR), vertical services, encryption key distribution, and inter-radio port controller (RPC). As in the case of the cellular switches, the switch in this type of architecture is not a PCS-only switch. This switch is an embedded public switched telephone network (PSTN) with PCS enhancements. In the RPC interface, the switch could be ISDN basic rate interface (BRI), primary rate interface (PRI), or analog lines with separate signaling.

The PCS-capable switch architecture has its advantages and disadvantages. One of the key advantages of this type of architecture is that an embedded PSTN switch can be upgraded to support the PCS. The next key advantage is that the PCS functionality is centralized in the switch. However, the key disadvantage of this architecture is that because the functions are bundled together, they are very hard to modify independently.

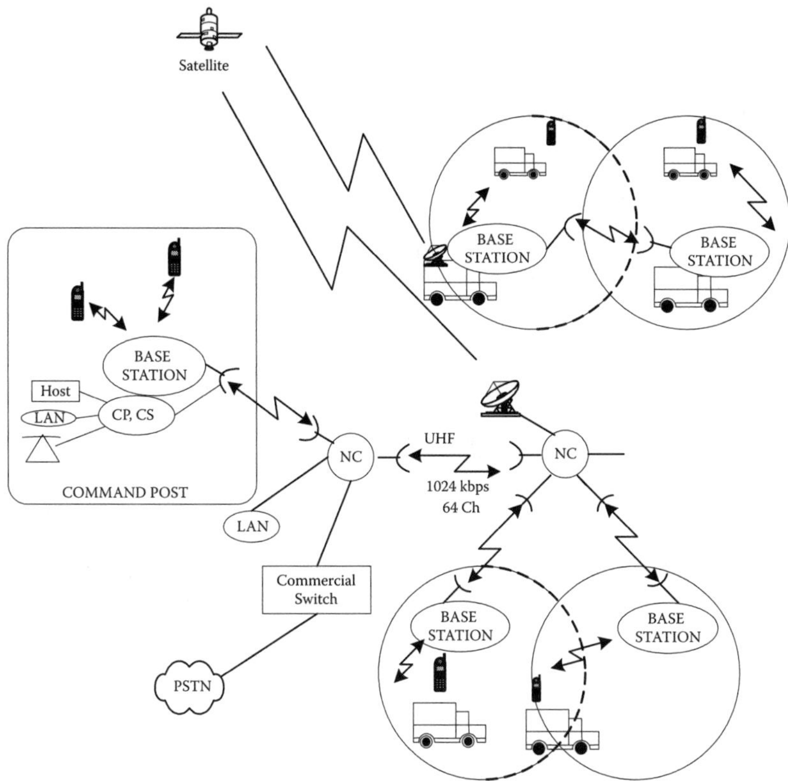

FIGURE 14.8 Adjusted MSE network model.

14.5 PCS STANDARD

The PCS standard covers many aspects of PCS. The main focus in the PCS wireless industry has been on communication with higher data throughput. However, in recent times, adequate attention is now being given to the set of applications requiring simple wireless connectivity with relaxed throughput and latency requirements. The standard has progressed from IEEE 802.15.4, 802.15.1 (Bluetooth), 802.11, 802.11b, 802.11g to 802.11a (HiperLAN). It has progressed continuously with increases in data rate.

14.5.1 THE PCS STANDARD COMMITTEES AND THEIR VARIOUS MODELS

There are many standard committees and models for PCS. Some of the major ones are discussed in this section. The Joint Technical Committee (JTC) has been responsible for developing standards for PCS in the United States. The JTC committee worked cooperatively with the TIA committee working on the TR-46 reference model and the ATIS committee working on the T1P1 reference model.

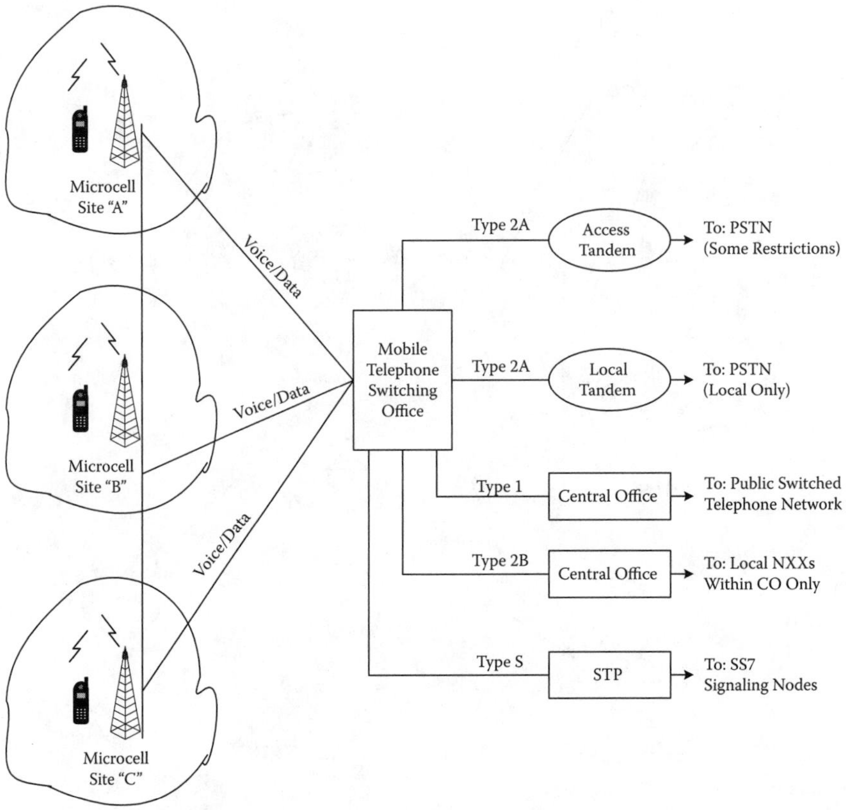

Microcell
Site "A"
Microcell
Site "B"
Microcell
Site "C"
Voice/Data
Voice/Data
Voice/Data
Mobile
Telephone
Switching
Office
Type 2A
Access
Tandem
To: PSTN
(Some Restrictions)
Type 2A
Local
Tandem
To: PSTN
(Local Only)
Type 1
Central Office
To: Public Switched
Telephone Network
Type 2B
Central Office
To: Local NXXs
Within CO Only
Type S
STP
To: SS7
Signaling Nodes

FIGURE 14.9 Personal communications network (PCN) concept.

(Note 1)
Base
Station
#1
(Note 3)
RASP #1
To PSTN
Fiber
Equipment
Fiber
Cable
Fiber
Node
(Note 2)
2-way Coaxial Cable
Base
Station
#1
(Note 1)
RASP #2
(Note 3)
(Note 4)
RAD
RAD
RAD Link #1
RAD
RAD
RAD Link #2
(Note 4)

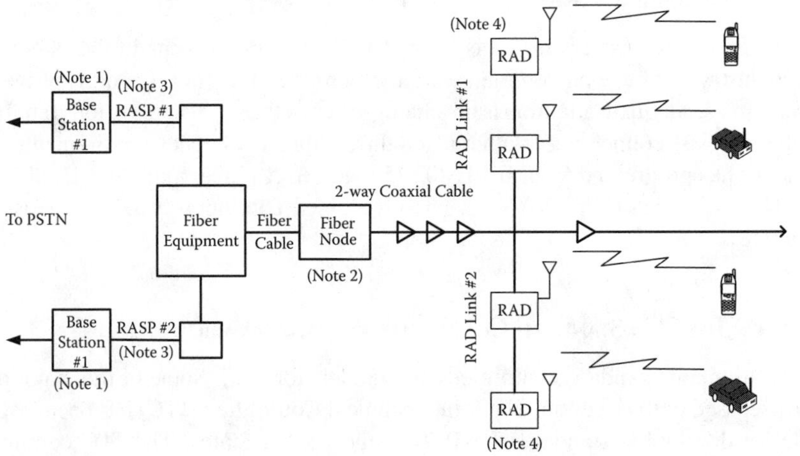

FIGURE 14.10 Generic cable architecture for PCS.

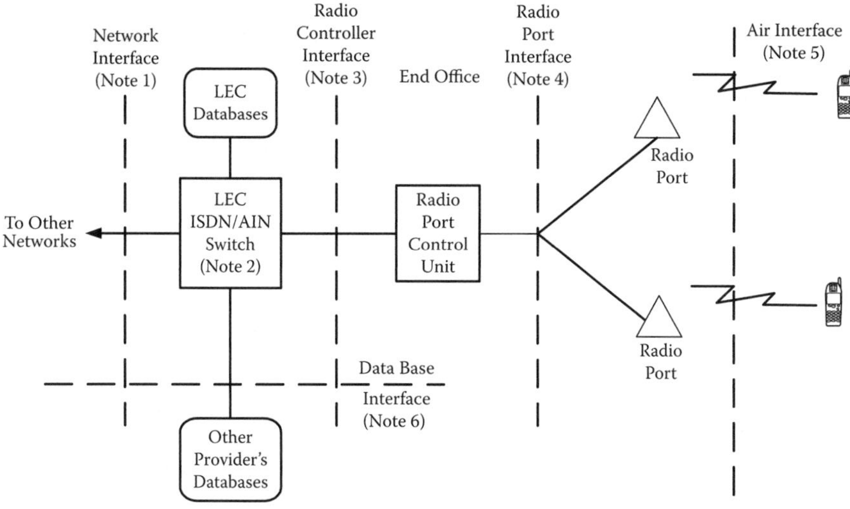

FIGURE 14.11 Simplified Bellcore proposed PCS architecture.

The TR-46 reference model has the following models:

- Personal station (PS): This can be a stand-alone device such as a PC or fax machine. PS terminates the radio path on the user side and allows the user to gain access to the network.
- Radio system (RS): This is often known as the base station. It terminates the radio path and connects to the personal communications switching center (PCSC).

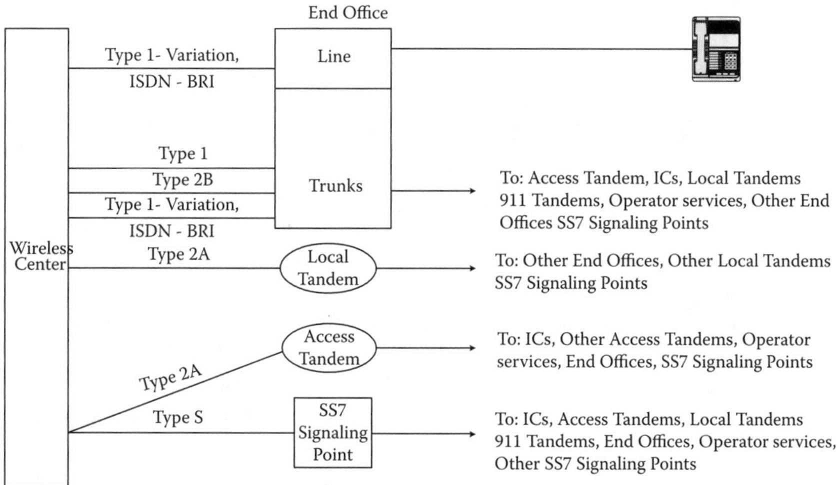

FIGURE 14.12 Interconnection examples overview.

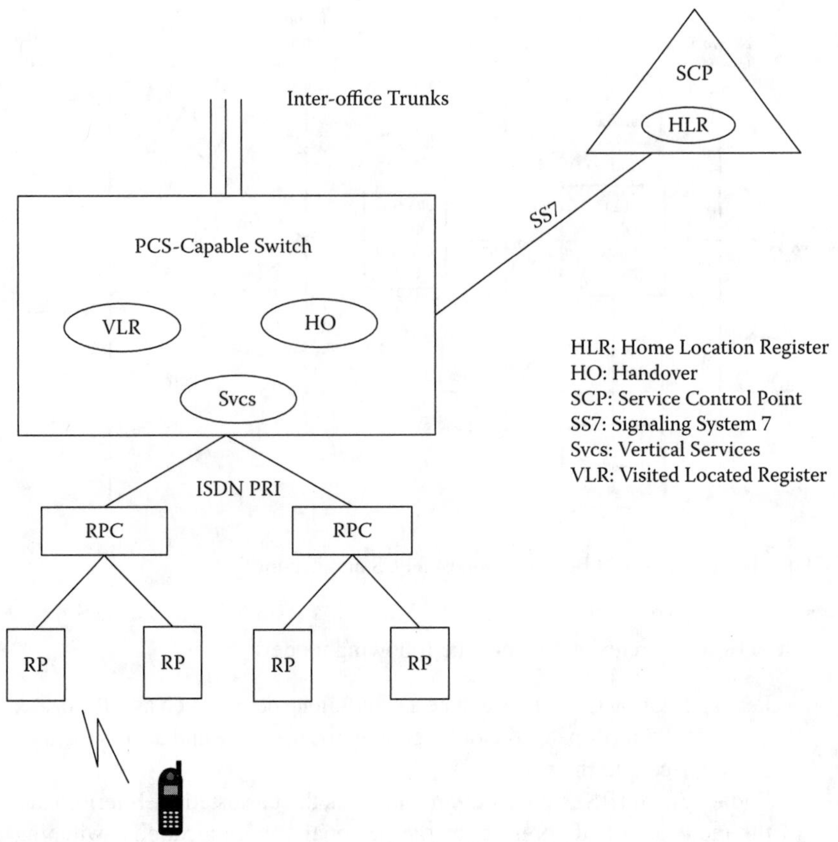

FIGURE 14.13 PCS-capable switch.

- Personal communication switching center (PCSC): This interfaces the user traffic from the wireless network to the wireline network.
- Home location register (HLR): This unit manages mobile subscribers. It may be part of PCSC or separate.
- Data message handler (DMH): This is used for billing.
- Visited location register (VLR): This stores subscriber information such as directory number (DN) and electronic serial number (ESN) obtained from the user's HLR.
- Authentication center: This manages authentication or encryption of information.
- Equipment identity register (EIR): This provides information about the PS.
- Operation system (OS): This takes care of the overall management of the wireless network.
- Interworking function (IWF): This interconnects PCSC to other networks and enables it to communicate with them.

Some interfaces are specified between the various elements of the system. Based on the TR-46 reference model, a call flows from a PS to an RS to the PCS switch to other network elements. The T1P1 PCS reference model is similar to the TR-46 model but with some differences.

With the TR-46 and T1P1 reference models, PCS has enough capabilities to support a wide range of services. These services include basic services such as call origination, call termination, call clearing, handoff, and emergence calls, and supplementary services such as automatic recall, automatic reverse charging, call hold and retrieve, call forwarding-default, call forwarding-busy, call waiting, call transfer, do not disturb, three-way calling, paging, email access, and voice message retrieval. It is evident that some of these services are similar to those of the wireline network or cellular network.

Satellites are instrumental in achieving global coverage and providing PCS services. Mobile satellite communications for commercial users is evolving rapidly toward PCS systems to provide basic telephone, fax, and data services virtually anywhere on the globe. Satellite orbits are being moved closer to the earth, improving communication speed and enabling PCS services. Global satellite systems are being built for personal communications. In the United States, the Federal Communications Commission (FCC) licensed five such systems: Iridium, Globalstar, Odyssey, Ellipso, and Aries. In Europe, ICO-Global is building ICO. Japan, Australia, Mexico, and India are making similar efforts. More about the use of satellite system for personal communication is discussed in Section 14.7.

Unlike GSM, PCS is unfortunately not a single standard but a mosaic consisting of several incompatible versions coexisting rather uneasily with one another. One major obstacle to PCS adoption in the United States has been the industry's failure to sufficiently convince customers of the advantages of PCS over AMPS, which already offers a single standard. This places the onus on manufacturers to inundate phones with features that attract market attention without compromising the benefits inherent in cellular phones. However, digital cellular technology enjoys distinct advantages. Perhaps the most significant advantage involves security because one cannot adequately encrypt AMPS signals.

There are four major problems currently facing PCS spectrum holders:

- Numerous obstacles to deploying the micro cell infrastructure, which requires more towers than cellular.
- Snags in voluntary surrender of the spectrum by current license holders. (Up-front auction costs of $7.7 billion for spectrum licenses and license holders are slow in surrendering them because they see gold. FCC did recognize this problem and moved to correct it by changing the rule.)
- Marketing, particularly being able to convince the users that PCS is better than cellular.
- Economics, particularly how the cost of using PCS compares with cellular.

The future growth and success of PCS services cannot be taken for granted. Like any new technology, the success of the PCS system will depend on a number of factors. These include initial system overall cost, quality, and convenience of the service provided, and cost to subscribers.

14.6 PCS SATELLITE SERVICES

As shown in Figure 14.8, satellite services play a significant role in the PCS implementations and also in the tactical system architectures of the feature. The emerging LEO satellite systems will continue to play greater roles in the future. Communication satellites have been operated in geostationary earth orbits (GEOs) in the past. This allows the ground antennas to remain fixed. As a result of this, only three or four satellites are required to cover the entire earth. By all measures, operationally this is a much simpler system. However, it requires a much more complicated satellite as a result of the satellite-to-ground large space loss.

There are two groups of categories for the LEOs. They are the little LEOs designed for non-voice services such as data and message transmission and the big LEOs designed to have much larger capacity and voice transmission capability than the little LEOs. The little LEO satellites are low in mass and small in size. They provide non-geostationary, non-voice services. The little LEO satellites were allocated frequencies at 137–138 MHz (space to earth) and 148–150.05 MHz (earth to space). The big LEO satellite systems were allocated frequencies at 2483.5–2500 MHz (space to earth) and 1610–1626.5 MHz (earth to space).

Some of the characteristics of the LEO satellite systems are

- propagation delay of 5–10 ms
- low elevation angle operation
- high replacement and investment cost for the space segment
- ground infrastructure is large
- propagation delay is very short
- handover requirements are very rapid

Some of the LEO systems now available include

- Globalstar
- Iridium
- Orbcomm
- Odyssey
- Ellipso
- Aries

These new LEO satellite systems provide services that are relevant to tactical users. The services include

- remote monitoring
- two-way paging
- voice and data communications
- email
- asset tracking
- search and rescue operations

TABLE 14.4
PCS Access Methods

	AMPS	N-AMPS	D-AMPS	GSM	CDMA
Standard	IS-91	IS-88	IS-136	GSM	IS-95
Channel Size	30 kHz	10 kHz	30 kHz	200 kHz	1.25 MHz
Total Channels	832	2412	832	1000 (MTA)	—
Users Per Channel	1	1	3	8	—
Voice Channels	Analog FDMA	Analog FDMA	Both TDMA	Digital TDMA	Digital CDMA
Control Channels	FDMA	FDMA	FDMA	TDMA	CDMA
Channel Reuse Factor	7	7	7 or 4	4 or 3	1
Channels per Cell Site	119	357	357 or 625	250 or 333	1
Capacity vs. AMPS	1	3	3	2.3 or 3.4	6–20

14.7 PCS ACCESS METHODS

PCS has several access methods, shown in Table 14.4. It covers the advanced mobile phone service (AMPS), narrowband advanced mobile phone service (N-AMPS), code division multiple access (CDMA), global system for mobiles (GSM-1900), and digital advanced mobile phone service (D-AMPS). Each has its various standards, channel sizes and total channels, users per channel, voice channels, control channels, channel reuse factor, channel per cell site, and, finally, channel versus AMPS.

SUMMARY

1. PCS began in the mid-1990s and is based on low earth orbiting (LEO) satellites.
2. PCS solutions provide a personalized set of communications services through an array of cellular communicators for various individuals.
3. The type of application that is considered for implementation, such as military or commercial, largely can determine the shape and type of any PCS architecture and/or environment.
4. Some of the key PCS architectures include Taiwan PCS architecture, tactical PCS system architecture, PCS architecture for cellular, generic cable and Bellcore systems, and PCS-capable switch architecture.
5. The standard has progressed from IEEE 802.15.4, 802.15.1 (Bluetooth), 802.11, 802.11b, 802.11g to 802.11a (HiperLAN).

REVIEW QUESTIONS

14.1 PCS stands for:
   a. Personal communications system
   b. Personal communications service
   c. Pacific communications service
   d. Personal commission system

14.2 PCS is based on low earth orbiting (LEO) satellites.
   a. True
   b. False

14.3 PCS normally operates on what range of frequency?
   a. 800 to 900 MHz
   b. 900 to 1800 MHz
   c. 1800 to 1900 GHz
   d. 1800 to 1900 MHz

14.4 The principal technology base for the PCS did not come from wireless analog cellular, the emerging and almost well-established digital wireless cellular systems.
   a. True
   b. False

14.5 The ability of a terminal to access telecommunication services from different locations and while in motion is known as:
   a. Terminal capability
   b. Personal mobility
   c. Terminal mobility
   d. Personal capability

14.6 The ability of the user to access the telecommunications services at any terminal on the basis of a unique personal number is known as:
   a. Personal mobility
   b. Personal capability
   c. Terminal capability
   d. Terminal mobility

14.7 The ability of a user to access and manipulate the user's service profile can be referred to as:
   a. Services capability management
   b. Services mobility management
   c. Services profile management
   d. Services profile mobility

14.8 In the United States, PCS operates between a spectrum of:
   a. 1.850 and 1.990 GHz
   b. 1850 and 1990 GHz
   c. 1.850 and 1.990 MHz
   d. 1.85 and 1.99 GHz

14.9 The 3G PCS operations were initiated by the European Union in 2000 by the International Mobile Telecommunications IMT band.
   a. True
   b. False

14.10 CDPD stands for what type of technology?
   a. Cellular data packet digital
   b. Cellular digital personal data
   c. Cellular data personal digital
   d. Cellular digital packet data

Answer: 14.1b, 14.2a, 14.3d, 14.4b, 14.5c, 14.6a, 14.7c, 14.8a, 14.9a, 14.10d

PROBLEMS

14.1 Define a PCS system. Why are "anywhere" and "anytime" some of the key characteristics of a PCS?
14.2 What are the basic features of a PCS?
14.3 Describe some of the key overall attributes that service providers must provide to satisfy their customers in using PCS.
14.4 Name at least four basic PCS architectures.
14.5 Describe at least two of the basic architectures of a PCS.
14.6 What are the differences between any of the architectures of a PCS?
14.7 In the tactical PCS architecture, what are the key driven requirements of the system architecture?
14.8 What are the major functions of the tactical PCS system architecture?
14.9 Why are standards important in the production of PCS systems?
14.10 What are the advantages and disadvantages of the PCS-capable switch architecture?
14.11 What are some of the characteristics of the LEO satellite systems?
14.12 Name some of the LEO systems now available.
14.13 What are some of the services of the LEO satellite systems?
14.14 Why are satellite PCS services important in telecommunications?
14.15 Develop a matrix of standards for PCS. Briefly describe each standard.
14.16 Why are standards important in the development of PCS?
14.17 Describe some of the access methods for PCS systems.
14.18 Compare and contrast cellular and PCS systems. What are the advantages and disadvantages of each?
14.19 List and describe the key frequency bands for PCS licensed operations.
14.20 List and describe the key frequency bands for PCS unlicensed operations.

15 Satellite Communication

Defeat is not the worst of failures. Not to have tried is the true failure.

—George E. Woodberry

Satellite-based communication has become a major facet of the telecommunication industry for at least two major reasons. First, it provides a means of broadcasting information to a large number of people simultaneously. Thus, satellite communication systems are an important ingredient in the implementation of a global communication infrastructure. Second, satellite communication provides a means of reaching isolated places on earth, where terrestrial telecommunications infrastructure does not exist or teledensity is low.

Satellite-based communication provides worldwide communications capability between two ground stations.

Satellite communication was first deployed in the 1960s and has its roots in military applications. Since the launch of the Early Bird satellite (first commercial communication satellite also known as Intersal I) by NASA in 1965 proved the effectiveness of satellite communications, satellites have played an important role in both domestic and international communications networks. They have brought voice, video, and data communications to areas of the world that are not accessible with terrestrial lines. By extending communications to the remotest parts of the world, virtually everyone can be part of the global economy.

Satellite communication is not a replacement of the existing terrestrial systems but rather an extension of wireless systems. However, satellite communication has the following merits over terrestrial communications:

- Coverage: Satellites can cover a much larger geographic area than the traditional ground-based systems. They have the unique ability to cover the globe.
- High bandwidth: A Ka band (27–40 GHz) can deliver a throughput rate of gigabits per second.
- Low cost: A satellite communications system is relatively inexpensive because there are no cable-laying costs and one satellite covers a large area.
- Wireless communication: Users can enjoy untethered mobile communication anywhere within the satellite coverage area.
- Simple topology: Satellite networks have simpler topology, which results in more manageable network performance.

TABLE 15.1
Advantages and Disadvantages of Satellites

Advantages	Disadvantages
Wide-area coverage	Propagation delay
Easy access to remote sites	Dependency on a remote facility
Costs independent of distance	Less control over transmission
Low error rates	Attenuation due to atmospheric particles (e.g., rain) can be severe at high frequencies
Adaptable to changing Network patterns	Continual time-of-use charges
No right-of-way necessary, earth stations located at premises	Reduced transmission during solar Equinox

- Broadcast/multicast: Satellites are naturally attractive for broadcast/multicast applications.
- Maintenance: A typical satellite is designed to be unattended, requiring only minimal attention by customer personnel.
- Immunity: A satellite system will not suffer from disasters such as floods, fire, and earthquakes and will therefore be available as an emergency service should terrestrial services be knocked out.

Of course, satellite systems do have some disadvantages. These are weighed with their advantages in Table 15.1. Some of the services provided by satellites include fixed satellite service (FSS), mobile satellite service (MSS), broadcasting satellite service (BSS), navigational satellite service, and meteorological satellite service.

This chapter explores the integration of satellites with terrestrial networks to meet the demands of highly mobile communities. After looking at the fundamentals of satellite communications, we will discuss the orbital and propagation characteristics and the various applications of satellite-based communications systems.

15.1 FUNDAMENTALS

A satellite communications system may be viewed as consisting of two parts: the space and ground segments. The space segment consists of the satellites and all their on-board tracking and control systems. The earth segment comprises the earth terminals, their associated equipment, and the links to terrestrial networks.

15.1.1 Types of Satellites—GEO, MEO, and LEO

There were only 150 satellites in orbit by September 1997. The number was expected to be roughly 1700 by the year 2002. With this increasing trend in the number of satellites, there is a need to categorize them. According to the height of their orbit and "footprint" or coverage on the earth's surface, they are classified as follows:

- Geostationary earth orbit (GEO) satellites: They are launched into a geostationary or geosynchronous orbit, which is 35,786 km above the equator. (Raising a satellite to such an attitude, however, requires a rocket, so that the achievement of a GEO satellite did not take place until 1963.) A satellite is said to be in geostationary orbit when the space satellite is matched to the rotation of the earth at the equator. A GEO satellite can cover nearly one-third of the earth's surface, i.e., it takes three GEO satellites to provide global coverage, as shown in Figure 15.1. Due to their large coverage, GEO satellites are ideal for broadcasting and international communications. [GEO are sometimes referred to as high earth orbits (HEO)]. Examples of GEO satellite constellations are Spaceway designed by Boeing Satellite Systems and Astrolink by Lockheed Martin. Another example is Thuraya, designed by Boeing Satellite Systems to provide mobile satellite services to the Middle East and surrounding areas. There are at least three major objections to GEO satellites. First, there is a relatively long propagation delay (or latency) between the instant a signal is transmitted and when it returns to earth (about 240 milliseconds). This is caused by speed-of-light

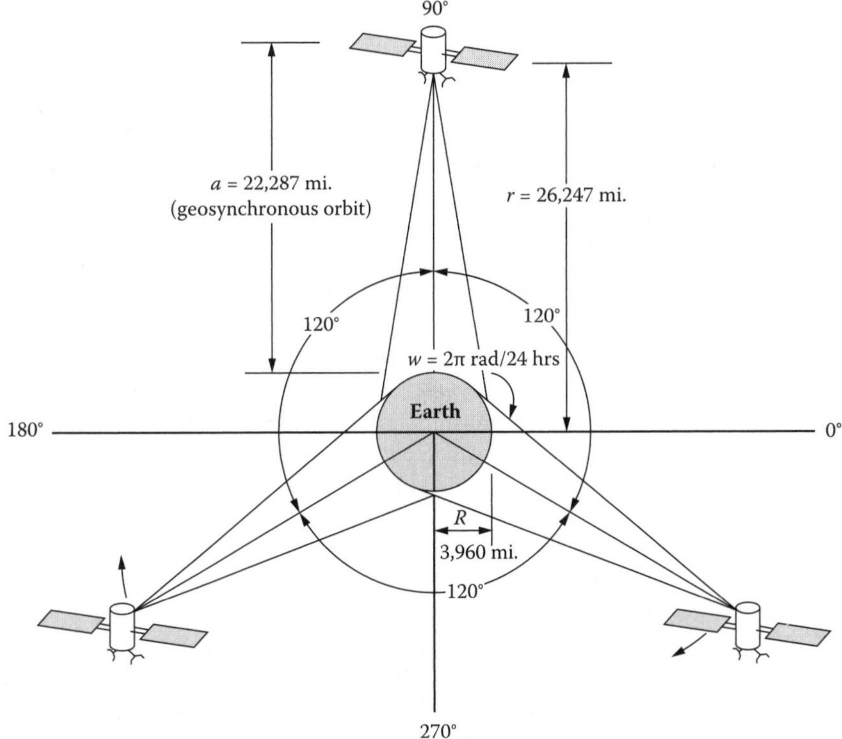

FIGURE 15.1 Three GEO satellites can provide full coverage of the earth with the exception of the polar caps.
Source: W. Hioki, *Telecommunications,* Prentice Hall, 4th ed., 2001, p. 307.

transmission delay and signal processing delay. This may not be a problem if the signal is going only one way. However, for signals such as data and voice, which go in both directions, the delay can cause problems. GEO satellites, therefore, are less attractive for voice communication. Second, there is lack of coverage at far northern and southern latitudes. This is unavoidable because a GEO satellite is below the horizon and may not provide coverage at latitudes as close to the equator as 45°. Unfortunately, many of the European capitals, including London, Paris, Berlin, Warsaw, and Moscow, are north of this latitude. Third, both the mobile unit and the satellite of a GEO system require a high transmit power. In spite of these objections, the majority of satellites in operation today are GEO satellites, but that may change in the near future.

- Middle earth orbit (MEO) satellites: They orbit the earth at 5,000 to 12,000 km. GEO satellites do not provide good coverage for places far north and satellites in inclined elliptical orbits are an alternative. Although the lower orbit reduces propagation delay to only 60 to 140 ms round trip, it takes 12 MEO satellites to cover most of the planet. MEO systems represent a compromise between LEO and GEO systems, balancing the advantages and disadvantages of each. [MEO are sometimes referred to as intermediate circular orbit (ICO).]
- Low earth orbit (LEO) satellites: They circle the earth at 500 to 3000 km. For example, the Echo satellite circles the earth every 90 minutes. To provide global coverage may require as many as 200 LEO satellites. Latency in an LEO system is comparable with terrestrial fiber optics, usually less than 30 ms round trip. LEO satellites are suitable for personal communication systems (PCS). However, LEO systems have a shorter life span of 5–8 years (compared with 12–15 years for GEO systems) due to the increased amount of radiation in low earth orbit. The LEO systems have been grouped as little LEO and big LEO. The little LEOs have less capacity and are limited to nonvoice services such as data and message transmission. An example is OrbComm designed by Orbital Corporation and consists of 36 satellites, each weighing 85 pounds. The big LEOs have larger capacity and voice transmission capability. An example is Loral and Qualcomm's Globalstar, which will operate in the L-band frequencies and employ 48 satellites organized in eight planes of six satellites each.

The arrangement of the three basic types of satellites is shown in Figure 15.2. The evolution from GEO to MEO and LEO satellites has resulted in a variety of global satellite systems. The convenience of GEO was weighed against the practical difficulty involved with it and the inherent technical advantages of LEO, such as lower delay and higher angles of elevation. While it has been conceded that GEO is in many respects theoretically preferable, LEO or MEO systems would be preferred for many applications. Although a constellation (a group of satellites) is required instead of only one for hemispheric coverage, the loss of individual satellites would cause only gradual degradation of the system rather than a catastrophic failure. A comparison of the three satellite types is given in Table 15.2.

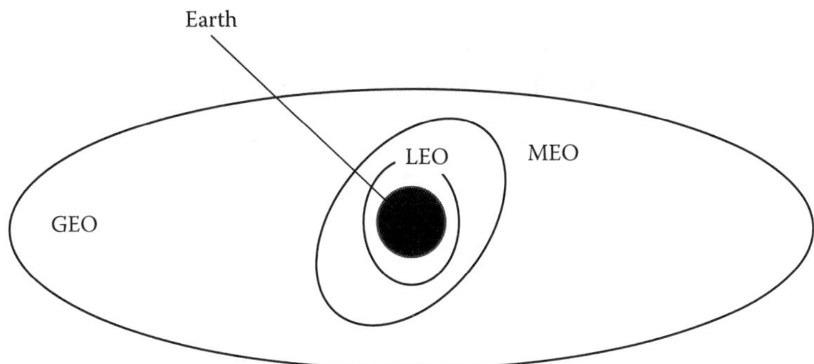

FIGURE 15.2 The three common types of satellites—GEO, MEO, and LEO.

15.1.2 Frequency Bands

Every nation has the right to access the satellite orbit and no nation has a permanent right or priority to use any particular orbit location. Without a means for nations to coordinate use of satellite frequency bands, the satellite services of one nation could interfere with those of another, thereby creating a chaotic situation in which neither country's signals could be received clearly.

To facilitate satellite communications and eliminate interference between different systems, international organizations govern the use of satellite frequency. The International Telecommunication Union (ITU) is responsible for allocating frequencies to satellite services. Since the spectrum is a limited resource, the ITU has reassigned the same parts of the spectrum to many nations and for many purposes throughout the world.

TABLE 15.2
Comparison of the GEO, MEO, and LEO

Type	Altitude	Coverage	Advantages	Disadvantages
LEO	300–1,000 km	Spot	Low path loss	Coverage is less
			High data rate	Need many satellites
			Low delay	Short orbital life
			Low launch cost	High Doppler
			Less fuel	Highly complex
MEO	1,000–10,000 km	Region	Moderate path loss	Multiple satellites
			Moderate launch cost	Moderate coverage
			Less fuel	Highly complex
GEO	36,000 km	Earth	Global coverage	High path loss
			Need few satellites	Long delay
			Long orbital life	Low data rate
			Low Doppler	High launch cost
			Less complex	Fuel for station keeping

TABLE 15.3
Satellite Frequency Allocations

Frequency band	Range (GHz)
L	1–2
S	2–4
C	4–8
X	8–12
Ku	12–18
K	18–27
Ka	27–40

The frequency spectrum allocations for satellite services are given in Table 15.3. Notice that the assigned segment is the 1–40 GHz frequency range, which is the microwave portion of the spectrum. As microwaves, the signals between the satellite and the earth stations travel along line-of-sight paths and experience free-space loss that increases as the square of the distance.

Satellite services are classified into 17 categories: fixed, intersatellite, mobile, land mobile, maritime mobile, aeronautical mobile, broadcasting, earth exploration, space research, meteorological, space operation, amateur, radio determination, radio navigation, maritime radio navigation, and standard frequency and time signal. The Ku band is presently used for broadcasting services, and also for certain fixed satellite services. The C band is exclusively for fixed satellite services, and no broadcasting is allowed. The L band is employed by mobile satellite services and navigation systems.

A satellite band is divided into separation portions: one for earth-to-satellite links (the uplink) and one for satellite-to-earth links (the downlink). Like a terrestrial microwave relay, a satellite must use separate frequencies for sending to the satellite (the uplink) and receiving from the satellite (the downlink), otherwise the powerful signal transmitted by the satellite would interfere with the weak incoming signal. Table 15.4 provides the general frequency assignments for uplink and downlink satellite frequencies. We notice from the table that the uplink frequency bands are slightly higher than the corresponding downlink frequency bands. This is

TABLE 15.4
Typical Uplink and Downlink Satellite Frequencies

Uplink frequencies (GHz)	Downlink frequencies (GHz)
5.925–6.425	3.700–4.200
7.900–8.400	7.250–7.750
14.00–14.50	11.70–12.20
27.50–30.0	15.70–20.20

to take advantage of the fact that it is easier to generate RF power within a ground station than it is onboard a satellite. In order to direct the uplink transmission to a specific satellite, the uplink radio beams are highly focused. In the same way, the downlink transmission is focused on a particular footprint or area of coverage. The most desired band frequency for satellite communication is 6 GHz for uplink and 4 GHz for downlink.

All satellite systems are constrained to operate in designed frequency bands depending on the kind of earth station used and service provided. The satellite industry, particularly in the United States, is subject to several regulatory requirements, domestic and international, depending upon which radio services and frequency bands are being proposed for use on the satellite. In the United States, the Federal Communications Commission (FCC) is the independent regulatory agency that ensures that the limited orbital/spectrum resources allocated to space radio communications services is used efficiently. After receiving an application for a U.S. domestic satellite, the FCC initiates the advance publication process. This is to ensure the availability of an orbit position when the satellite is authorized. FCC does not guarantee international recognition and protection of satellite systems unless the authorized satellite operator complies with all coordination requirements and completes the necessary coordination of its satellites with all other administrations whose satellites are affected.

15.1.3 Basic Satellite Components

Every satellite communication involves the transmission of information from a ground station to the satellite (the uplink), followed by a retransmission of the information from the satellite back to the earth (the downlink). Hence the satellite system must typically have a receiver antenna, a receiver, a transmitter antenna, a transmitter, some mechanism for connecting the uplink with the downlink, and a power source to run the electronic system. These components are illustrated in Figure 15.3 and explained as follows:

- Transmitters: The amount of power required by a satellite transmitter to send out depends on whether it is a GEO or LEO satellite. The GEO satellite is about 100 times farther away than the LEO satellite. Thus, a GEO satellite would need 10,000 times as much power as an LEO satellite. Fortunately, other parameters can be adjusted to reduce this amount of power.
- Antennas: The antennas dominate the appearance of a communication satellite. The antenna design is one of the more difficult and challenging parts of a communication satellite project. The antenna geometry is constrained physically by the design and the satellite topology. A major difference between GEO and LEO satellites is their antennas. Since all the receivers are located in the coverage area, which is relatively small, a properly designed antenna can focus most of the transmitter power within that area. The easiest way to achieve this is simply to make the antenna larger. This is one of the ways the GEO satellite makes up for the apparently larger transmitter power it requires.

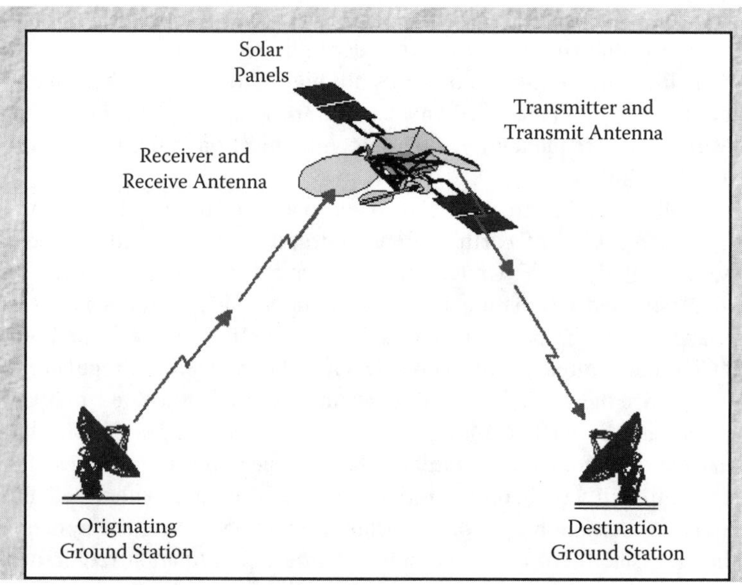

FIGURE 15.3 Basic components of a communication satellite link.
With permission of Regis Leonard, NASA Lewis Research Center.

- Power generation: The satellite must generate all of its own power. The power is often generated by large solar cells, which convert sunlight into electricity. Since there is a limit to how large the solar panels can be, there is also a practical limit to the amount of power that can be generated. Satellites must also be prepared for periods of eclipse, when the earth is between the sun and the satellite. This necessitates having batteries on board that can supply power during eclipse and recharge later.
- Transponders: These are the devices each satellite must carry. A transponder is equipment that receives a weak signal at one frequency, amplifies it, and changes its frequency to another for transmission to another earth station. The block diagram of a typical transponder is shown in Figure 15.4. For example, a GEO satellite may have 24 transponders with each assigned a pair of frequencies (uplink and downlink frequencies).
- Ground stations: The ground (or earth) stations form the ground segment of the satellite communication system. The ground station is responsible for interacting and communicating with the satellites. Most ground or earth stations simply transmit and receive signals with a fixed antenna. At least one ground station must perform the task of controlling and monitoring the satellite. In a transmitting ground station, the information signal (voice, video, or data) is processed, amplified, and transmitted. In a receiving ground station, the reverse process takes place. In the past, ground stations were massive, expensive, and owned by common carriers and the military. Now, earth stations are small, less expensive, and owned

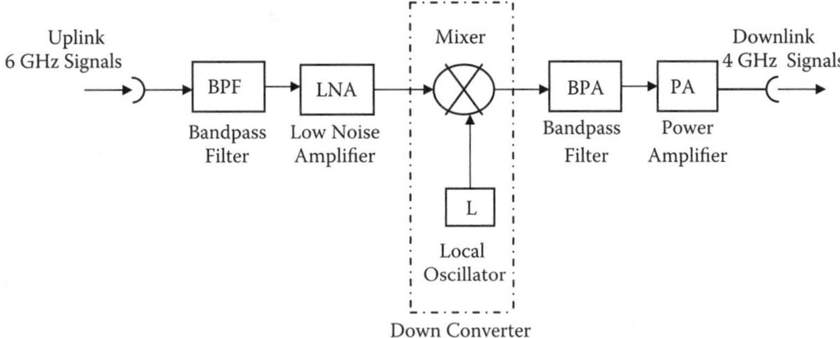

FIGURE 15.4 A simplified block diagram of a typical transponder.

or leased by private organizations. The antenna is a vital component of the ground station and its size varies considerably from a 1-m diameter parabolic reflector used for TV programs at home to a 64-m diameter reflector used in a deep space network.

- Telemetry and control: This is partly implemented by the ground stations and partly by the satellite. Telemetry allows the ground station to manage the satellite. Telemetry, tracking, and command (TTC) are all used to monitor and control the satellite while in orbit. Telemetry is the means by which measurement made is transmitted to an observer at a distance. Tracking is collecting data to monitor the movement of an object. Command is the process of establishing and maintaining control. Tracking and command are used by the terrestrial control station to determine the position of the satellite and predict its future location. The satellite contains telemetry instrumentation that continuously gathers information and transmits it to the ground station.

15.2 ORBITAL CHARACTERISTICS

Since a satellite is a spacecraft that orbits the earth, an intuitive question to ask is: "What keeps objects in orbit?" The answer to the question in found is the orbital mechanical laws governing satellite motion. Satellite orbits are essentially elliptical and obey the same laws of Johannes Kepler that govern the motion of planets around the sun. The three Kepler's laws are stated as follows:

- First law: The orbit of each planet follows an elliptical path in space with the sun serving as the focus.
- Second law: The line linking a planet with the sun sweeps out equal areas in equal time.
- Third law: The square of the period of a planet is proportional to the cube of its mean distance from the sun.

Besides these laws, Newton's law of gravitation states that any two bodies attract each other with a force proportional to the product of their masses and inversely proportional to the square of the distance between them, i.e.

$$\mathbf{F} = -\frac{GMm}{r^2}\mathbf{a}_r \tag{15.1}$$

where M is the mass of one body (earth), m is the mass of other body (satellite), $\mathbf{F}$ is the force on m due to M, r is the distance between the two bodies, $\mathbf{a}_r = \mathbf{r}/r$ is a unit vector along the displacement vector $\mathbf{r}$, and $G = 6.672 \times 10^{-11}$ Nm/kg^2 is the universal gravitational constant. If M is the mass of the earth, the product $GM = \mu = 3.99 \times 10^{14}$ m^3/s^2 is known as Kepler's constant.

Kepler's laws in conjunction with Newton's laws can be used to completely describe the motion of the planets around the sun or that of the satellite around the earth. Newton's second law can be written as

$$\mathbf{F} = m\frac{d^2r}{dt^2}\mathbf{a}_r \tag{15.2}$$

Equating this with the force between the earth and the satellite in Equation (15.1) gives

$$\frac{d^2r}{dt^2}\mathbf{a}_r = -\frac{\mu}{r^2}\mathbf{a}_r \tag{15.3}$$

or

$$\ddot{\mathbf{r}} + \frac{\mu}{r^3}\mathbf{r} = 0 \tag{15.4}$$

where $\ddot{\mathbf{r}}$ is the vector acceleration. The solution to the vector second-order differential Equation (15.4) is not simple but it can be shown that the resulting trajectory is in the form of an ellipse given by

$$r = \frac{p}{1 + e\cos\theta} \tag{15.5}$$

where r is the distance between the geocenter and any point on the trajectory, p is a geometric constant (the latus rectum), e ($0 \le e < 1$) is the eccentricity of the ellipse, and θ (known as the true anomaly) is the polar angle between r and the point on the ellipse nearest to the focus. These orbital parameters are illustrated in Figure 15.5. The eccentricity e is given by

$$e = \sqrt{1 - \left(\frac{a}{b}\right)^2} \tag{15.6}$$

The point on the orbit where the satellite is closest to the earth is known as the perigee, while the point where the satellite is farthest from the earth is known as the

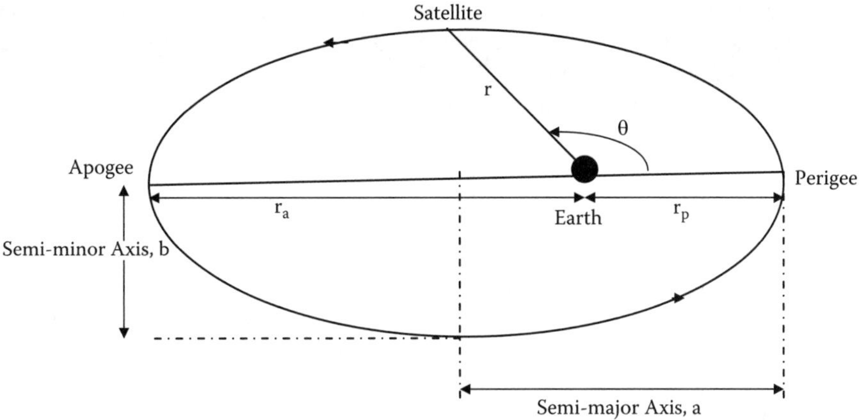

FIGURE 15.5 Orbital parameters.

apogee. The fact that the orbit is an ellipse confirms Kepler's first law. If a and b are the semimajor and semiminor axes (see Figure 15.5), then

$$b = a\sqrt{(1-e^2)} \tag{15.7a}$$

$$p = a(1-e^2) \tag{15.7b}$$

Thus, the distance between a satellite and the geocenter is given by

$$r = \frac{a(1-e^2)}{1+e\cos\theta} \tag{15.8}$$

Note that the orbit becomes a circular orbit when $e = 0$.

The apogee height and perigee height are often required. From the geometry of the ellipse, the magnitudes of the radius vectors at apogee and perigee can be obtained as

$$r_a = a(1+e) \tag{15.9}$$

$$r_p = a(1-e) \tag{15.10}$$

To find the apogee and perigee heights, the radius of the earth must be subtracted from the radii lengths.

The period T of a satellite is related to its semimajor axis a using Kepler's third law as

$$T = 2\pi\sqrt{\frac{a^3}{\mu}} \tag{15.11}$$

For a circular orbit to have a period equal to that of the earth's rotation (a sidereal day 23 hours, 56 minutes, 4.09 s), an attitude of 35,803 km is required. In this equatorial plane, the satellite is "geostationary."

The velocity of a satellite in an elliptic orbit is obtained as

$$v^2 = \mu\left(\frac{2}{r} - \frac{1}{a}\right) \tag{15.12}$$

For a synchronous orbit ($T = 24$ h), $r = a = 42{,}230$ km, and $v = 3074$ m/s or 11,070 km/h. The closer the satellite is to the earth, the stronger is the effect of gravity, which constantly pulls it toward the earth, and so the greater must be the speed of the satellite to avoid falling into the earth.

A constellation is a group of satellites. The total number N of satellites in a constellation depends on the earth central angle γ and is given by

$$N \approx \frac{4\sqrt{3}}{9}\left(\frac{\pi}{\gamma}\right)^2 \tag{15.13}$$

Example 15.1

A satellite has the following orbital parameters:

semimajor axis: 42,168 km
mean anomaly: 28.4°
eccentricity: 0.00034

Calculate (a) the maximum and minimum distances the satellite could be from the center of the earth during each revolution, (b) the orbital period in hours, and (c) the mean orbital velocity.

Solution

(a) Maximum distance $r_a = a(1 + e) = 42168(1 + 0.00034) = 42{,}182.3$ km; minimum distance $r_p = a(1\text{-}e) = 42168(1 - 0.00034) = 42{,}153.7$ km

(b) $$T = 2\pi\sqrt{\frac{a^3}{\mu}} = 2\pi\sqrt{\frac{42{,}168^3}{398613.52}} = 86174.42 \text{ s}$$

$$= 23.94 \text{ hr}$$

(c) $$r = \frac{a(1-e^2)}{1+e\cos\theta} = \frac{42{,}168(1-0.00034^2)}{1+0.00034\cos 28.4^\circ}$$

$$= 42{,}155.381 \text{ km}$$

$$v^2 = \mu\left(\frac{2}{r} - \frac{1}{a}\right) = 398613.52\left(\frac{2}{42,155.38} - \frac{1}{42,168}\right)$$

$$= 9.4586$$

or $v = 3.0755$ km/s

15.3 PROPAGATION CHARACTERISTICS

There are two major effects space has on satellite communications. First, the space environment, with radiation, rain, and space debris, is harsh on satellites. The satellite payload, which is responsible for the satellite communication functions, is expected to be simple and robust. Traditional satellites, especially GEOs, serve as bent pipes and act as repeaters between communication points on the ground, as shown Figure 15.6a. There is no on-board processing (OBP). However, new satellites allow OBP, including decoding, demodulation, transponder, beam switching, and routing, as in Figure 15.6b, where a network of satellites is connected by intersatellite links (ISL).

The second effect is that of wave propagation. Light-of-sight and attenuation due to atmospheric particles (rain, ice, dust, snow, fog, etc.) are not significant at L, S, and C bands. Above 10 GHz, the main propagation effects are

- tropospheric propagation effects: attenuation by rain and clouds, scintillation, and depolarization
- effects of the environment on mobile terminals: shadowing, blockage, and multipath caused by objects in the surrounding of the terminal antenna

The troposphere can produce significant signal degradation at the Ku and Ka bands, particularly at lower elevation angles. Most satellite systems are expected to operate at an elevation angle above roughly 20°. Rain constitutes the most fundamental obstacle encountered in the design of satellite communication systems at frequencies above 10 GHz. The resultant loss of signal power makes for unreliable

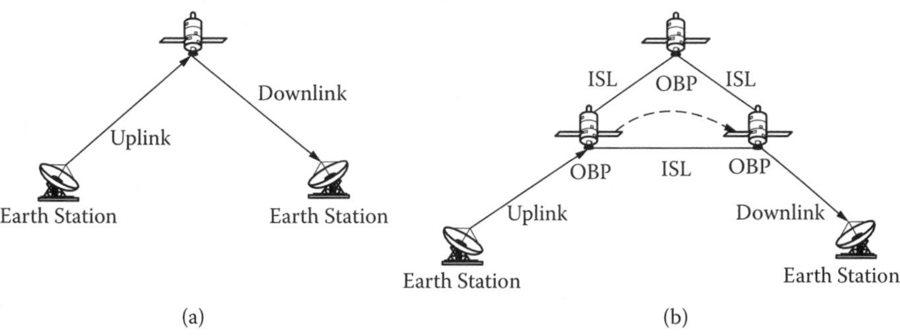

FIGURE 15.6 Satellite configuration types: (a) bent pipe, (b) on-board processing (OBP) switching and routing.

TABLE 15.5
Attenuation Coefficients

Frequency f (GHz)	a	b
8.5–25	$4.21 \times 10^{-5}(f)^{2.42}$	$1.41(f)^{-0.0779}$
25–54	$4.21 \times 10^{-5}(f)^{2.42}$	$2.63(f)^{-0.272}$
54–100	$4.09 \times 10^{-2}(f)^{0.699}$	$2.63(f)^{-0.272}$

transmission. Based on empirical data, the specific attenuation per unit α (dB/km) is related to rain intensity or rain rate R (in mm/hr) as

$$\alpha = aR^b \tag{15.14}$$

where a and b are frequency-dependent coefficients. The approximate expressions for $a(f)$ and $b(f)$ given in Table 15.5 are suitable for engineering purposes. The total attenuation loss in dB is given by

$$A = \alpha L_{eq} \tag{15.15}$$

where L_{eq} is the equivalent length, which is determined by the height of the freezing level and it depends on the rain rate R and elevation angle θ. It is given by

$$L_{eq} = [7.413 \times 10^{-3} R^{0.766} + (0.232 - 1.803 \times 10^{-4} R)\sin\theta]^{-1} \tag{15.16}$$

The rain rate R is given by the Rice-Holmberg model. The percent of an average year for which the rain rate exceeds R at a medium location is given by the Rice-Holmberg distribution

$$P(R) = ae^{-0.03R} + be^{-0.258R} + ce^{-1.63R} \tag{15.17}$$

where

$$a = M\beta/2922, \quad b = M(1 - \beta)/438.3, \quad c = 1.86\beta \tag{15.18}$$

M is the total mean yearly rainfall in millimeters and β is the ratio of thunderstorm rain accumulation to total accumulation. Attenuation due to other hydrometeors such as oxygen, water vapor, and fog is not as significant as that due to rain.

To determine the amount of power received on the ground due to satellite transmission, we consider the power density

$$\Psi = \frac{P_t}{S} \tag{15.19}$$

where P_t is the power transmitted and S is the terrestrial area covered by the satellite. The value of P_t is a major requirement of the spacecraft. The coverage area is given by

$$S = 2\pi R^2(1 - \cos\gamma) \tag{15.20}$$

where $R = 6378$ km is the radius of the earth. S is usually divided into a cellular pattern of spot beams, thereby enabling frequency reuse. The effective area of the receiving antenna is a measure of the ability of the antenna to extract energy from the passing electromagnetic wave and is given by

$$A_e = G_r \frac{\lambda^2}{4\pi} \tag{15.21}$$

where G_r is the gain of the receiving antenna and λ is the wavelength. The power received is the product of the power density and the effective area. Thus

$$P_r = \Psi A_e = \frac{P_t}{4\pi r^2} G_t A_e \tag{15.22}$$

or

$$P_r = G_r G_t \left(\frac{\lambda}{4\pi r}\right)^2 P_t \tag{15.23}$$

where G_t is the gain of the transmitting antenna and r is the distance between transmitting and receiving antennas. This is known as Friis equation, relating the power received by one antenna to the power transmitted by the other. We first notice from this equation that for a given transmitted power P_t, the received power P_r is maximized by minimizing the coverage area S. Second, mobile terminals prefer having nondirectional antennas, thereby making their gain G_r fixed. Therefore, to maximize P_r encourages using as long a wavelength as possible, i.e., as low a frequency as practicable within regulatory and technical constraints.

The path loss accounts for the phenomenon that occurs when the received signal becomes weaker as the distance between the satellite and the earth increases. In free space, the strength of the radiated signal diminishes as the square of the distance it travels, so the received power density is inversely proportional to the square of the distance. The path "free-space" loss (in dB) is given by

$$L_p = 92.45 + 20\log_{10} f + 20\log_{10} r \tag{15.24}$$

where f is the frequency (in GHz) and r is the distance (in km).

The noise density N_o is given by

$$N_o = kT_o \tag{15.25}$$

where $k = 1.38 \times 10^{-23}$ Ws/°K is Boltzmann's constant and T_o is the equivalent system temperature, which is defined to include antenna noise and thermal noise generated at the receiver. Shannon's classical capacity theorem for the maximum error-free transmission rate in bits per second (bps) over a noisy power-limited and bandwidth-limited channel is

$$C = B\log_2\left(1+\frac{P_r}{N_o}\right) \tag{15.26}$$

where B is the bandwidth and C is the channel capacity.

Example 15.2

Calculate the free space loss in dB between a satellite and earth station separated by a distance of 26,000 km for a frequency of 22 GHz.

Solution

$$\begin{aligned} L_p &= 92.45 + 20\log_{10} f + 20\log_{10} r \\ &= 92.45 + 20\log_{10}(22) + 20\log_{10}(26{,}000) \\ &= 207.6 \text{ dB} \end{aligned}$$

15.4 BROADBAND COMMUNICATIONS TO LAND AND SEA LOCATIONS

There is a need for applying broadband communications to land and sea locations. This section explores some of the key areas that can be considered in the process.

15.4.1 Earth Station Engineering at High Frequencies

The basic satellite earth station link for earth station engineering at high frequencies is shown in Figure 15.7. The outdoor unit consists of the following:

- Antenna: This is a device used for transmitting and receiving radio waves, usually designed to focus the waves to or from one direction. It is often referred to as a dish. The antenna also contains the feedhorn. This feedhorn is a piece of antenna hardware, located at the focal point of the parabolic reflector, that radiates RF energy toward the antenna reflector and collects (received) RF energy from the antenna reflector.
- Power amplifier: This is a device that amplifies a specific band of frequencies such as the Ku and Ka bands by a large amount, sufficiently large to enable the antenna to beam them up to the satellite.
- Up converter: This is a device that increases the frequency of the carrier, typically from intermediate frequency (IF) to radio frequency (RF).
- Low noise amplifier: This is a device used to amplify the received modulated carrier while minimizing noise.
- Down converter: This is a device that lowers the frequency of the carrier, typically from radio frequency (RF) to intermediate frequency (IF).

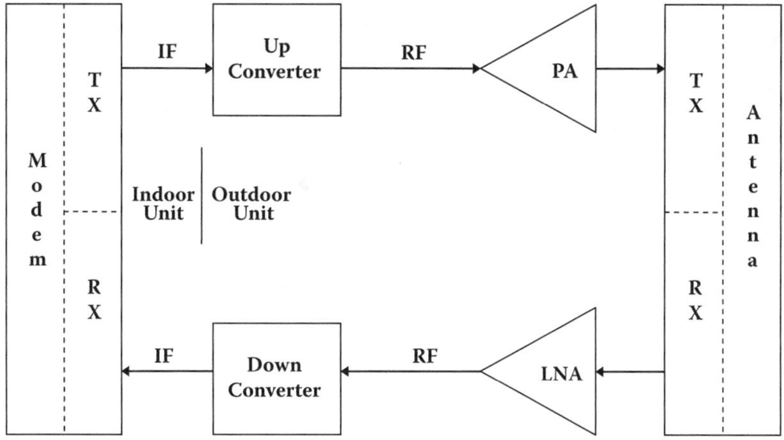

PA: Power Amplifier
LNA: Low Noise Amplifier
IF: Intermediate Frequency
TX: Transmitter
RX: Receiver
RF: Radio Frequency

FIGURE 15.7 The basic satellite earth station link.

In earth station engineering, an earth station is used to transmit/receive satellite transmissions. An earth station dish antenna is typically about 0.66 m–1.8 m in diameter. It is mounted on a roof, wall, or ground. This antenna, along with the attached low-noise blockdownconverter (LNB), which receives satellite signals, and the transmitter, which sends signals, make up the VSAT outdoor unit (ODU). This is one of the two components of a VSAT earth station. The second component of a VSAT earth station is the indoor unit (IDU). The IDU is a small desktop box or PC that contains receiver and transmitter boards and an interface to communicate with the user's existing in-house equipment—LANs, servers, PCs, TVs, kiosks, etc. This IDU is connected to the ODU with a pair of cables.

The satellite earth stations can be installed anywhere within the satellite footprint as long as they have an unobstructed view of the satellite. They are capable of sending and receiving video, data, and audio content at the same high speed regardless of their distance from terrestrial switching offices and infrastructure. Figure 15.8 shows an example of the satellite network star architecture, which has a hub and the key in-house equipment.

The indoor unit consists of the following:

- Receiver electronics: This is the equipment that allows a specific satellite signal to be separated from all others being received by an earth station, and converts the signal format into a format for video, voice, or data. This equipment consists of a modem and possibly a router.

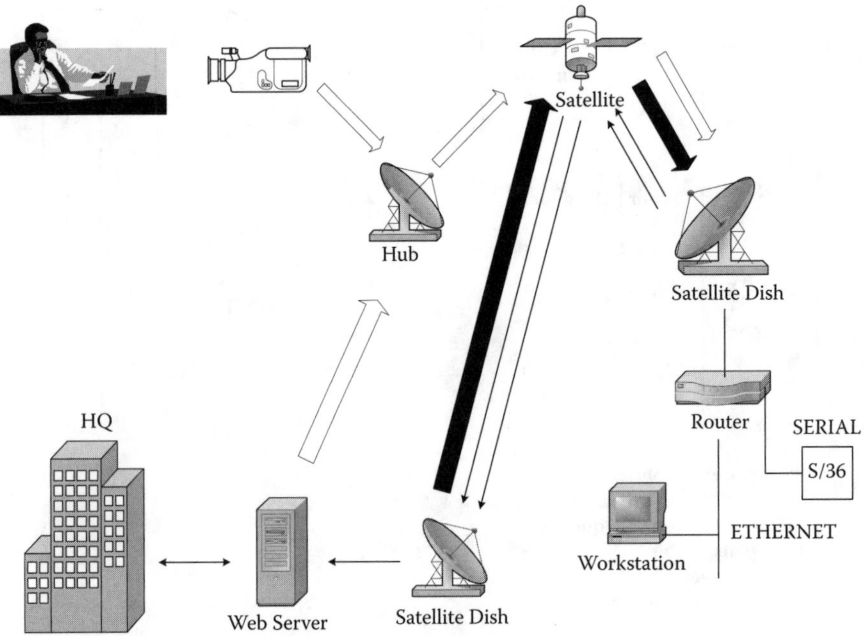

FIGURE 15.8 Satellite network star architecture.

- Modulator/demodulator (modem): This is a device that converts binary data streams, such as those from a PC, to communicate over an analog transmission medium such as telephone lines or on a carrier wave for wireless transmissions. The satellite modem can either connect to the computer serial port, or to a network device via an Ethernet connection.
- Router: This is a network layer device that determines the optimal path along which network traffic should be forwarded. Routers forward packets from one network to another based on network layer information.

In the implementation of a satellite earth station's efficiency, the satellite dish efficiencies are of importance. Most satellite dishes use a parabolic curve design, which permits the focusing and amplifying of the satellite signal. The ability of the parabolic antenna to amplify signals is directly related to the exactness of this parabolic curve as well as good antenna assembly techniques practiced during installation.

The teleport is another important earth station satellite engineering component. A teleport comprises one or many telecommunication hubs. It is a satellite communication center capable of switching voice, data, and image transmissions to and from any location. It provides customers with access to the terrestrial world of services and applications.

Teleports provide customers with the ability to share among many the costs of teleport buildings, land, telecommunication equipment, redundancy, and engineering support, thereby, achieving economies of scale. Teleports offer customers flexibility. They provide better service monitoring, equipment redundancy, standby power, signal protection, and network management, all from one central location.

The satellite's inherent strength is point-to-multipoint broadcasting of data, video, or audio. The satellite can simultaneously deliver information to a virtually unlimited number of end-user locations—at high speeds up to 45 Mbps. This avoids duplicate transmissions and maximizes the efficiency of infrastructure and bandwidth.

The satellite networks are cost effective to deploy, maintain, and operate. The individual remote earth stations can be relatively inexpensive (about the same cost as a router in a frame relay network) and can be quickly and easily installed by a field technician. Hub and satellite costs are shared among thousands of customer sites, so the per-site cost of equipment, maintenance, and management is low—and gets lower as more sites are added to the network. As such, one remote earth station can aggregate all of a community's telecommunication user requirements and achieve improved economics as a result.

15.4.2 State of the Art in Broadband VSATs: Standardized Architecture and Capabilities

There has been an increasing demand for broadband mobile communication systems. This has led to the successful deployment of both terrestrial and satellite wireless networks with standardized architectural capabilities. Besides the high data rates, current broadband VSAT wireless architectural networks can be inexpensive, support reconfigurability, and provide time-varying and space-varying coverage at low cost. However, in parallel with these two well-established methods for providing wireless communication services, another alternative has attracted the attention of the telecommunication community in recent years.

This other alternative is called high-altitude platforms (HAPs) or stratospheric platforms (SPEs), and is located 17–22 km above the earth's surface. It is based on quasistationary aerial platforms operating in the stratosphere. The most important advantages of the HAP systems are

- easy and incremental deployment
- flexibility/reconfigurability
- low-cost operation
- low propagation delay
- high elevation angles
- broad coverage
- broadcast/multicast capability
- broadband capability
- ability to move around in emergency situations

However, there are also key disadvantages such as

- monitoring of the station
- immature airship technology
- stabilization of the on-board antenna

It is worth noting that for the same bandwidth allocation, terrestrial systems need a huge number of base stations to provide the needed coverage, while GEO satellites face limitations on the minimum cell size projected on the earth's surface and LEO satellites suffer from handover problems. Therefore, HAPs seem to be a very good design compromise.

15.5 BROADBAND AIRBORNE MOBILE SATELLITE COMMUNICATION

Airborne mobile satellite communications is another key area where broadband communications has been found to be needed. In this section, we explore some of the various important considerations in broadband airborne mobile satellite communication.

15.5.1 Customers and Installations

Initially, the customer installations of VSAT at Ku and Ka bands are very pricey.

The payoff comes from much power costs on large data that moves the large VSAT pipe that are always operated at speeds sometimes measured in Mbps. Suppliers have been able to compete based on the ability to multiplex many signals through the open pipeline. Due to the high cost (circa $60,000) of the antenna, VSAT may cost upwards of $4,000 per month, with payoff resulting from moving large chunks of data, video, fax, and voice through the big pipe.

15.5.2 Services and Applications

The key component for VSAT applications is the high-power amplifier (HPA).

Classes of power required for both Ku- and Ka-band frequencies are

- 1 W power amplifier (+30 dBm)
- 2 W power amplifier (+33 dBm)
- 4 W power amplifier (+36 dBm)

Providers of VSAT provide applications to go with the hardware and access to the landside teleports that bring the connectivity to home offices. For example, in the cruise industry, the package includes Internet cafes and Wi-Fi hubs, with capabilities to bill passenger accounts. In the offshore business, the solution providers may offer shortcuts for automating the entry of seismic data and linkages to various logging devices. An important corollary is the ability to assist in handling ship's business—linking purchasing and maintenance systems to the shoreside office or providing video conferencing and access to electronic chart updates.

15.5.3 L-, C-, Ku-, and Ka-Band Advantages and Disadvantages

The commonly used satellite frequency bands are L, C, and Ku. The most common frequencies for fixed satellite services such as video, data, and service are the C and Ku bands. L is the most common frequency for mobile satellite services such as voice and data. The Ka band is used for video and data services. The frequency and power output of a satellite signal determine the size of the earth station antenna. When the frequency increases, the wavelength decreases. As the wavelength increases, larger antennas are necessary to gather the signal. As an example, C-band satellite transmissions are at a low frequency in the 4–6 GHz range with larger wavelengths than Ku band or Ka band. These larger wavelengths mean that a larger satellite antenna is required to gather the minimum signal strength typically a 1.8 m to 3.7 m for C band, whereas Ka and Ku are approximately 0.66 m to 1.2 m respectively. We are assuming in this case that the bandwidth is constant in this relativity statement.

Atmospheric particles can negatively affect satellite communications, particularly in the higher satellite bands such as Ku and Ka. Rain is the major atmospheric particle that affects satellite communications. Other atmospheric conditions such as clouds, snow, and fog can also affect satellite communications. Ka-band satellite links operating at moderate to high elevation angles are affected by the following propagation factors:

- Tropospheric scintillation: Amplitude fluctuations that are produced by refractive inhomogeneities present in the lower part of the troposphere are called tropospheric scintillations. The scintillation can occur with or without fading on the path; the former is known as the dry scintillation and the latter as wet scintillation since it is accompanied by rain on the path.
- Rain and ice depolarization: Some satellite systems employ frequency re-use by means of orthogonal polarizations. Such satellite systems may suffer from interference through coupling between wanted and unwanted polarizations. These kinds of coupling are a result of antenna imperfections and atmospheric depolarization caused by precipitation particles. Rain and ice depolarization is a function of the polarization angle, elevation angle, frequency, and attenuation.
- Rain attenuation: At Ka-band frequencies and even at Ku-band frequencies, rain attenuation is the dominant propagation impairment. Rain attenuation is a function of the polarization angle, elevation angle, frequency, rain intensity, raindrop size distribution, and raindrop temperature. Fade durations and rates are closely correlated with the rain type. Raindrop size distribution and the rain temperature largely determine the frequency scaling of the rain attenuation.
- Melting layer attenuation: We can define the melting layer as the region around the 0°C isotherm where snow and ice particles from aloft melt to form rain. Under light rain and low elevation angle conditions, melting layer attenuation may become a significant factor to the total path attenuation.
- Cloud attenuation: Clouds containing liquid water can produce both signal attenuation and amplitude scintillation at Ka-band frequencies. In general, ice clouds do not produce these effects.
- Gaseous absorption: Gaseous absorption arising from oxygen and water vapor present in the atmosphere is relatively small compared to other absorptive effects. Absorption due to oxygen is nearly constant and that due to water vapor varies slowly with time in response to variations in the water vapor content in the atmosphere. As a result of this, the gaseous absorption increases with the relative humidity as well as the temperature.

15.5.4 Challenges of Aeronautical Broadband Satellite Communication

Satellites are long-term devices, a fact that most designers and operators tend to ignore. Aeronautical broadband satellite communication systems are no exceptions to the problem that it takes some time from the time a satellite is conceptualized to the time you actually get it into space. If you place an order today for a complex new satellite, it will typically take about three years or more before it is in orbit. The typical lifetimes are 8 to 15 years. Satellites can be designed to cover a wide range of markets and may have many requirements to fulfill. Is this good or bad? It is a big challenge that leads to very interesting debates in conceptualizing what type of satellite to design. This debate is normally between management who wants to cover all of the markets, content provision partners who are understandably interested in just their service, the bankers trying to minimize their risk exposures, and the engineers trying to optimize the design that has ever-changing requirements.

Some other key challenges of aeronautical broadband communication are

- Frequency band and spectrum regulation: To satisfy the increasing data rate requirements and to overcome restricted bandwidth availability at lower frequencies, broadband services have to move to higher frequency bands. Frequencies up to Ka band and above are suitable for aeronautical communication systems. However, in the short/medium term, the future broadband aeronautical broadband satellite communications (AirCom) will most likely lead to the use of available transponders and systems and "cheaper" antenna technology, thereby giving some preference to C and Ku bands.
- Aircraft and flight characteristics: In this case, the interests are in
  - geometric/geographic aspects of single flights
  - global/regional flight route statistics/data
  - geometric implications from body structure

These three together affect the key issues in antenna and system design and dimensioning.

- Satellite constellation issues: The first option to consider for the near-term future and any other evolutionary approaches towards aeronautical multimedia communications is a broadband network based on geostationary satellites. The main critical issues in a GEO system are the coverage deficiencies at higher latitudes and the extreme antenna steering requirements at the lowest elevation angles (i.e., again highest latitudes) in light of the important near-polar flight routes in the northern hemisphere. These problems could be avoided with an appropriate non-GEO satellite constellation. In the case of an LEO or MEO solution, potential system capacity limitations and latency for real-time communications could be reduced. However, networking complexity tends to increase while moving to lower orbits. Satellite handover will become a major issue, and intersatellite links may be necessary at least for LEO constellations to provide connectivity over large ocean areas.

- Aeronautical channel implications: The channel for aeronautical transmission is very important for the design of high-quality services. Since the spectrum at lower frequencies is too limited for multimedia applications, frequencies at Ku band, K/Ka band, and above are likely to be used. Using a steered Cassegrain antenna with a beam width of 4.2°, the aeronautical channel in K band has been about 18.685 GHz. The radio signal at millimeter wavelength is subject to atmospheric attenuation due to vapor and oxygen. These effects can be calculated statistically using the absorption and rain models of the International Telecommunication Union (ITU). The attenuation depends on the flight altitude, the region, and the weather conditions. Shadowing and diffraction in aeronautical multimedia services will cause the radio connection to drop since the satellite systems will most likely not provide exhaustive link margins to compensate for deep power fades.
- Terminal antenna design: The outdoor part of the antenna design has to be of minimal protuberance because of aerodynamic issues. Therefore, a minimized antenna size is of great advantage because the antenna size decreases for a fixed gain at higher frequencies. High data rates will require the use of high-gain antennas with small beam widths. Such antennas have to be pointed toward the satellite throughout all flight and on-ground maneuvers of the aircraft in order to enable a continuous communication link. An aircraft is a very agile, attitude-changing vehicle. In the aeronautical application, the absolute values of the attitude angles are a technical challenge, while in a land mobile scenario, the angular change rates are the dominating factor for the antenna beam agility design. This antenna beam agility is one of the most important technical issues for the realization of a broadband aeronautical communication link. The aircraft's attitude changes, antenna technologies, antenna diversity, and the optimal satellite constellation are some of the key issues of importance to be considered.
- Network architecture and system design: The network architecture and system design are critical issues that must be taken into consideration from the earlier conception of the AirCom network architecture. One interesting area is the type of LAN that will be used in the aircraft. Some of the basic choices include the option of having one integrated network such as all-IP LAN versus several tailored subnetworks for the various service and application categories. The gigabit Ethernet is an integrated solution that could be based on wireless ideas such as Hyperlan II or IEEE 802.11b. In this case, any wireless solution has the potential of simplifying cabin or seat reconfiguration and reduces the weight of the communication infrastructure, but introduces possible electromagnetic interference. Relevant specifications and guidelines must be taken into account. The capacity dimensioning process is another important area in the AirCom system design. A lot of information from various earlier design steps, such as market/terminal figures, service profile and service bit rate specifications, flight route statistics, satellite constellation, satellite spot beam antenna design, etc. are needed.

- Market entry options and business case implications: In the early stages of the AirCom system design, one of the challenging decisions is deciding the different market entry options and reference business cases. The frequency band used and the type of technology are driven by the type of market entry options. The evolutionary path leads through C/Ku bands and existing GEO transponders, whereas the "revolutionary" path may target from the beginning advanced Ka-band technology and the design of a tailor-made, potentially non-GEO system. Where business cases are concerned, earlier decisions must be made about how far the AirCom market must go.

15.6 APPLICATIONS

Satellite communication services are uniquely suited for many applications involving wide area coverage. Satellites provide the key ingredient in the development of broadband communications and information processing infrastructure. Here, we consider five major applications of satellite communications: the use of very small aperture terminals (VSATs) for business applications; fixed satellite service (FSS), which interconnects fixed points; mobile satellite (MSAT) service (MSS), which employs satellites to extend cellular networks to mobile vehicles; satellite radio, which continuously provides entertainment to listeners; and satellite-based Internet, which enables IP-over-satellite connectivity.

15.6.1 VSAT Networks

A very small aperture terminal (VSAT) is a dish antenna that receives signals from a satellite. (The dish antenna has a diameter that is typically in the range of 1.2 m to 2.8 m but the trend is toward smaller dishes, not more than 1.5 m in diameter.) A VSAT may also be regarded as a complete earth station that can be installed on the user's premises and provide communication services in conjunction with a larger (typically 6–9 m) earth station acting as a network management center (NMC), as illustrated in Figure 15.9.

VSAT technology brings the features and benefits of satellite communications down to an economical and usable form. VSAT networks have become mainstream networking solutions for long-distance, low-density voice and data communications because they are affordable to both small and large companies. Other benefits and advantages of VSAT technology include lower operating costs, ease of installation and maintenance, ability to manage multiple protocols, and ability to bring locations where the cost of leased lines is very expensive into the communication loop.

Satellite links can support interactive data applications through two types of architectures: mesh topology (also called point-to-point connectivity) and star topology (also known as point-to-multipoint connectivity). Single-hop communications between remote VSATs can be achieved by full-mesh connectivity. Although the mesh and star configurations have different technical requirements, it is possible to integrate the two if necessary.

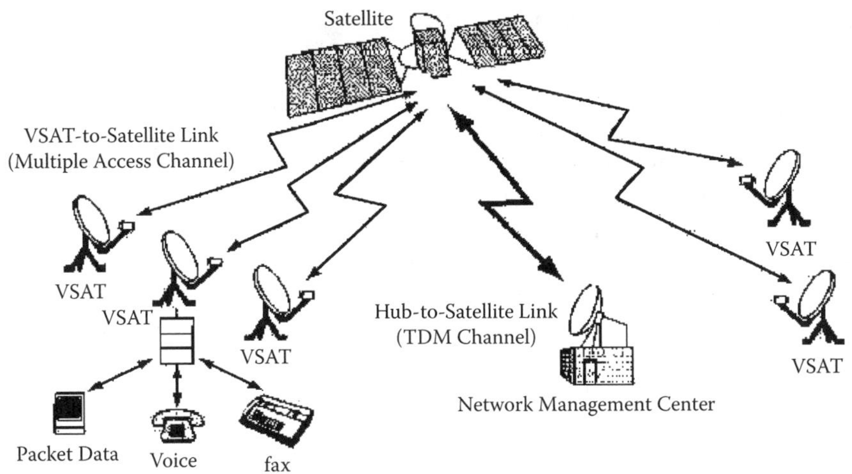

FIGURE 15.9 A typical VSAT network.

The star network employs a hub station. The hub consists of an RF terminal, a set of baseband equipment, and network equipment. A VSAT network can provide transmission rates of up to 64 kbps. As common with star networks, all communication must past through the hub. That is, all communication is between a remote node and the hub; no direct node-to-node information transfer is allowed in this topology. This type of network is highly coordinated and can be very efficient. The point-to-multipoint architecture is very common in modern satellite data networks and is responsible for the success of the current VSAT.

A mesh network is more versatile than a star network because it allows any-to-any communications. Also, the star network can provide transmission rates up to 64 kbps per remote terminal, whereas the mesh network can have its data rates increased to 2 Mbps or more. Mesh topology was used by the first satellite networks to be implemented. With time, there was a decline in the use of this topology but it remains an effective means of transferring information with least delay. Mesh topology applies to either temporary connections or dedicated links to connect two earth stations. All full-duplex point-to-point connectivities are possible and provided, as typical of a mesh configuration. If there are N nodes, the number of connections is equal to the permutation of N(N-1)/2. Mesh networks are implemented at C and Ku bands. The transmission rate ranges from about 64 kbps to 2.048 Mbps (E1 speed). Users have implemented 45 Mbps.

Several types of VSAT networks are now in operation, both domestically and internationally. There were over 1000 VSATs in operation at the beginning of 1992. Today, there are over 100,000 two-way Ku-band VSATs installed in the United States and over 300,000 worldwide. Almost all of these VSATs are designed primarily to provide data for private corporate networks, and almost all two-way data networks with more than 20 earth stations are based on some variation of an ALOHA protocol for access. The price of a VSAT started around $20,000 and dropped to around $6,000 in 1996.

15.6.2 Fixed Satellite Service

Several commercial satellite applications are through earth stations at fixed locations on the ground. The international designation for such an arrangement is fixed satellite service (FSS). FSS provides communication service between two or more fixed points on earth, as opposed to mobile satellite services (MSS) (to be discussed later), which provides communication for two moving terminals. Although ITU defined FSS as a space radio communication service covering all types of satellite transmissions between given fixed points, the borderline between FSS and broadcasting satellite service (BSS) for satellite television is becoming more and more blurred. FSS applies to systems that interconnect fixed points such as international telephone exchanges. It involves GEO satellites providing 24-hour-per-day service.

Table 15.6 shows the WARC (World Administrative Radio Conference) frequency allocations for FSS. The table only gives a general idea and is in no way comprehensive.

The FSS shares frequency bands with terrestrial networks in the 6/4 GHz and 14/12 GHz bands. Thus, it is possible that a terrestrial network could affect a satellite on the up-link or that a terrestrial network may be affected by the down-link from a satellite.

As exemplified by Intelsat, FSS has been the most successful part of commercial satellite communications. Early applications were point-to-point telephony and major trunking uses. Current applications of the FSS can be classified according to frequency (from about 3 MHz to above 30 GHz), the lowest frequency being the HF band. They include high-frequency (HF) service, private fixed services, auxiliary broadcasting (AUXBC) services, cable relay service (CARS), and federal government fixed services.

Although the telecommunications industry as a whole is growing rapidly, the FSS industry is not. The market trend is toward the replacement of long-haul microwave system with fiber. Fiber provides much greater capacity than microwaves.

15.6.3 Mobile Satellite Service

There is a need for global cellular service in all geographic regions of the world. The terrestrial cellular systems serve urban areas well, but they are not economical for rural or remote areas where the population or teledensity is low.

TABLE 15.6
Frequency Allocations for FSS (below ~30 GHz)

Downlinks (in GHz)	Uplinks (in GHz)
3.4–4.2 and 4.5–4.8	5.725–7.075
7.25–7.75	7.9–8.4
10.7–11.7	
11.7–12.2 (Region 2 only)	12.75, 13.25, and 14.0–14.5
12.5–12.75 (Region 1 only)	
15.7–21.2	27.5–31.0

Mobile satellite (MSAT) systems complement the existing terrestrial cellular networks by extending communication coverage from urban to rural areas.

Mobile satellite services (MSS) are not limited to land coverage but include marine and aeronautical services. Thus, the coverage of mobile satellite is based on geographic and not on population coverage as in terrestrial cellular systems, and could be global.

MSAT or satellite-based PCS/PCN is being developed in light of the terrestrial constraints. The low cost of installation makes satellite-based PCS simple and practical. The American Mobile Satellite Corporation (AMSC) and Telesat Mobile of Canada are designing a geosynchronous MSAT to provide PCS to North America. The concept of MSAT is illustrated in Figure 15.10.

Satellite communication among mobile earth stations is different from cellular communication. First, the cells move very rapidly over the earth, and the mobile units, for all practical purposes, appear stationary—a kind of inverted cellular telephone system. Second, due to different designs, use of a handheld is limited to the geographic coverage of a specific satellite constellation and roaming of handheld equipment between different satellite systems is not allowed. With personal communication systems (PCS), there will be a mix of broad types of cell sizes: the pico cell for low-power indoor applications; the micro cell for lower-power outdoor pedestrian applications; the macro cell for high-power vehicular applications; and the supermacro cell with satellites. For example, a micro cell of a PCS has a radius of 1 to 300 m.

There are two types of constellation design approaches to satellite-based PCS. One approach is to provide coverage using three GEO satellites at approximately 36,000 km above the equator. The other approach involves using the LEO and MEO

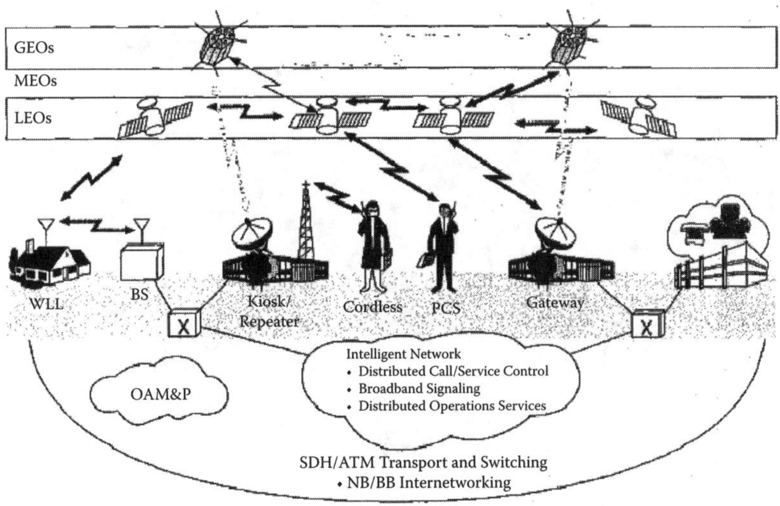

FIGURE 15.10 MSAT concept.

satellites at approximately 500 to 1500 km above the earth's surface. Thus, MSS are identified as either GEO or nongeostationary orbit (NGSO) satellites.

The main purpose of MSAT or MSS is to provide data and/or voice services to a fixed or portable personal terminal, close to the size of today's terrestrial cellular phones, by means of interconnection via satellite. LEO and MEO satellites have been proposed as an efficient way to communicate with these handheld devices. The signals from the handheld devices are retransmitted via a satellite to a gateway (a fixed earth station) that routes the signals through the public switched telephone network (PSTN) to its final destination or to another handheld device.

Satellite systems designed for personal communications include the Iridium, Globalstar, and ICO systems. All are global systems covering everywhere on earth. Each of these is characterized by two key elements: a constellation of non-geosynchronous satellites (LEO or MEO) arranged in multiple planes and a handheld terminal (handset) for accessing PCS, as illustrated in Figure 15.11.

Iridium (www.iridium.com), which began in 1990, is the first mobile satellite telephone network to offer voice and data services to and from handheld telephones anywhere in the world. It uses a network of intersatellite switches for global coverage and GSM-type technology to link mobile units to the satellite network. Several modifications have been made to the original idea, including reducing the number of satellites from 77 to 66 by eliminating one orbital plane. (The name Iridium was based on the fact that Iridium is the element in the periodic table whose atom has 77 electrons.) Some of the key features of the current Iridium satellite constellation are

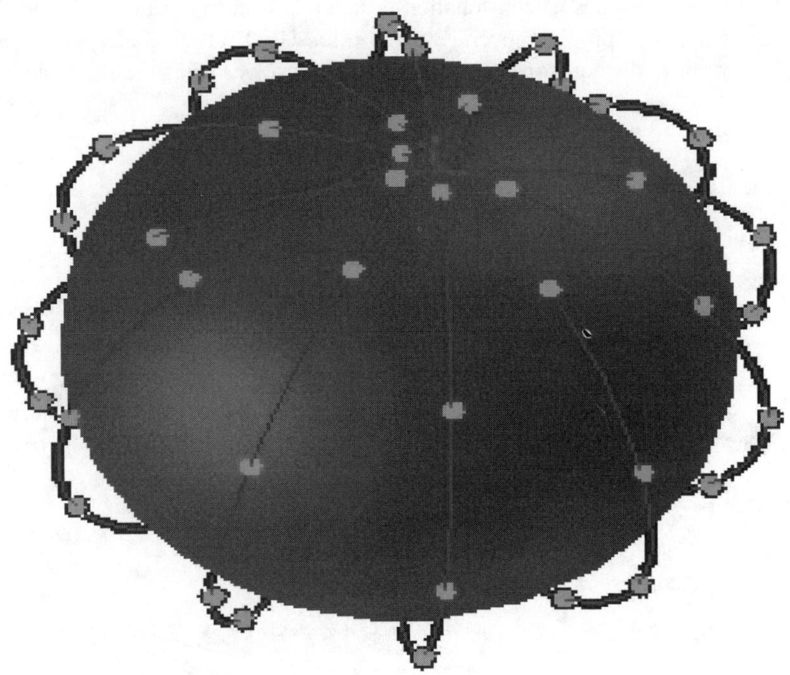

FIGURE 15.11 A typical satellite constellation.

- number of (LEO) satellites: 66 (each weighing 700 kg or 1500 lb)
- number of orbital planes: 6 (separated by 31.6° around the equator)
- number of active satellites per plane: 11 (uniformly spaced, with one spare satellite per plane at 130 km lower in the orbital plane)
- altitude of orbits: 780 km (or 421.5 nmi)
- inclination: 86.4°
- period of revolution: 100 minutes
- design life: 8 years

In spite of some problems expected of a complex system, Iridium is already at work. Its 66 LEO satellites have been fully commercial since November 1, 1998. But on August 13, 1999, Iridium filed for bankruptcy and was later bought by Iridium Satellite LCC. Vendors competing with Iridium include Aries, Ellipso, Globalstar, and ICO.

The second system is Globalstar (www.globalstar.com), which is a satellite-based cellular telephone system that allows users to talk from any place in the world. It serves as an extension of terrestrial systems worldwide except for polar regions. The constellation is capable of serving up to 30 million subscribers. Globalstar is being developed by the limited partnership of Loral Aerospace Corporation and Qualcomm with ten strategic partners. A functional overview of Globalstar is presented in Figure 15.12. The key elements are

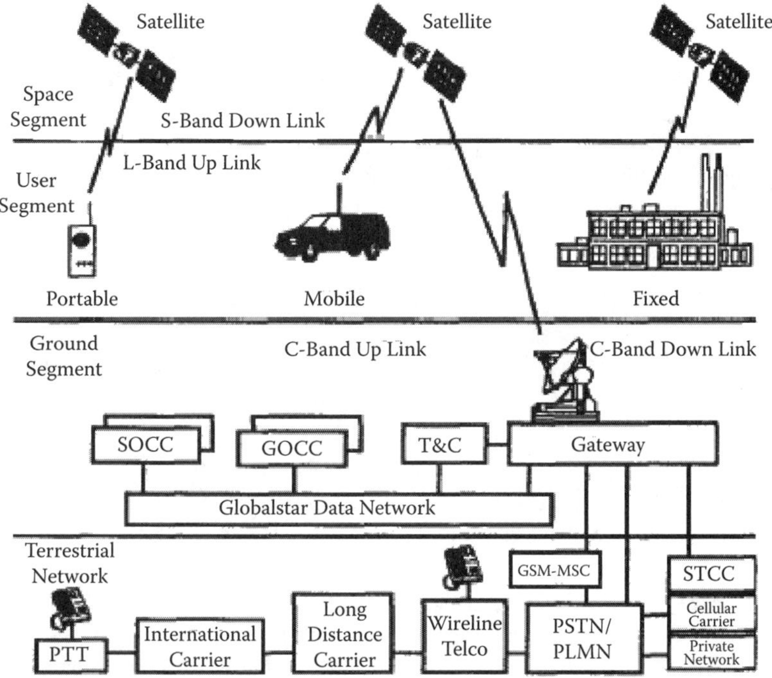

FIGURE 15.12 Globalstar system architecture.

- Space segment: This consists of a constellation of 48 active LEO satellites located at an altitude of 1414 km and equally divided in 8 planes (6 satellites per plane). The satellite orbits are circular and are inclined at 52° with respect to the equator. Each satellite illuminates the earth at 1.6 GHz L-band and 2.5 GHz S-band with 16 fixed beams with service links, assignable over 13 FDM channels.
- User segment: This includes mobile and fixed users.
- Ground segment: This consists of gateways (large ground station), ground operations control center (GOCC), satellite operations control center (SOCO), and Globalstar data network (GDN). The gateway enables communications to and from handheld user terminals (UTs), relayed via satellite, with public switched telephone network (PSTN). A gateway with a single radio channel transmits on a single frequency.
- The Globalstar satellites employ "bent pipe" transponders with the feeder link at C band. Each satellite weighs about 704 pounds and has a capacity of 2800 full-duplex circuits. It covers the earth with only 16 spots beams.

Since Globalstar plans to serve the military with commercial subscriptions, it employs signal encryption for protection from unauthorized calling party. Unlike Iridium, which offers a global service, Globalstar's business plan calls for franchising its use to partners in different countries.

The third system is the ICO system (originally called Inmarsat-P), which was built by Hughes Space and Communications (now Boeing Satellite Systems). ICO constellation is made of

- 10 operational MEO satellites with five in each of the two inclined circular orbits at an attitude of 10,355 km
- one spare satellite in each plane, making 12 total launched
- each satellite employs 163 spot beams
- each satellite will carry an integrated C- and S-band payload
- 12 satellite access nodes (SANs) located globally

The inclination of the orbits is 45°, making it the lowest of the systems described. Although this reduces the coverage at high latitudes, it allows for the smallest number of satellites. The ICO system (www.ico.com) is designed to provide the following services:

- global paging
- personal navigation
- personal voice, data, and fax

The three constellations are compared in Table 15.7.

As more and more customers sign up for mobile satellite service (MSS), information will flow more freely and the world will grow smaller, and the global economy will be stimulated. The dawning age of global personal communication will bring the world community closer together as a single family.

TABLE 15.7
Characteristics of Satellite PCS Systems

Parameter	Iridium	Globalstar	ICO
Company	Motorola	Loral/Qualcomm	ICO-Global
Number of satellites	66	48	10
Number of orbit planes	6	8	2
Altitude (km)	780	1,414	10,355
Weight (lb)	1,100	704	6050
Bandwidth (MHz)	5.15	11.35	30
Frequency Up/down (GHz)	30/20	5.1/6.9	14/12
Spot beams/satellite	48	16	163
Carrier bit rate (kps)	50	2.4	36
Multiple access	TDMA/FDMA	CDMA/FDMA	TDMA/FDMA
Cost to build ($billion)	4.7	2.5	4.6
Service start date	1998	1999	2003

15.6.4 Satellite Radio

Satellite radio is broadcasting from satellite.

With satellite radio, one can drive from Washington DC to Los Angeles, CA without changing the radio station and without static interference. Satellite eliminates localization, which is the major weakness of conventional radio. It transforms radio from a local medium into a national one. Satellite radio will permanently change radio just as cable changed television. It is regarded as radio beyond AM, beyond FM, or radio to the power of X.

Figure 15.13 displays a typical architecture of satellite radio. Satellite radio is based on digital radio, which produces a better sound from radio than analog radio. Digital radio systems are used extensively in communication networks. Digital radio offers CD-quality sound, efficient use of the spectrum, more programming choice, new services, and robust reception even under the most challenging conditions.

Satellite radio is both a new product and a service. As a product, it is a new electronic device that receives the satellite signal. As a service, it will provide consumers with 100 national radio stations, most of which will be brand-new, comprising various music, news, sports, and comedy stations.

Satellite radio service is being provided by two companies: XM Satellite Radio (also known as XM Radio), based in Washington, and Sirius Satellite Radio, based in New York. The two companies obtained FCC licenses to operate digital audio radio service (DARC) systems coast-to-coast throughout the continental USA. To avoid competition with terrestrial radio broadcasters, both satellite broadcasters will carry advertisement of nationally branded products.

XM Satellite Radio is made possible by two satellites, officially named "Rock and Roll," placed in geostationary orbit, one at 85° W longitude and the other at 115° W longitude. Rock and Roll were built by Boeing Satellite Systems. The satellites

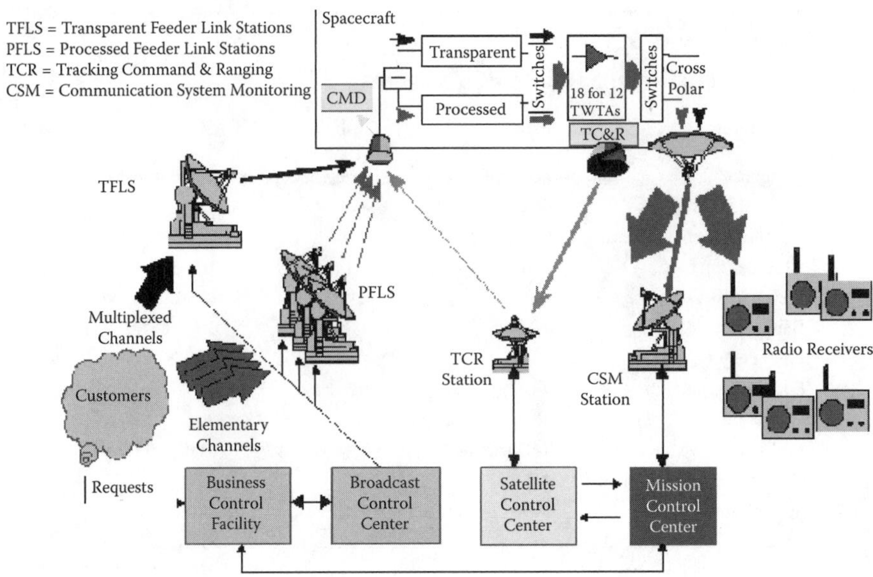

FIGURE 15.13 A typical architecture of satellite radio.
Source: EBU Technical Review.

will be positioned above the United States. In September, 2001, XM Satellite Radio started to broadcast. Subscribers will pay as little as \$9.95 per month after they purchase an AM/FM/XM radio.

Sirius Satellite Radio, on the other hand, does not use GEO satellites. Rather, it is flying three satellites that are equally spaced in an elliptical 47,000 × 24,500 km orbit that takes 24 hours to complete. This ensures that each satellite spends about 16 hours a day over the continental United States, with at least one satellite over the country at any time. It also means that Sirius will be higher in the sky than XM, which is at the zenith only at the equator. Sirius charges \$12.95 a month for their service. The systems of both satellite radio companies are compared in Table 15.8.

TABLE 15.8
Comparison of XM and Sirius Systems

	XM	**Sirius**
Constellation	2 satellites	3 satellites
Satellite type	Boeing 702	SS/Loral FS-3000
Terrestrial repeaters	1500 in 70 areas	105 in 46 areas
Satellite costs	\$439m	\$120m
Transmission rate	4 Mbps	4.4 Mbps
Uplink frequencies	7.05–7.075 GHz	7.06–7.0725 GHz
Downlink frequencies	2.3325–2.345 GHz	2.32–2.3325 GHz

Of paramount importance to the two broadcasters are early agreements from major U.S. car manufacturers. General Motors, Acura, American Honda Motors, Saab, Suzuki, and Isuzu have agreed to install XM radio receivers in some of their car and truck models. Sirius has exclusive alliances to install AM/FM/XM radios in Ford, Chrysler, Mercedes, BMW, Jaguar, and Volvo cars.

Both companies need the support of radio receiver manufacturers. XM-ready radio receivers produced by Pioneer are now available for sale by some retailers such as Best Buy, Circuit City, Sears, Tweeter, and Radio Shack. XM Satellite Radio is now extending its market to airlines. The Sirius system is being developed by manufacturers such as Panasonic, Visteon, Kenwood, Clarion, and Jensen.

Besides XM Satellite Radio and Sirius Satellite Radio, which operate in the United States, WorldSpace is another radio satellite broadcasting company already broadcasting in Africa and Asia. With a constellation of three satellites (AfriStar to cover Africa and the Middle East, AmeriStar to serve Latin America and the Caribbean, and AsiaStar to serve nearly all Asia), WorldSpace intends to touch all or parts of the four continents, especially those areas of the world that most conventional radio stations cannot reach.

As a new technology, satellite radio is not without its own peculiar problems. First, people are not yet used to paying for radio programming. If the programming of the satellite broadcasters is not better than what people are getting free from regular, terrestrial radio, they will be reluctant to pay. So the real question is: How many people are going to subscribe? Second, satellite broadcasting requires a near-omnidirectional receive antenna for cars, which in turn requires a powerful signal from the satellite. Third, some believe that the two companies will face a big hurdle in transforming radio from a local medium into a national one. The pie-in-the-sky companies are faced with great risks ahead of them.

Satellite radio will transform the radio industry, which has seen little technological change since the discovery of FM, some 40 years ago. Receiving digital-quality music from radio satellite is a major technical milestone. It is as revolutionary to the entertainment industry as was the invention of radio itself. The future of radio-by-satellite is exciting but uncertain.

15.6.5 Satellite-Based Internet

The Internet is becoming an indispensable source of information for an ever-growing community of users. The thirst for Internet connectivity and high performance remains unquenched. This has led to several proposals for integrating satellite networks with terrestrial ISDN and the Internet.

Several factors are responsible for this great interest in IP-over-satellite connectivity. First, satellites cover areas where land lines do not exist or cannot be installed. Satellites can serve as an access link between locations separated by great distances. Second, developments in satellite technology allow home users to receive data directly from a geostationary satellite channel at a rate 20 times faster than that of an average telephone. With more power transponders utilizing wider frequencies, commercial satellite links can now deliver up to 155 Mbps. Third, the unique positioning of satellites between sender and receivers lends itself to new applications such as IP multicast, streaming

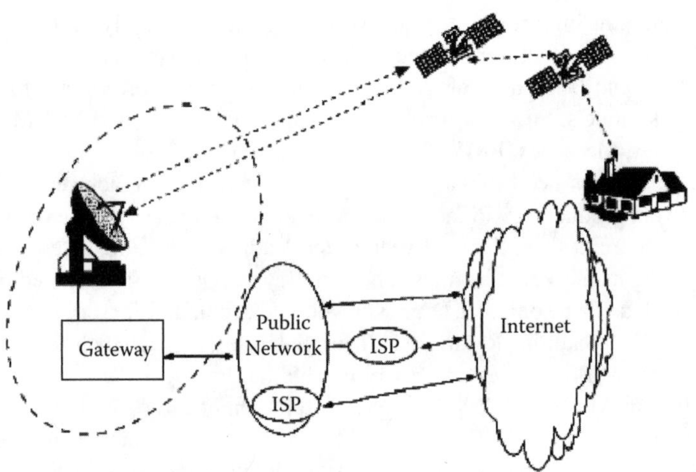

FIGURE 15.14 A typical configuration for satellite-based Internet.

data, and distributed web caching. Fourth, satellite connectivity can be rapidly deployed because trenches and cable installation are unnecessary. Moreover, satellite communication is highly efficient for delivering multimedia content to businesses and homes.

As an inherently broadcast system, a satellite is attractive to point-to-multipoint and multipoint-to-multipoint communications especially in broadband multimedia applications. The asymmetrical nature of Web traffic suggests a good match to VSAT systems since the VSAT return link capacity would be much smaller than the forward link capacity.

A typical network architecture for a satellite-based Internet service provider (ISP) is shown in Figure 15.14, which has been simplified to focus on the basic functionality. It includes its own satellite network and a network of ground gateway stations. The ground gateway stations interface with the public network through which access to the Internet is gained. The number of satellites may vary from dozens to hundreds, and they may be GEO, MEO, or LEO. Thus, the satellite-based Internet has several architectural options due to the diverse designs of satellite systems, orbit types, payload choice, and intersatellite links designs.

There is ongoing research into various aspects of implementation and performance of TCP/IP over satellite links. Related issues include the slow start algorithm, the ability to accommodate large bandwidth-delay products, congestion control, acknowledgment, and error recovery mechanisms.

SUMMARY

1. Satellites have the capability to provide coverage over large geographic areas.
2. Satellites are classified as geostationary earth orbit (GEO), middle earth orbit (MEO), and low earth orbit (LEO) according to their distance from the earth. Most communications satellites use geostationary orbits, which are so far from earth that they cause a noticeable delay.

3. Satellite orbit motion is governed by Kepler's laws in conjunction with Newton's laws.
4. In earth station engineering, an earth station is used to transmit/receive satellite transmissions.
5. Atmospheric particles can negatively affect satellite communications, particularly in the higher satellite bands such as Ku and Ka.
6. The antenna system is one of the most important performance factors in a mobile satellite communication system and even in a HAP configuration.
7. Several tracking methods are used in helping to point the antenna in any desired direction.
8. Link budget determines the maximum allowed path loss of a wireless connection.
9. Applications of satellite communications include very small aperture terminals (VSATs) for business applications; fixed satellite service (FSS), which interconnects fixed points; mobile satellite (MSAT) service, which employs satellite to extend cellular network to mobile units; satellite radio, which continuously provide entertainment to listeners; and satellite-based Internet, which enables IP-over-satellite connectivity.

REVIEW QUESTIONS

15.1 Which is not one of the advantages of satellite?
- a. Wide-area coverage
- b. Easy access to remote sites
- c. Low error rates
- d. Propagation delay

15.2 A satellite that is located at 1000 km above the equator is classified as a:
- a. GEO
- b. MEO
- c. LEO
- d. HEO
- e. ICO

15.3 Which of the following is not an advantage of GEO satellites?
- a. Noticeable latency
- b. Three of them provide global coverage
- c. They remain almost stationary with respect to an earth station
- d. Effects of Doppler shift are negligible

15.4 The component of the satellite system that monitors and controls the satellite while in orbit is called:
- a. Antenna
- b. Transponder
- c. Telemetry
- d. Payload

15.5 Which component of the satellite receives a weak signal at one frequency, amplifies it, and changes its frequency to another for transmission to another earth station?

a. Antenna
b. Transponder
c. Telemetry
d. Payload

15.6 The orbital motion of a satellite is governed by:
a. Maxwell's equations
b. Euler equations
c. Kepler's laws
d. Newton's laws

15.7 The fixed satellite service (FSS) industry is not growing because the market trend is toward the replacement of long-haul microwave system with fiber.
a. True
b. False

15.8 Which of the satellite services does not involve mobile units?
a. VSAT
b. FSS
c. MSAT
d. Satellite radio

15.9 Which global mobile satellite system has the largest number of satellites?
a. Globalstar
b. ICO
c. Ellipso
d. Iridium

15.10 The future of satellite radio is exciting but uncertain.
a. True
b. False

Answers: 15.1d, 15.2c, 15.3a, 15.4c, 15.5b, 15.6c,d, 15.7a, 15.8a,b, 15.9d, 15.10a

PROBLEMS

15.1 What basic need is a satellite designed to meet?

15.2 What advantages do satellite-based communications have over terrestrial communications?

15.3 What are the benefits of using GEO satellites for commercial purposes? Why are there objections to GEO satellites?

15.4 How do LEO satellites differ from GEO satellites? What are the advantages of the former?

15.5 What are typical uplink and downlink frequencies? Why do they differ?

15.6 State Kepler's laws of planetary motion.

15.7 A constellation is designed such that the earth central angle is 18.44°. How many satellites does the constellation consist of?

15.8 A satellite is in an orbit with semimajor axis 42,164 km and eccentricity 0.001. Determine (a) the period of the satellite, (b) the apogee height, and (c) the perigee height.

15.9 A satellite is orbiting in the equatorial plane with a period from perigee to perigee of 14 h. Given that the eccentricity is 0.003, calculate the semimajor and semiminor axes.

15.10 At 40 GHz, calculate the attenuation due to rain rate of 50 mm/hr.

15.11 Determine, for a frequency of 14 GHz, the rain attenuation for a rain rate of 20 mm/hr.

15.12 A satellite transmitter has a power of 20 W. An earth station antenna located 36,000 km away has an effective area of 2 m^2. Calculate the power received. Assume that the transmission is in all directions (isotropic).

15.13 Name and describe the various components of the satellite earth station link outdoor unit.

15.14 Name and describe the various components of the satellite earth station link indoor unit.

15.15 What are some of the characteristics of the satellite teleports?

15.16 Why should any business enterprise choose satellite network?

15.17 What are some of the most important advantages of the HAP satellite system?

15.18 What are some of the most important disadvantages of the HAP satellite system?

15.19 What are the spectrum allocations for the Ku band and Ka band in the transmit and receive modes?

15.20 What are the advantages of the L, C, Ku, and Ka band frequencies?

15.21 What are the disadvantages of the L, C, Ku, and Ka band frequencies?

15.22 What are the propagation factors affecting Ka band satellite links operating at moderate to high elevation anlgles?

15.23 Compare and contrast the mesh and star topologies of VSAT network.

15.24 What are the major limitations of VSAT networks?

15.25 Mention some applications of fixed satellite services.

15.26 Discuss briefly the need for an MSS or MSAT system and compare it with the existing terrestrial cellular networks.

15.27 Explain the operation of Iridium.

15.28 Discuss Globalstar system and its limitations.

15.29 How is ICO different from Iridium and Globalstar?

Part 5

Network Management and Security

16 Network Management

An expert is a person who has made all the mistakes that can be made in a very narrow field.

—Niels Bohr

Today's information technology (IT) spans from local area networks to wide area networks that support numerous applications locally and remotely. The job of a network manager is complicated and he/she needs to manage the heterogeneous network locally and remotely. Network management today is mostly a combination of local and remote configuration and management with software. A collection of management applications that provides an interface to network management protocols [e.g., simple network management protocol (SNMP)] is typically called a network management system (NMS). The management device is called a management station and the managed device is called the agent. A management station provides the required values of different parameters from its virtual store called the management information base (MIB). A management application can query an agent's MIB and present the result in readable form on the management station. There are two types of network management: centralized and distributed network management system (CDNMS). There are many issues for network management. The International Organization for Standardization (ISO) has formally defined five network management categories. They are performance management, fault management, configuration management, security management, and accounting management.

This chapter overviews briefly network management architecture, simple management information protocols, management information base (MIB), remote network monitoring (RMON), and a summary of the subsequent sections.

16.1 NETWORK MANAGEMENT ARCHITECTURE

A network management system depends on the network topology. This can be viewed as a collection of software, hardware, and some procedures. The scope of network management extends from the local network to the wide area network. It is complex in nature due to the hybrid/heterogeneous nature of the networks. A network management system uses different software/tools to process a network parameters' values, and monitor and display the parameter values in a readable format to manage and configure the networks. Numerous commercial software packages are sold that provide users with collections of network management tools such as facilities to diagram the network, to isolate problems, and to maintain logs of network performance statistics and events.

A general architecture or framework is usually used for network management. NMS may be referred to as integrated network management systems (INMS) or enterprise network management systems (ENMS). Two definitions we need to understand the architecture are (1) manager: a node that is the host of the controlling network management software and (2) intelligent devices/managed device: all other nodes such as front-end processors, remote work station, etc. The agent can be any application software such as Oracle, Microsoft SMS, and Lotus Notes. Therefore, an agent is not only hardware but could be application software running on the network. Typically an agent compiles information about its device and stores it in the database and finally interacts with the NMS via simple network management protocol (SNMP) and common management information protocol (CMIP). Management proxies provide information about other entities of the network. Typical network management architecture is shown in Figure 16.1. A generic network management configuration is shown in Figure 16.2.

Each host/agent (Figure 16.2) has a network management entity (NME) that performs a number of tasks. These are collecting information and storing it in a database, interacting with the manager and providing support to the manager for on-demand tests for performance measurement or fault detection, etc. To support multiplatform networks, several ENMS architectures are developed. Among these, simple network management protocol (SNMP) is well known and suited for managing TCP/IP networks. Others are the AT&T unified network management architecture (UNMA), and

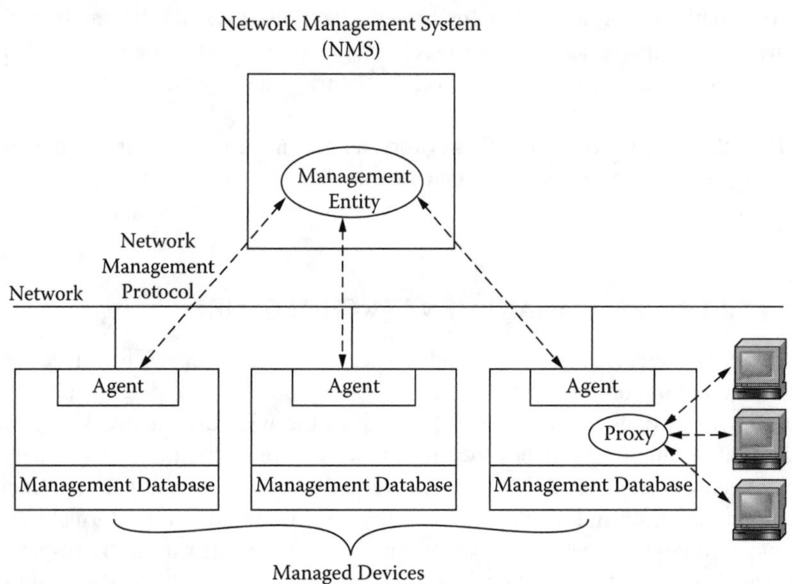

FIGURE 16.1 Typical network management architecture.
Source: Cisco Web site.

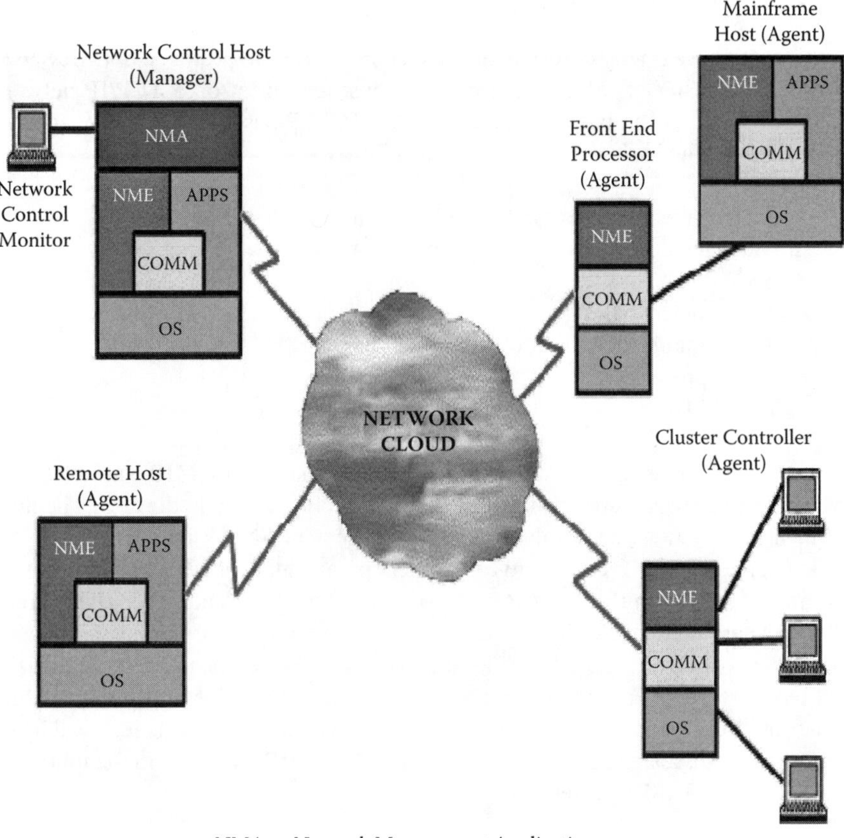

FIGURE 16.2 Enterprise network management system.
Source: Network Management Architectures and Protocols In-Depth by AA Verstraete © Revised: April 8, 1998.

the ISO's common management interface protocol (CMIP). For telecommunications network management, ITU's standard is telecommunications management network (TMN). Also IEEE 801.1 protocol includes management functions for LANs.

There are many NMS packages sold on the market. The choice depends on the manager's operating system, data architecture type, and the ENMS protocols CMIP/SNMP. Some well-known commercial packages are

- IBM's Systemview
- Hewlett-Packard's OpenView
- Novel's NetWare Management System and
- AT&T's Accumaster

16.2 NETWORK MANAGEMENT PROTOCOLS

In this section we briefly describe the most common ENMS protocol: SNMP. SNMP allows the transfer of information between host and agent on a TCP/IP network. Figure 16.3 shows the basic components of a TCP/IP-based network management environment. The basic components are as follows:

1. Management application (hosted by management station)
2. Management process
3. SNMP
4. Management MIB
5. A communication architecture
6. Agent process
7. Agent MIB

In general, management applications control the agent's MIB variables. The management process supports the management application to do the task. The management applications pass values to the management SNMP module. An SNMP packet contains SNMP commands and others parameters' values and it is passed to the transport layer in the TCP/IP protocol stack. This layer and lower layers create other headers that enable the formation of a frame and then it is sent to the agent over the network. On the other side, the agent's SNMP extracts the MIB variable and passes it to the appropriate agent process. The agent process also returns MIB values of its current database to the agent SNMP, which eventually is received by the management station and stored in the management's MIB database. In the following, we discuss the SNMP packet.

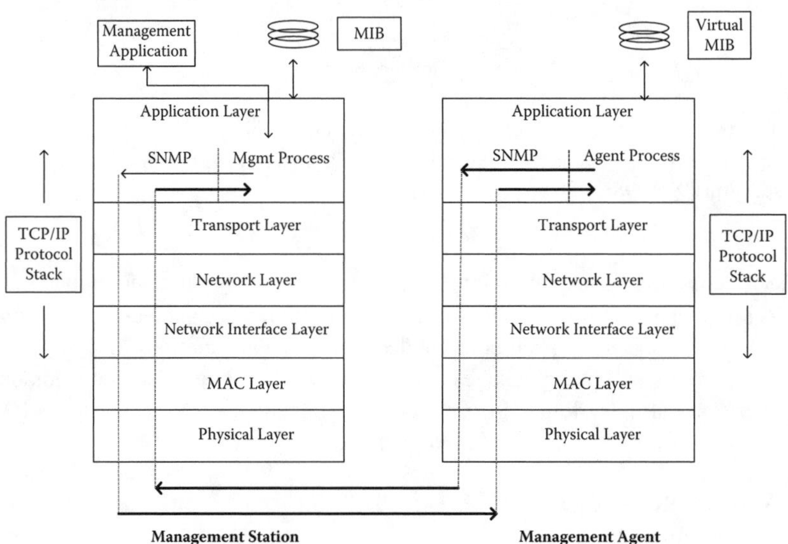

FIGURE 16.3 Components of a TCP/IP based network management environment.

16.3 SIMPLE NETWORK MANAGEMENT PROTOCOL (SNMP)

Figure 16.4 shows the fields of an SNMP packet. Figure 16.5 shows the fields of the protocol data unit (PDU) of SNMP. We briefly described the fields here.

- Version is the value used to indicate the version number of SNMP that is being used in management and agent stations. The same version of SNMP is used for the agents and the management station. Version "0" is used to indicate SNMPv1 packet.
- Community name is a password used by the management station and the agent. Both use the same password. This allows the management station to control the agent's MIB elements. This field is not encrypted and raises the question of security. SNMPv3 gives attention to this matter.
- Protocol data unit (PDU) is a common name for an SNMP message. Network-management systems and agents communicate using messages. SNMPv1 supports five messages: GetRequest, SetRequest, GetNextRequest, GetResponse, and Trap. The general format of the PDU is shown in Figure 16.5.
- Request ID is an integer used to identify the message type, e.g., Get-Request, Set-Request, Get-Next-Request, Get-Response, and Trap. The numbers are given in parentheses:
  - Get-Request (0) requests a value or set of values from the management agent.
  - Get-Next-Request (1) is used by the network-management system to retrieve the next value in a table or a list within an agent.
  - Get-Response (2) informs the management station of the results of a Get-Request or SetRequest by returning an error indication and a list of variable/value bindings.
  - Set-Request (3) is used by the network-management system to set the values within a device.
  - Trap (4) message is initiated by the agents for an alarm or event to be taken care of by the network manager. The trap command PDU is shown in Figure 16.6.

Version	Community Name	PDU

FIGURE 16.4 The SNMPv1 Packet.

Request ID	Error Status	Error Index	VarBindList

FIGURE 16.5 Fields of PDU.

Enterprise	Agent Address	Generic Trap Number	Specific Trap Number	Time Stamp	VarBindList

FIGURE 16.6 SNMP trap PDU.

Variable ID	Variable Value

FIGURE 16.7 VarBindList pairs.

  - Enterprise contains an object identifier that creates the trap message.
  - Agent address is the IP address of the device.
  - Generic trap number is an integer number (0–7) that shows the trap type defined by the RFC 1157.
  - Specific trap number is a code of proprietary MIB information of the agent used by the management station to interpret the trap number.
  - Time stamp contains the time in 1/100th of a second since the agent on the device was initialized.
  - VarBindList contains troubleshooting information.
- Error status has the value 0 for the Get-Request message. For the Get-Response message used by the agent there are six possible values: noError = 0; tooBig = 1: PDU size is big; noSuchName = 2: no object with this name; badValue = 3: PDU type is bad; readOnly = 4: incorrect implementation of SNMP; genErr = 5: errors of other types.

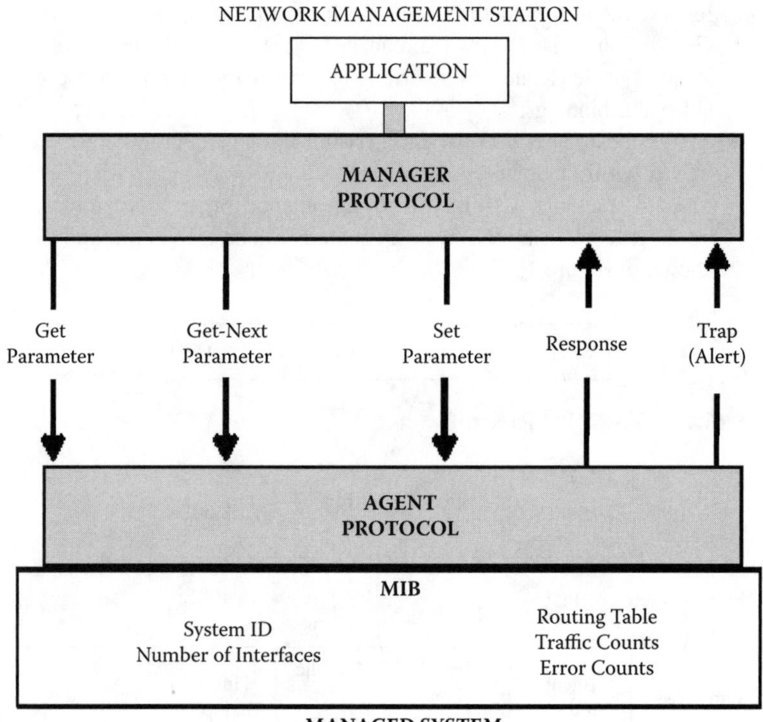

FIGURE 16.8 Messages sent between a manager and an agent.

- Error index is an integer. The number represents the respective variable of VarBindList of the Get-Response message that caused the error code to appear in the error status field. When the management station requests the error status, the error index fields show a value of ϕ.
- VarBindList is a list of variable IDs and variable value pairs are as shown in Figure 16.7. The list may contain several pairs. Variable ID is the object identifier defined by the structure of management information (SMI). The object identifier defines the path to the object in the MIB tree as shown in Figure 16.9. Variable value contains the value of the variable in the form of an integer/octet string/IP address.

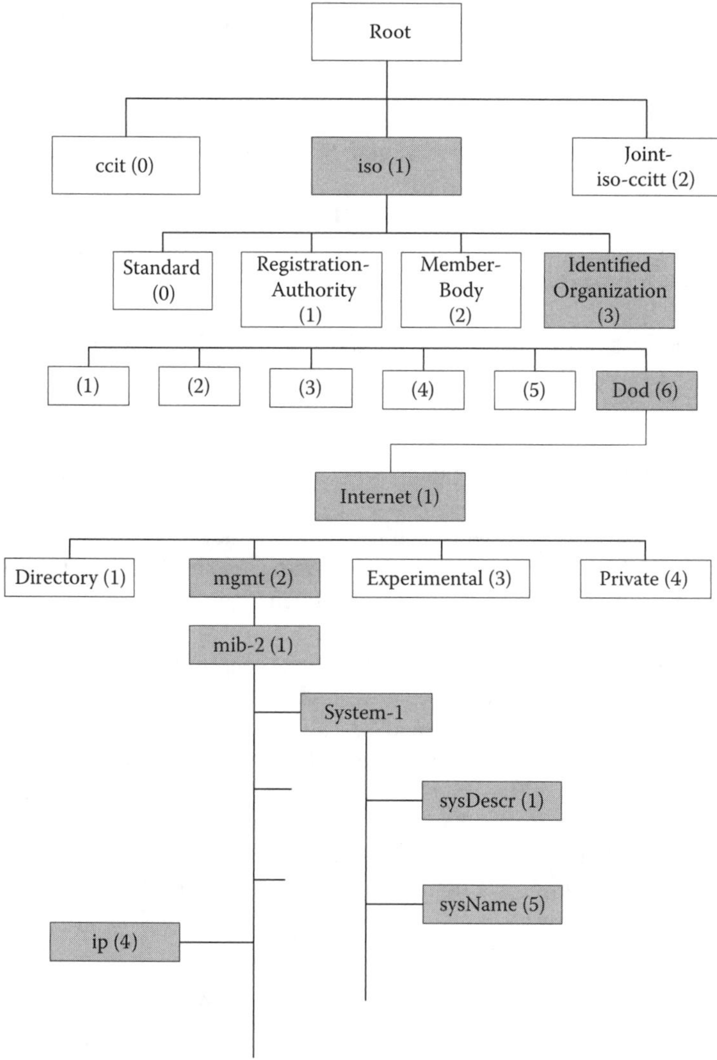

FIGURE 16.9 Management information base tree.

Figure 16.8 shows the interaction between the management station and the agent using SNMP. Please check the Internet references given in the reference section for detailed information on SNMP and SNMP2.

16.4 MANAGEMENT INFORMATION BASE (MIB)

The MIB is a logical database stored on a network device. This database may be a combination of hardware configurations, and status and statistical information. SNMP standards use this logical database information to manage the network. The internal structure of these MIBs is not important for the manager, but the accessibility of the database is very important. There are various variables we can keep in the MIB, but for standardization, the SNMP community uses the following approach:

- define groups of useful parameters
- fine-tune the fields after several months of experience (remove parameters and add new ones)
- committees of experts to define MIB variables for special technologies (e.g., bridges, token-ring interfaces)
- add vendor-specific extension

MIB 1 is the first version developed based on the TCP/IP suite. It includes the following objects/variables:

- system description
- the number of networking interfaces that a system has (Ethernet adapters, serial ports)
- IP address associated with each network interface
- counter that counts the numbers of incoming and outgoing datagrams
- table of information about active TCP connections

ANS.1 (abstract syntax notation 1) data type language is used to define MIB variables. It has the ability to create a data structure with more capabilities than a conventional language such as C and Pascal. Basic encoding rule (BER) is used for translations of the value of ASN.1 variables into a fixed transmission format. ISO and CCIT developed the ISO/IEC OSI standards for the management of the information as a tree structure. This provides a modular framework that allows adding new branches to the management information tree without any changes of protocol. If we want to perform an action we just define the variable that causes the action when it is set to 1. A new action is just a new variable in the MIB.

16.5 STRUCTURE OF MANAGEMENT INFORMATION (SMI)

SMI describes how to manage information on a TCP/IP network. The SNM community uses the term object instead of variable and an object is associated with a name, attributes, and a set of operations that can be performed on the object. SNMP describes managed objects that hold the network information. A MIB is a set of managed objects. ISO and CCITT use the idea of structuring information into a global naming tree and assigning an identifier to each object. Figure 16.9 shows the management information tree. The root has three main nodes: CCITT (0), ISO (1), and joint-ISO-CCITT (2). Since the Internet Activity Board (IAB) decided to base the structure of management information on the Internet, we are only interested in nodes under the ISO node. However, the nodes under joint-ISO-CCITT are important for the ISO management structure.

The management node MIB-2 is assigned under the Internet node. Hence the prefix to all MIB-2 objects is iso.org.dod.internet.mgmt.mib-2. The Internet node is managed by the IAB.mib-2 (1) under the management (2) node which has the Internet standard MIB objects. To understand how an information object is defined the readers should read the ISO/IEC standard 8824, part 1-4.

16.6 REMOTE NETWORK MONITORING (RMON)

RMON is used to monitor the remote network and its concept is based on the MIB. It is an MIB element developed by the IETF (Engineering Task Force) that helps to monitor and analyze the IEEE-specified LAN protocols. Its main purposes include network fault diagnosis, tuning the network's performance, and planning and design. Hence it is a part of the network management functions. To use RMON products an agent must have an SNMP NME (network management entity).

The original RMON was developed for LAN technology. It captured traffic statistics from the MAC (media access control) layer of the remote LAN segment. The current RMON can serve FDDI and token ring technology too. The second-generation RMON can support the OSI-7 layer concepts. The added upper-layer functionality allows net managers to know not only who is talking to whom, but also what applications are being used.

RMON is similar to SNMP. It defines some variables and captures the data from the network segments, which it stores in its database. A management application retrieves that information for the network manager. The network manager manages the network using this information. This helps reduce on-site workloads and keeps the network running smoothly. For more details concerning RMON and RMON2, see 3COM Report on RAMON (http://www.3com.com/nsc/501305.html).

SUMMARY

1. Information technology (IT) spans from local area networks to wide area networks that support numerous applications locally and remotely.
2. Network management is a complicated issue that involves managing heterogeneous networks locally and remotely.
3. Network management today is mostly a combination of local and remote configuration and management with software. A collection of management applications that provides interface to network management protocol [e.g., simple network management protocol (SNMP)] is typically called a network management system (NMS).

4. The management device is called a management station and the managed device is called the agent.
5. SNMP protocol is one of the main protocols used for TCP/IP suite for interactions between the management station and other manageable devices (agents).
6. In monitoring remote networks, RMON is used. The concept is based on MIB.

REVIEW QUESTIONS

16.1 Network management involves managing heterogeneous networks locally and remotely.
   a. True
   b. False

16.2 A network management system does not depend on the network topology.
   a. True
   b. False

16.3 A general architecture or framework is usually used for network management.
   a. True
   b. False

16.4 INMS means:
   a. Industrial Network Management Systems
   b. Individual Network Management Systems
   c. Integrated Network Management Systems
   d. Immobile Network Management Systems

16.5 Some well-known commercial packages are:
   a. IBM's System view
   b. Hewlett-Packard's OpenView
   c. Novel's NetWare Management System
   d. AT&T's ACCUMASTER
   e. All of the above

16.6 A collection of management applications that provides an interface to network management protocol, e.g., simple network management protocol (SNMP), is typically called a network management system (NMS).
   a. True
   b. False

16.7 SNMP protocol is not one of the main protocols used for the TCP/IP suite for interactions between the management station and other manageable device agents.
   a. True
   b. False

16.8 SMI describes how to manage information on the TCP/IP network.
   a. True
   b. False

16.9 MIB 1 is the first version developed based on the TCP/IP suite.
  a. True
  b. False

16.10 ANS.1 (abstract syntax notation 1) data type language is used to define which variables?
  a. MIB variables
  b. NMS variables
  c. SNMP variables
  d. SNM variables
  e. All of the above

Answer: 16.1a, 16.2b, 16.3a, 16.4c, 16.5e, 16.6a, 16.7b, 16.8a, 16.9a, 16.10a

PROBLEMS

16.1 Draw a diagram of a typical network management architecture and describe how it operates.

16.2 Name and describe some of the well-known commercial NMS packages used and sold in the market today.

16.3 What does the choice of some of the well-known commercial NMS packages depend upon? Describe each of them.

16.4 Name basic components of a TCP/IP-based network management environment.

16.5 Describe the basic components of a TCP/IP-based network management environment.

16.6 Name and describe the fields of a SNMP packet.

16.7 What kinds of fields are required of the protocol data unit (PDU) of SNMP?

16.8 Briefly describe the fields of the PDU of SNMP.

16.9 MIB 1 is based on the TCP/IP suite. Briefly describe the objects/variables.

16.10 What is RMON? State the purposes of RMON.

16.11 Using the request ID integer describe the PDU that could be used to identify the message type, e.g., Get-Request, Set-Request, Get-Next-Request, Get-Response, and Trap in any SNMP.

16.12 Using an SNMP management station, how can you monitor a system that runs an SNMP agent at your site?

16.13 Name the different areas of network management and describe each.

16.14 Performance management has goals. What are they?

16.15 What are the goals of configuration management in any network management system?

16.16 Security is very essential in the management of network systems. What are the security management goals?

16.17 Questions 16.17–16.19 deal with simple network management protocol (SNMP) management infrastructure. Define the following terms: managing entity, managed device, management agent, MIB, and network management protocol.

16.18 Use a diagram to depict the management infrastructure from part 16.7 above and show how these parts are related to one another.

16.19 Name four of the six analytical approaches useful in network management problem solving, and briefly describe how each works.

16.20 Name and briefly define the five areas of network management defined by the ISO. Name and briefly define the four key stages of network management and provide a simple diagram showing how the stages flow into one another.

REFERENCES

For more general information on network management:

SNMP & CMIP: an introduction to network management (http://www.inforamp.net/~kjvallil/t/snmp.html)

Network Management: What it is and what it isn't. (http://netman.cit.buffalo.edu/Doc/DStevenson/)

The "Simple" Web: Information on Network Management (http://snmp.cs.utwente.nl/)

Network Management Server (archive base for comp.dcom.net-management) (http://netman.cit.buffalo.edu/index.html)

Networks and Network Management (http://www.wmpenn.edu/PennWeb/Academic/CompSci/NM210/default.htm)

SNMP Version 3 Working Group of the IETF http://www.ietf.org/html.charters/snmpv3-charter.html

SNMP Agent Extensibility (AgentX) Protocol ftp://ietf.org/internet-drafts/draft- ietf-agentx-ext-pro-02.txt

The Simple Times: SNMP Users Publication http://www.simple-times.org/pub/simple-times/

For information on CMIP and Web-based Network Management:

IBM's Webbin' CMIP (and SNMP) http://misa.zurich.ibm.com/Webbin/

ISO's CMIP Management Standards http://wwwsnmp.cs.utwente.nl/iso/standards.html

17 Network Security

A successful man is one who makes more money than his wife can spend. A successful woman is one who can find such a man.

—Lana Turner

We use computer networks for everything from banking and investing online to communicating with others through email or chat programs. Due to the value of information on these networks, it has become a lucrative target for unauthorized users or criminals. To protect both the network and the information on it, network managers must consider taking some security measures. Security measures help you to stop unauthorized users from accessing any part of your computer network. As more and more people become "wired" or "connected" and the world becomes more tightly interconnected, network security becomes more important.

Network security is the process of preventing and detecting unauthorized use of your network.

We begin this chapter by looking at the network security requirements. We then discuss the types and sources of threats there are against computer networks. We go on to consider access control methods, cryptography, firewalls, intrusion detection, and security standards.

17.1 SECURITY REQUIREMENTS

Security is an important issue for networks for at least two reasons. First, information that resides on the networks is increasingly being used for education, commerce, health care, national defense, and many other endeavors. Such information may be security sensitive (e.g., tactical military information). Second, whether a computer is used at home, university, business, or government agency, if it is connected via a network to other computers, its resources are at risk. A network-connected computer can reach tens of millions of users in every part of the globe. The vast connectivity poses monumental risks. Since anyone can reach out to the network, almost anyone can reach in.

Before information exchange using computer network became widespread, several methods, such as signatures, notarization, identification, and dating, were used to give the parties involved confidence about the genuineness of the documents.

In the same way, documents in electronic form must possess some form of integrity and authenticity. To achieve this, the following security requirements must be met:

- Confidentiality: This prevents the disclosure of transmitted data to unauthorized parties. Information should be available only to those who should rightfully have access to it. Network transmission of sensitive information, such as strategic or tactical military information, requires confidentiality. A common way of keeping transmitted data confidential between the sender and the receiver is to encrypt the data before transmission and decrypt it upon receipt. This can also be used for integrity, authentication, and nonrepudiation.
- Integrity: This involves detecting modification, insertion, or deletion of transmitted data. Integrity guarantees that a message being transferred is not corrupted. Information should be modified only by those who are authorized to do so. Information that cannot be trusted is worthless. A way to ensure information integrity is to use cryptographic algorithms.
- Availability: Information should be accessible to those who need it when they need it. Information that is not readily available when needed is not useful. Availability ensures the survivability of the network services despite attacks.
- Nonrepudiation: This is preventing either the sender or receiver in a communication network from denying their participation. In other words, nonrepudiation ensures that the originator of a message cannot deny having sent the message.
- User authentication: This is demonstrating that the identity of a user or system is valid. The most common method of user authentication is the use of passwords. This method is seldom secure since the password, which is changed infrequently, has to flow over the wire. If an eavesdropper detects the password, he or she can masquerade as the legitimate user and gain access to all resources on the network.
- Access control: This is guarding against unauthorized use of resources. Even if users are authenticated, access control allows them to access only the information and network resources to which they are allowed.

These requirements apply to home Internet users just as much as they would to any corporate or government network. In this age of laptops, wireless communication, desktop modems, and handheld computing gadgets, the security parameter is convoluted.

17.2 NETWORK THREATS

Network intruders (also referred to as hackers, attackers, or crackers) are always discovering new vulnerabilities to exploit in computer software. The most common methods used by intruders to gain control over computer networks and what can be done to protect your network against various threats are discussed here.

- Denial-of-service: The primary goal of this form of attack is to deny the victim(s) access to a particular resource or prevent legitimate users of a service from using that service. For example, an attack may be made to "flood" a network, thereby preventing legitimate network traffic. Denial-of-service (DoS) attacks are perhaps the nastiest, because they are very easy to launch and difficult to track. This type of attack causes your computer to crash or to become so busy processing data that you are unable to use it. Denial-of-service attacks come in a variety of forms and aim at a variety of services. For example, in February 2001, the public access of the sites of Amazon, eBay, Yahoo!, and other dot-coms was cut using DoS.

- Unauthorized access: The primary goal of this form of attack is to access some resource on your machine. When you attempt to log on to a computer network, you are presented with a logon dialog. This dialog prompts for the user name and password. This information is used for authentication purposes. A remote user can access certain information on the network without authentication. Criminals may use different tactics to gain unauthorized access and perform privileged actions on the network.
- Email spoofing: Spoofing is generally used by spammers as a first defense against people finding out who they are. Email spoofing is the act of changing your name in email so that it looks like the email came from someone else. It is usually an attempt to trick the user into releasing sensitive information. Email spoofing is commonly used by computer viruses that attempt to deceive the recipient into believing the email is from someone other than the actual sender. Although spoofing can be a legitimate and helpful tool for someone with more than one email account, spoofing anyone but yourself is absolutely illegal.
- Email-borne viruses: A virus cannot exist in an email text message. Viruses must be attached to and infect an executable program (.exe, .com). Viruses and other types of malicious code are often spread as attachments to emails. Email worms and viruses can reach your system and infect users through harmful attachments. Be sure you know the source of the attachment before you open an attachment. Never open a suspicious attachment.
- Packet sniffing: A packet sniffer is a wire-tap device that plugs into computer networks and eavesdrops on the network traffic. It could also be a program that captures data from information packets as they travel over the network. The data may include user names, passwords, bank account or credit card information, and proprietary information. With that data, intruders can launch widespread attacks on systems. The best defense against packet sniffing is to encrypt your data, so that while they can sniff it, they cannot read it.

17.3 ACCESS CONTROL METHODS

Security measures are often distributed throughout a computer network. These measures can be implemented in host computers, terminals, and special security devices. When users have common access to a computer network, it is necessary that we control access to the system. Access control decisions consist of two main steps: authentication of the user's digital identity and authentication of the user's right to perform the desired action. In current computer networks, authentication of the user's digital identity utilizes three types of access control methods: passwords, tokens, and biometrics. These three types indicate that access control schemes apply something an individual either knows (passwords), has (tokens), or is (biometrics).

- Passwords: Passwords are the most common form of securing a network. They are usually required by the host computer, which asks a user for identification. If the user is unable to provide this, the machine refuses to grant access. Because your password is like a key to your account, you need to safeguard it. Methods of ensuring that password-based systems are secure include assigning passwords to users instead of letting them choose their passwords, requiring that passwords be at least six characters, requiring that users change their password on a regular basis, allowing a password to be used only once, and denying multiple simultaneous logins. A major drawback of passwords is carelessness on the part of many users. Some users write down passwords where they can be easily discovered and others select a familiar character (like 123456), which can be guessed.
- Tokens: A token is a small hardware device that is used for authentication purposes. Most token-based cards require both a username and PIN. They use a technique known as two-factor authentication, which requires some personal, unique information from the user, plus some data generated by the hardware, to authenticate the user. In contrast to password authentication, which relies solely on the use of a single password, two-factor authentication incorporates a personal identification number (PIN) in addition to a hardware or software token device. Smart cards (credit-card-sized devices that are preprogrammed with unique passwords) are an example of hardware tokens. Software tokens are generated by software installed on a computer. After being activated by a user entering a PIN, a software token provides a unique password for authentication. Since the time-based token expires every 60 minutes, intruders are prevented from using the passcode in replay attacks. A drawback of the hardware-based token mechanism is that the user cannot access the system if the token is lost or forgotten.
- Biometrics: Biometrics are automated methods of identifying a person or verifying the identity of a person based on a physiological or behavioral characteristic. A biometric is basically a measurable, robust, and distinct physical characteristic or personal trait of an individual that can be used to identify the individual or verify the user's claimed identity. The various biometrics commonly used include fingerprints, offline/online signatures, iris recognition, retina, voice, palm scan, gait, thumb prints, voice recognition, facial geometry, or other physical feature. They are each in their own way unique to an individual. A biometric device uses some characteristic of the user for authentication. Biometric-based authentication systems and applications are expected to support multiple biometric devices and biometric data. Because biometrics are not secrets, systems must not use them like passwords, otherwise, biometric-based authentication will reduce security rather than enhance it.

17.4 CRYPTOGRAPHY

People have been trying to protect information from unauthorized access since they learned to write. Why do we use envelopes for regular mail rather than send a postcard? Are we trying to hide something? There are many ways to protect sensitive information.

Since the Internet is a global network, no secret of any value can be communicated over it without the help of cryptography (data encryption). Data encryption is used in various applications and environments. In fact, cryptography has been of great value to the military, spies, and diplomats since antiquity. When sensitive information is encrypted with a powerful algorithm, the information can be protected from eavesdroppers.

Cryptography is derived from the Greek words for "secret writing." The field of cryptography deals with the technique for conveying information or a message securely. The block diagram of a cryptographic system is shown in Figure 17.1. The message in its original form (in intelligible form) is called plaintext. The transmitter will encrypt the plaintext in order to hide its meaning. This encrypted message is called ciphertext. The algorithm used in encrypting the message is known as a cipher. The process of converting from plaintext to ciphertext is called encryption, while the reverse process is called decryption. Most cryptographic algorithms make use of a secret value called the key. The process of breaking ciphers when the key is unknown is called cryptanalysis.

> **Cryptography** is a process associated with scrambling plaintext into ciphertext (a process called encryption), then back again (known as decryption).

Cryptography is generally used to protect data while it is being transferred from one point to another or while it is stored in a medium vulnerable to physical theft.

Crytographic algorithms (or ciphers) can be classified using several criteria. According to one criterion, ciphers are either symmetric or asymmetric. In conventional cryptography, the sender and receiver of a message use the same secret key. This method is known as a secret key or symmetric cryptography. The main challenge is getting the sender and receiver to agree on the secret key without anyone else finding out. The cryptography is asymmetric (or public key) if the sender and receiver use different keys. In this system, each person gets a pair of keys, one called the public key and the other called the private key. The public key is published, while the private key is kept secret. The two types of cryptosystems

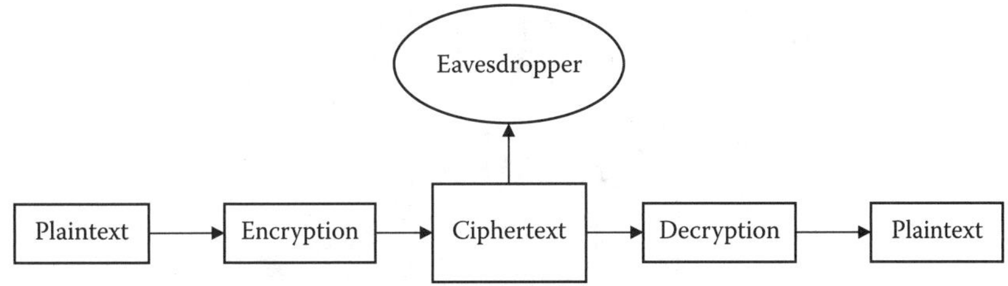

FIGURE 17.1 Block diagram of a cryptographic system.

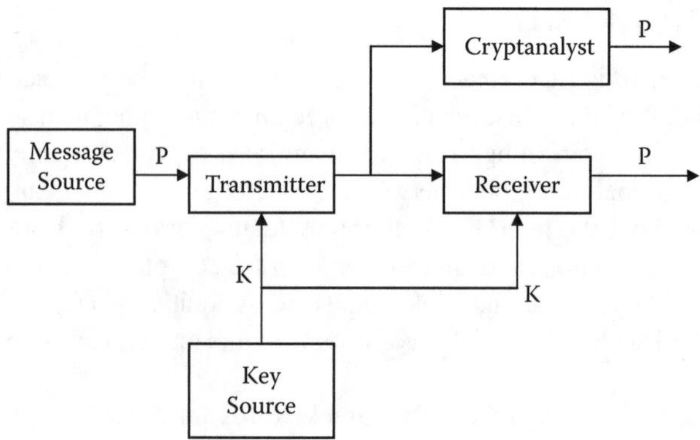

FIGURE 17.2 Conventional (symmetric) cryptographic system.

are shown in Figure. 17.2 and Figure 17.3. The most important modern symmetric encryption algorithm is that contained in data encryption standard (DES), to be discussed fully in Section 17.7. Public key cryptography provides advantages over symmetric cryptography. First, the problem of managing secret keys is greatly reduced. With a symmetric or conventional cipher, it is necessary to transfer secret keys to both communicating parties before secure communication can begin. As shown in Figure 17.2, the key is distributed over a secure channel. Public keys may be transferred over a public channel. Second, public key cryptography provides a digital signature. Authentication can be provided with public key cryptography. Digital signatures are set to replace handwritten signatures in many applications.

Over the years several encryption algorithms have been developed. Here, we will consider only the substitution ciphers, transposition ciphers, and Rivest-Shamir-Adleman (RSA) public key scheme. The data encryption standard (DES) will be considered in Section 17.7.

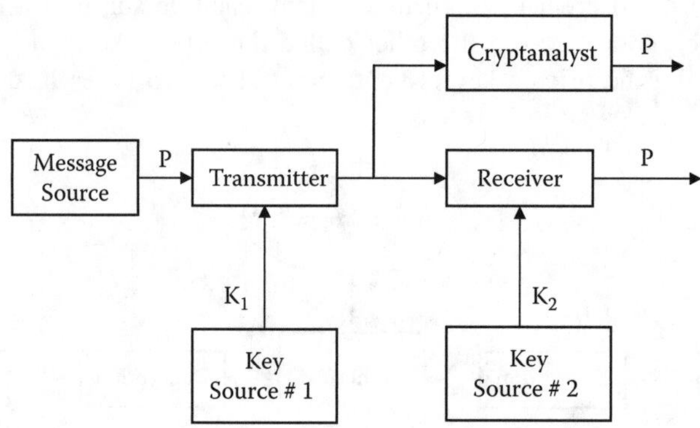

FIGURE 17.3 Public key (asymmetric) cryptographic system.

17.4.1 Substitution Cipher

Simple (monoalphabetic) substitution ciphers replace each letter in the plaintext with another letter to produce the ciphertext. Although this class of ciphers is easy to implement and use, they are not difficult to break. An example is a substitution cipher known as shifted alphabet. One was used by Julius Caesar to send messages to his generals. Using Caesar cipher, we replace each letter in the alphabet with the letter three places further down the alphabet. The alphabet is wrapped around so that A follows Z. Thus, the plaintext letter A becomes D, B becomes E, …, Z becomes C. By this method, we obtain

plain: a b c d e f g h i j k l m n o p q r s t u v w x y z
cipher: d e f g h i j k l m n o p q r s t u v w x y z a b c

For example,

plain: T H I S I S F U N
cipher: W K L V L V I X Q

If a numerical equivalent is assigned to each letter (a = 1, b = 2, …, z = 26), then the encryption algorithm becomes

$$E(\lambda) = (\lambda + 3) \bmod (26) \tag{17.1}$$

where λ stands for an English letter. Since the shift could be any amount, the general Caesar algorithm is

$$E(\lambda) = (\lambda + k) \bmod (26) \tag{17.2}$$

where k varies from 1 to 25.

The ciphertext alphabet used in the Caesar cipher is orderly. Some substitution ciphers use a scrambled alphabet that has no apparent order. For example, A might be substituted with T, B with U, and C with Q. However, all simple substitution ciphers can be easily broken using frequency analysis. In the English language, E is the most frequent letter, followed by T, O, A, N, I, R, S, H, and so forth. Z is the least frequent letter. Polyalphabetic substitution ciphers use multiple alphabets to conceal the single letter frequency distribution of the plaintext letters in the ciphertext.

Example 17.1

We may use message reversal for encryption. In this case, the plaintext is written backwards to produce the cipher. For example,

plaintext: we must attend the party
cipher: ytrap eht dnetta tsum ew

Obviously, this can easily be decoded and the message is not secure.

TABLE 17.1
Matrix Used to Create Transposed Ciphertext

A	T	R	A
N	S	P	0
S	I	T	I
O	N	C	I
P	H	E	R

17.4.2 Transportation Ciphers

This family of ciphers requires that the letters in the plaintext be rearranged or transposed. For this reason, the ciphertext contains the same letters as the plaintext. Hence, the frequency analysis does not provide any useful information in an attempt to produce the plaintext.

A common method of transportation is to insert the plaintext into a matrix in some known way and form the ciphertext by reading the letters in another known way. For example, the plaintext can be inserted into the matrix by rows and the ciphertext can be obtained by reading the matrix by columns.

An example of transportation cipher is shown in Table 17.1. The ciphertext is formed by writing the plaintext into the rows of a 5×4 matrix. We then read the text out by the columns.

Plaintex: A T R A N S P O S I T I O N C I P H E R
Ciphertext: A N S O P T S I N H R P T C E A O I I R

17.4.3 RSA

The RSA scheme was developed by Ronald Rivest, Adi Shamir, and Len Adelman. It is based on the fact that it is easy to generate two large prime numbers and multiply them together, but it is much more difficult to factor the result. (Try multiplying 13,107 and 73,351 or try factoring 536,560,557 by hand.) The product can therefore be made public as part of the enciphering key without compromising the factors that effectively constitute the deciphering key.

The RSA scheme involves the following five steps:

- Step 1: Select two large prime numbers p and q, each about 100 decimal digits long. Multiply them together to get n = pq.
- Step 2: Obtain $\phi = (p - 1)(q - 1)$. Choose a random integer E (an encryption key), between 3 and $\phi - 1$ which has no common factors with ϕ.
- Step 3: Compute the decryption key D, such that

$$D = E^{-1} \bmod \phi \tag{17.3}$$

which means that D and E are multiplicative inverses of each other mod ϕ. The public information consists of E and n. All other quantities here are kept secret.

- Step 4: Given a plaintext message P which is an integer between 0 and n – 1 and the public encryption number E, we form the ciphertext integer.

$$C = P^E \bmod n \tag{17.4}$$

This means that we raise P to power E, divide the result by n, and let C be the remainder.

- Step 5: Using the secret decryption number D, we recover the plaintext P from the ciphertext as

$$P = C^D \bmod n \tag{17.5}$$

With RSA, users can create both a public key and a secret key. RSA is ideally suited for electronic mail systems. The RSA algorithm can also be used for a kind of unforgettable digital signature. In this application, the secret key is used to encrypt a message, which can then itself be decrypted by anyone possessing the public key. Using the private key for encryption means that an author can uniquely "seal" information with his or her own signature. Authentication using digital signatures not only assures the receiver that electronic message was indeed sent by, say, Mellon Bank, but also that the bank's message ("Transfer $3 million from account X to account Y.") has not been manipulated.

Example 17.2

As an example of the RSA scheme, let p = 5 and q = 11. Then

$$n = pq = 55, \qquad \phi = (p-1)(q-1) = 40$$

Suppose E = 7, then D = 23 (7 × 23 = 161 = 1 mod 40). In other word, $D = E^{-1} \bmod \phi = 7^{-1} \bmod 40$ since ED mod 40 = 7 × 23 mod 4 = 1.

If the plaintext P = 2, then

$$C = 2^7 \bmod 55 = 18$$

and

$$P = 18^{23} \bmod 55 = 18^1\ 18^2\ 18^4\ 18^{16} \bmod 55 = 18\ 49\ 36\ 26 \bmod 55 = 2$$

which is the original plaintext. (You may use MATLAB to check this.)

17.5 FIREWALLS

Firewalls can help to mitigate some of the risks present on the Internet without authentication or encryption. They have become a critical component of today's computer networks because they provide a very powerful way to protect your system from most Internet security threats. Firewalls in networks keep file damage on one

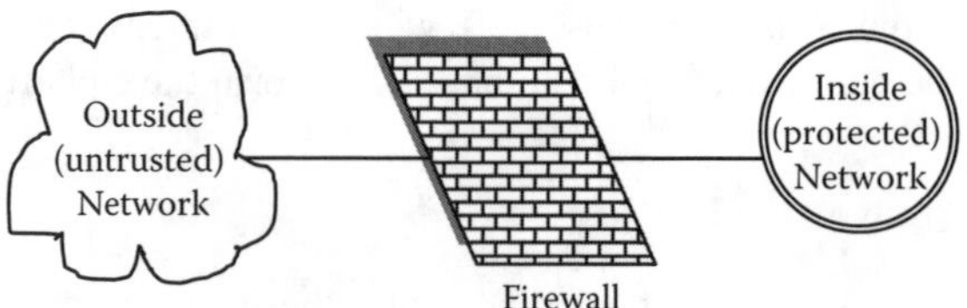

FIGURE 17.4 A firewall separates a protected network from an outside network.

part of the network from spreading to the rest of the network. Whether data is entering or leaving a domain, it is protected from eavesdropping (passive wiretapping).

A **firewall** is a protective device designed to block unwanted network traffic while allowing other traffic to pass.

It is a system that enforces an access control policy between two networks, as shown in Figure 17.4. It may also be regarded as a pair of mechanisms: one which exists to block traffic, and the other which exists to permit traffic. Some firewalls permit only email traffic through them, thereby protecting the network against any attacks other than attacks against the email service.

For example, if a firewall is installed on your computer, it will absolutely isolate your computer from the Internet. The firewall has total power over your computer's receipt of packets or information from the Internet. Since every arriving packet must contain the correct IP address of the sender's machine, the firewall can be selective about which packets are admitted and which are dropped. Its "permit" or "deny" decisions are only based upon the source and destination addresses. But the real power of a firewall is derived from its ability to be selective about what it passes and what it blocks.

There are several types of firewalls. Three popular types are

- Packet-filtering firewall: This is the simplest firewall to implement. It involves using only a packet-filtering router. Routers can be configured to filter packets based on comparison of packet content with predefined specification criteria. IP addresses, TCP or UDP port numbers, etc. can be used as criteria for access or denial. Packet-filtering firewalls examine endpoint identifiers in packets passing through a link to determine if each packet should be allowed to proceed.
- Application-based firewall: This involves a software application package that allows or denies access across networks. An application-based firewall does not allow any direct connection between internal and external hosts. An application-based firewall works by allowing some applications to access the Internet while denying others. While it may not prevent a denial-of-service attack on your connection or block more sophisticated hacking activities, it will prevent most trojans or viruses.
- Proxy firewall: This type of firewall acts as a mediator between two devices attempting to communicate through a firewall. Proxy firewalls

> provide an intermediate connection between the servers on the Internet and your internal servers. They terminate a flow at one side of the firewall, provide some proxy service to examine that data within the flow, and then recreate the flow in the network at the other side of the firewall. (A proxy is a process that resides between client and server processes.) For incoming packets, the proxy acts as a server to clients within your internal network. For outgoing packets, the proxy acts as a client sending data out to servers on the external network.

Firewalls may be hardware devices, software programs, or combinations of hardware and software.

Firewalls are certainly a key component of a good security system. They have several advantages. First, a firewall is likely to be more secure than an average host. Second, protocols and services that are not necessary or cannot be secured can be protected from attack. Third, services are located on the firewall system as opposed to being distributed on many systems. Fourth, extended logging of network traffic may be focused on one system. Finally, software and logging are not distributed on many hosts. They are located on the firewall system.

Firewalls have some disadvantages. First, firewalls can be circumvented or penetrated via social engineering methods. Second, firewalls do not protect a network against attacks that do not go through the firewall. For example, computers using dial-up modems outside the network can provide a back-door entry into the network. Third, firewalls do not protect from internal threats from employees and/or contractors. A high percentage of attacks do originate internally from disgruntled employees. Fourth, firewalls cannot protect very well against things like viruses.

Keep in mind that not everyone approves of firewalls. On the one hand, proponents say that many computer systems and networked applications are not secure and could use firewalls. Also, they are widely accepted, readily available, and easily justifiable to purchasing managers. On the other hand, opponents claim that firewalls generate a false sense of security that leads to laxity in enforcing security measures. Also, a firewall does not offer any protection from inside threats.

17.6 INTRUSION DETECTION

Many organizations currently secure their networks with the use of firewalls in conjunction with an intrusion detection system. They realize that a firewall is not enough to protect their networks. A firewall separates a protected network from an unprotected one. Intrusion detection systems function as a secondary system designed to back up systems such as firewalls, encryption, and authentication. A few years ago, the solution to a security problem would have been to cut off Internet access, but with e-commerce and e-business driving the market, that is simply not feasible now. The necessity of maintaining Internet connectivity has made intrusion detection a top priority.

Intrusion detection is important because it is impossible to keep up with the rapid pace of potential threats to computer systems. The Internet is changing rapidly day by day and businesses are expanding exponentially using the Internet.

The possibilities and opportunities of using the Internet are limitless; unfortunately, so too are the risks and chances of malicious intrusions. Intrusion detection is a proactive process that requires the constant attention of network administrators.

An **intrusion** is any set of actions that attempt to compromise the integrity, confidentiality, or availability of a resource.

An intruder (also known as a "hacker" or "cracker") is somebody who attempts to break into or misuse your computer networks. There are two types of potential intruders:

- **Outside intruders:** These are "hackers" coming in over the Internet to break into your system.
- **Inside intruders:** Studies show that 80% of intrusions and attacks come from within organizations. An insider knows the layout of your network and where the valuable data are. An inside intruder misuses privileges (such as the Social Security employee who marked someone as being dead because they didn't like that person).

A mechanism is needed to detect *both* types of intruders.

As shown in Figure 17.5, intruders may also be classified as passive or active.

- Passive intruder: The passive intruder passively observes the data traffic between the sender and the receiver. The confidentiality of data in transmission is threatened by a passive attack. The feasibility of passive attacks depends on the physical transmission media. Radio, satellite links, metallic conductors such as twisted pair are easy to intercept. But the tapping of optical fiber is more difficult.
- Active intruder: The traffic between the sender and the receiver is fully controlled by the intruder. The active intruder can modify, extend, delay, destroy, or copy data units. The integrity or availability of the data in transmission is threatened.

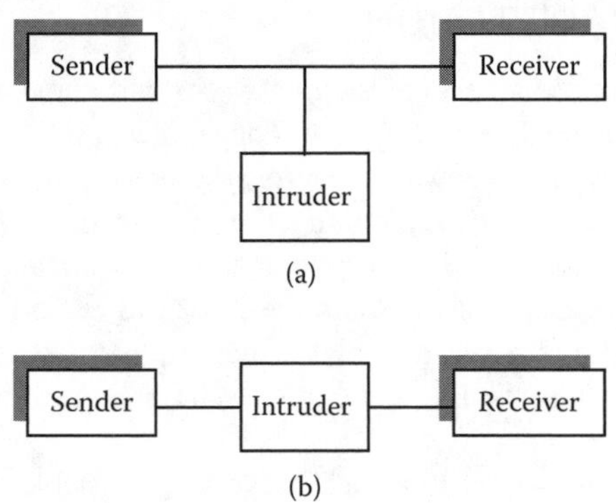

FIGURE 17.5 (a) Passive intruder, (b) active intruder.

Passive attacks are hard to detect but can easily be prevented, whereas active attacks are easy to detect but hard to prevent.

Who are these intruders? The vandals probing networks for security vulnerabilities may be curious teenagers, disgruntled employees, or corporate criminals from rival companies.

How do intruders break into the network? System flaws that exist in software are the root cause of network intrusions. Software always has bugs. System administrators and programmers can never track down and eliminate all possible holes. Intruders have only to find one hole to break in.

Intrusion may be regarded as the problem of identifying individuals who are using a computer system without authorization and those who have legitimate access to the network but are abusing their privileges. Intrusions can be classified into two major groups:

- **Anomaly** intrusions, which are based on observations of deviations from normal system usage patterns. They are detected by building up a profile of the system being monitored, and detecting significant deviations from this profile. A basic assumption of anomaly intrusion is that attacks differ from normal behavior. However, anomalous activities that are not intrusive may be flagged as intrusive. Also, intrusive activities that are not anomalous result in false negatives.
- **Misuse** intrusions, which are well-defined attacks on known weak points of a network. They can be detected by looking for certain actions being performed on certain objects. Misuse detection systems essentially define what is wrong. The idea behind misuse detection schemes is that there are ways to represent attacks in the form of a pattern or a signature so that even variations of the same attack can be detected.

If the activities of network users are monitored, it is possible to find intrusion attempts by looking at activities that infer a certain intrusion scenario.

There are also two types of detection: statistical-anomaly and rules-based.

- Anomaly detection: Such detection tries to detect security breaches by analyzing audit-log data for abnormal user and system behavior. It detects misuse by measuring a norm over time. This requires that the organization first build profiles of normal user and network behavior to serve as a statistical base for intrusion detection. The main advantage of anomaly detection is that it does not require prior knowledge of security flaws in networks to detect possible intrusion. It can detect previously unknown attacks. However, anomaly detection systems are difficult to train in highly dynamic environments.
- Rules-based detection: Rules-based system searches for attack signatures in the network traffic. In this form of detection, organizations form models of high-level system state changes or audit-log events that occur during attacks. These models form rules bases. Thus, rules-based detection systems monitor logs and resources and search for models that match an attack profile. This requires that security experts or administrators regularly update the rules base to reflect newly discovered attack methods.

Vendors have developed hybrid products that combine the advantages of statistical-anomaly and rules-based intrusion-detection systems while avoiding some of their weaknesses.

An intrusion detection system (IDS) is a system for detecting such intrusions.

An **Intruction detection system** (IDS) is any hardware, software, or combination that monitors a network for malicious activity.

IDSs are based on the premise that an intruder's behavior will be noticeably different from that of a legitimate user and that many authorized actions are detectable. The goal of IDS is to identify unauthorized use, misuse, and abuse of the network by both internal and external penetrators.

IDSs monitor network state looking for unauthorized usage, denial of service, and anomalous behavior. If there are attacks on a network, we would like to detect them as soon as possible and take appropriate action. Just like a burglar alarm, an IDS remains dependent on a human operator to act on the data collected. An IDS does not usually take preventive measures when an attack is detected; it is a reactive rather than proactive device. An IDS plays the role of an informant rather than a police officer. An intrusion detection system (IDS) will run constantly on the system, working away in the background, and only notifying you when it detects something it considers suspicious or illegal. From the observations made over the system usage over time, the IDS will compute metrics about the system's overall state, and decide whether an intrusion is currently occurring. An IDS may also perform its own system monitoring. However, the task is difficult. In fact, IDS does not detect intrusions at all—it only identifies evidence of intrusions, either while they are taking place or after the fact.

The most common way to detect intrusions has been by using the audit data generated by the operating system, as illustrated in Figure 17.6. An audit trail is a record of activities on a system that are logged to a file in chronologically sorted order. Audit trails are particularly useful because they can be used to establish guilt of attackers, and they are often the only way to detect unauthorized user activity.

There are two types of IDSs: network IDS (NIDS) and host IDS (HIDS).

- Host-based IDS: HIDS monitors attacks at the operating system or application level. It has access to audit logs, error messages, service and application rights, and any resources available to the monitored host. It bases its decisions on information obtained from a single host. Since it relies on

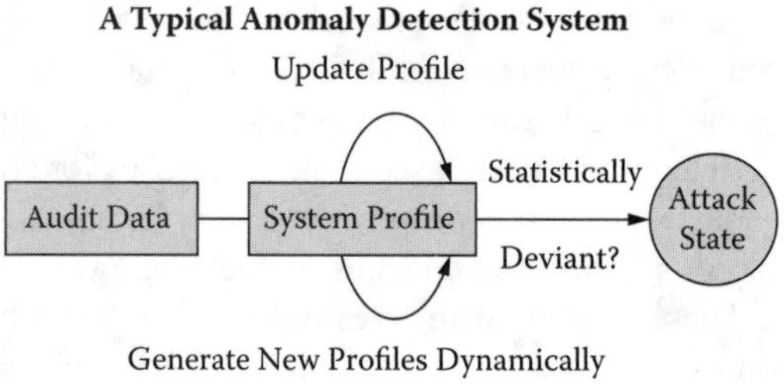

FIGURE 17.6 A typical anomaly detection system.

audit logs and error messages, it is basically detecting attacks after they have occurred.

- Network-based IDS: NIDS is usually placed at a strategic point in the network and monitors traffic as it flows to other hosts. It obtains data by monitoring the traffic information in the entire network. An NIDS is more cost effective than an HIDS because with one device, an NIDS can protect a large chunk of the network. The popularity of NIDSs has outpaced the acceptance of HIDSs.

Network security administrators and vendors now look beyond the traditional intrusion detection technology, which detects problems after they have occurred, to a more active approach: intrusion prevention. This is an emerging field and it is unclear which of many intrusion-prevention techniques will become most popular.

Example 17.3

The following scenarios are examples of intrusion[1]

- A student gains access to his/her academic records.
- A person leans over your shoulder to observe your password.
- An employee browses through his/her boss's employer reviews.
- A user exploits a flaw in a file server program to gain access to and then corrupt another user's file.
- An intruder uses a script to "crack" the passwords of other users on the network.
- An intruder installs a program on a computer to inspect network traffic.
- An intruder modifies the router table in a network to prevent the delivery of messages to a particular destination.
- Government intrusion on the rights of individuals.
- Invasion of individual rights by the intelligence community.

17.7 SECURITY STANDARDS

Network security is a never-ending battle between the network administrators and engineers and the computer hackers. In various attempts to keep the hackers out, engineers have developed several encryption algorithms, authentication protocols, and intrusion detection systems. Because security systems vary so widely, many believe that standards have become important. There are two popular standards for computer network security: IPSec and DES.

17.7.1 IPSec

IPSec (IP SECurity) refers to a set of standards developed by the Internet Engineering Task Force (IETF). It consists of network-layer protocols that allow two or more hosts

---

[1] Adapted from: N. J. Puketza et al., "A methodology for testing intrusion detection systems," *IEEE Trans. Software Engineering*, vol. 22, no. 11, Oct. 1996, pp. 719–729.

to communicate in a secure manner. IPSec solves two problems that have plagued the IP protocol suite for years: host-to-host authentication and encryption.

IPSec provides security for transmission of sensitive information over unprotected networks such as a LAN or the Internet. It acts at the network layer as shown in Figure 17.7, protecting and authenticating IP packets between participating IPSec devices.

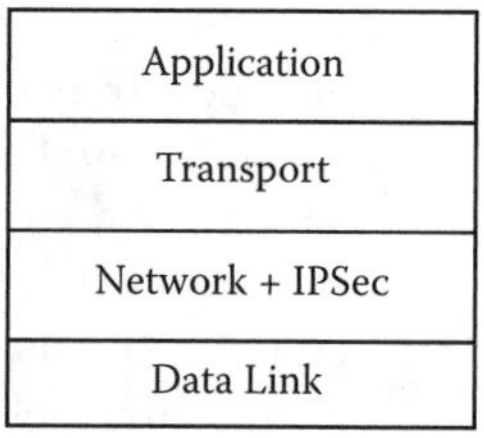

Application
Transport
Network + IPSec
Data Link

FIGURE 17.7 IPSec stack layering.

IPSec (IP Security) is a set of network layer protocols developed by the IETF to support secure exchange of packets at the IP layer.

IPSec protocols aim to address several problems: password sniffing, IP spoofing, session hijacking, and denial-of-service. IPSec provides the following optional network security services. In general, local security policy will dictate the use of one or more of these services:

- Data confidentiality: The IPSec sender can encrypt messages before transmitting them across a network.
- Data integrity: The IPSec receiver can authenticate messages sent by the IPSec sender to ensure that the data has not been altered during transmission.
- Data origin authentication: The IPSec receiver can authenticate the source of the IPSec messages sent. This service depends on the data integrity service.
- Anti-replay: The IPSec receiver can detect and reject replayed messages.

IPSec provides two choices of security service: authentication header (AH), which essentially allows authentication of the sender of data, and encapsulating security payload (ESP), which supports both authentication of the sender and encryption of data as well. The specific information associated with each of these services is inserted into the packet in a header that follows the IP packet header.

As mentioned earlier, IPSec consists of two sub-protocols:

- Authentication header (AH), which protects the IP packet header from third-party interference and spoofing, by computing a cryptographic checksum and hashing the IP packet header fields with a secure hashing function. This is then followed by an additional header that contains the hash, to allow the information in the packet to be authenticated. AH provides proof-of-data origin on received packets, data integrity, and antireplay protection. It defines an algorithm-independent mechanism for providing cryptographic authentication without encryption to IPv4 and IPv6 packets.
- Encapsulated security payload (ESP), which protects the IP packet data from third-party interference, by encrypting the contents. ESP provides all that AH provides in addition to optional data confidentiality and limited traffic flow confidentiality. It defines an algorithm-independent mechanism for providing encryption to IPv4 and IPv6 packets.

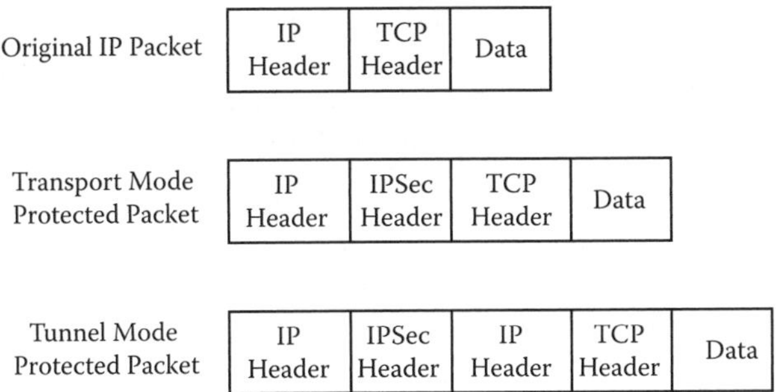

FIGURE 17.8 IP packets protected by IPSec in transport and tunnel modes.

ESP and AH can either be used together or separately, depending on the environment. In other words, you can configure both of them, or decide to use only one.

The IPSec protocols (AH and ESP) can be used to protect either an entire IP payload or the upper-layer protocols of an IP payload. This distinction results in two different modes of IPSec: transport and tunnel. The transport mode is used to protect upper-layer protocols. In this mode, an IPSec header is inserted between the IP header and the upper-layer protocol header, as shown in Figure 17.8. The tunnel model is to protect whole IP datagrams. In this mode, the entire IP packet is encapsulated in another IP datagram and the IPSec header is inserted between the inner and outer IP headers.

A big advantage of IPSec is that security arrangements can be handled without requiring changes to individual user computers. With standard-based IPSec, data can be transmitted across a LAN, private and public networks, and the Internet without fear of observation, modification, or spoofing. This enables applications such as virtual private networks (VPNs), including intranets, extranets, and remote user access.

17.7.2 DES

The data encryption standard (DES) was published by the U.S. National Bureau of Standards in 1977 (it has since been renamed the National Institute of Standards and Technology or NIST). This publication provides a complete description of a mathematical algorithm for encrypting (enciphering) and decrypting (deciphering) binary coded information. It is both a U.S. standard and a private sector standard. It has also been a worldwide standard through ISO.

DES is a symmetric algorithm since the same key is used by both the encryption and the decryption. DES is also a block cipher since it operates on blocks (64-bit wide) of data at a time. It acts on a fixed-length block of plaintext and converts it into a block of ciphertext of the same size by using the secret key. This structure leads naturally to the block diagram in Figure 17.9.

A key for the DES algorithm consists of 64 binary digits (0s and 1s) of which 56 bits are randomly generated and used directly by the algorithm. The other 8 bits, which are not used by the algorithm, are used for error detection. Hence the effective key length is only 56 bits. Authorized users of encrypted computer data must have

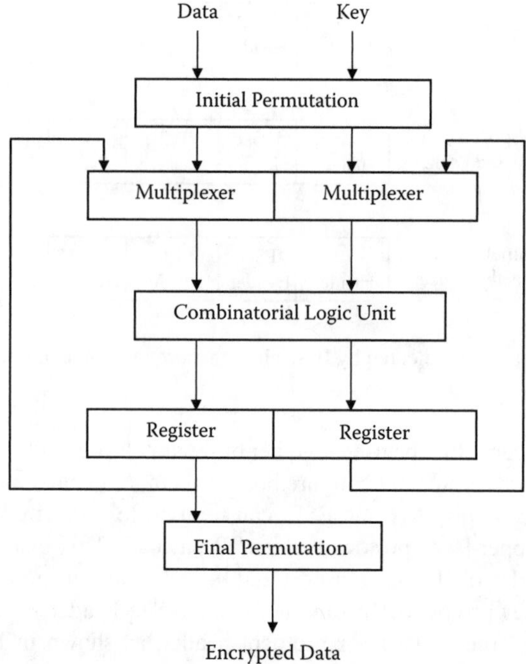

FIGURE 17. 9 DES block diagram.

the key that was used to encipher the data in order to decrypt it. Data can be recovered from the cipher only by using exactly the same key used to encipher it. To break the cipher would require calculating on average 2^{55} encryptions before the correct one is found and that would take forever.

The algorithm specified in this standard may be implemented in software, firmware, hardware, or any combination thereof. The specific implementation may depend on several factors such as the application, the environment, the technology used, etc. Since the publication of DES, NIST validates approximately three implementations each year. Some exports of cryptographic modules implementing this standard and technical data regarding them must comply with federal regulations and be licensed by the U.S. Department of State.

DES is no longer secure and 3DES (or triple-DES) is a temporary replacement. 3DES is a cryptosystem that can encrypt and decrypt data using a single secret key. In 3DES, we apply three stages of DES with a separate key for each stage, as shown in Figure 17.10, so the key length in 3DES is 168 bits. 3DES is much faster than public key cryptography methods like the RSA method. This is one of the main advantages of using a system like 3DES. However, the transmission of the secret key between users is unsafe. This is where public key cryptography excels.

As mentioned before, DES is no longer secure and 3DES is only a temporary arrangement. Hence NIST published the Advanced Encryption Standard (AES) in 2001. The AES now formally replaces 3DES as the encryption standard.

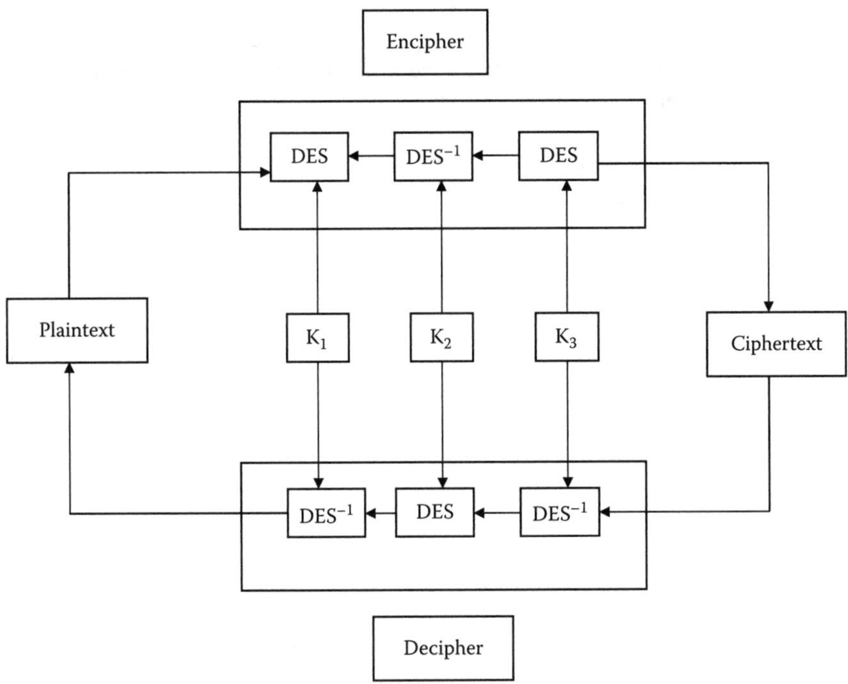

FIGURE 17.10 Triple-DES encryption and decryption.

SUMMARY

1. Network security is authorizing access to information on the network. It is necessary to protect resources on the network from losses of confidentiality, integrity, or availability.
2. Access control methods apply what one knows (password), one has (token), or one is (biometrics).
3. Cryptography involves encryption and decryption of messages to protect them. There are symmetric (or secret key) cryptography and asymmetric (or public key) cryptography.
4. Firewalls maintain privacy and ensure the authenticity of data communications that pass through their domain's boundaries.
5. Intrusion detection is a process that must be executed by system administrators in order to maintain secure networks. The goal of intrusion detection systems (IDSs) is to detect intrusion.
6. IPSec provides a standard and robust mechanism that helps to provide security to IP and upper-layer protocols (e.g., TCP and UDP).
7. The data encryption standard (DES) is an NIST standard for encryption. It has been replaced by the advanced encryption standard (AES).

REVIEW QUESTIONS

17.1 An email appears to have originated from one source when it was actually sent from another source. This is called:
  a. Email spamming
  b. Email spoofing
  c. Email hacking
  d. Email sniffing

17.2 A security requirement that ensures that certain information is never disclosed to unauthorized entities is:
  a. Confidentiality
  b. Integrity
  c. Authentication
  d. Nonrepudation

17.3 A security requirement that ensures that a message has not been modified in transit is:
  a. Confidentiality
  b. Integrity
  c. Authentication
  d. Nonrepudation

17.4 Firewall offers some protection from inside threats.
  a. True
  b. False

17.5 At all seven OSI layers, cryptographic mechanisms can be applied.
  a. True
  b. False

17.6 System flaws that exist in software are the root cause of network intrusions.
  a. True
  b. False

17.7 IPSec was developed by:
  a. ITU
  b. ANSI
  c. IEEE
  d. IETF

17.8 IPSec operates at what layer?
  a. Application
  b. Transport
  c. Network
  d. Data Link

17.9 It is impossible to build a system which is absolutely secure.
  a. True
  b. False

17.10 DES is a public or asymmetric encryption algorithm.
  a. True
  b. False

Answer: 17.1b, 17.2a, 17.3b, 17.4b, 17.5a, 17.6a, 17.7d, 17.8c, 17.9a, 17.10b

PROBLEMS

17.1 Explain what we mean by confidentiality, integrity, and authentication.

17.2 Discuss three methods used by intruders to gain access to computer networks.

17.3 Explain how authentication will thwart the man-in-the-middle attack.

17.4 How would you use something one knows, one has, or one is to authenticate a user's identity?

17.5 (a) Using the general Caesar algorithm with k = 3, obtain the cipher corresponding to the following plaintext:

MR WATSON I NEED YOUR HELP

(b) Repeat part (a) if k =5 .

17.6 Using the Caesar algorithm with k = 4, decrypt the following message:

MNYWXKSXQEVVMIH

17.7 Use the transposition matrix to encrypt the message:

F O O D I S G O O D F O R Y O U

17.8 Using modular arithmetic, evaluate the following:
a. 154 mod 13
b. 6571 mod 51
c. 32^5 mod 35
d. 22^{11} mod 65

17.9 Solve for X in each of the following:
a. 5X mod 13 = 1
b. 3X mod 17 = 2
c. 11X mod 46 = 1

17.10 Let p = 5 and q = 13, and use the RSA algorithm to encrypt P = 22, and then decrypt the encoded message. Select E = 11.

17.11 Explain the three types of firewalls.

17.12 Discuss the advantages and limitations of firewalls.

17.13 What is intrusion detection?

17.14 Explain the two major types of IDS.

17.15 What is IPSec designed to accomplish?

17.16 Why is a commercial data encryption standard necessary?

17.17 How does 3DES work?

17.18 IDEA, the International Data Encryption Algorithm, was developed in Zurich, Switzerland. Using Web resources, write a half-page summary of IDEA.

17.19 What is a digital signature? Using Web resources, find out about digital signature standard (DSS).

17.20 Using Web resources, write a half-page summary of AES (Advanced Encryption Standard).

18 Network Testing, Fault Tolerance, and Analysis

Science without religion is lame, religion without science is blind.

—Albert Einstein

Today's broadband network is heterogeneous in nature. The core network, access network, and the customer premises network are of different kinds and we follow the open system interconnection (OSI) model for interoperability and reliability issues. Interoperability is the key issue of internetworking for current service providers and vendors. Every vendor confirms interoperability of its product before it brings any product to market. These interoperability tests require rigorous testing, which requires workers and specific testing tools; and, hence, the budget for testing takes a large percentage of product development costs. Automating the testing process therefore helps to reduce cost. For each type of system, it is usually the practice that a forum is initiated and most of the vendors' members join the forum to finalize a standard testing report. Although it takes months and even years to make a standard, this is a necessary part of product development. Otherwise, developed products from different vendors become interoperable if not standardized.

Network analysis is another issue for service providers. The health of the network has ups and downs because of the high demands of bandwidth. It also depends on adding new services to the current network system. Network health depends almost all the time on the number of active lines (users). Before implementing any network, the network professionals analyze the current and future demands and propose network topology. They also have a plan for how the network will evolve with future growth of demand.

Fault tolerance is an important issue in broadband communication networks. The majority of broadband communication applications, from credit card transactions to cellular conversations, assume that a reliable network will always be available. Therefore, designing and sustaining a fault-tolerant broadband network covering several failure scenarios becomes a major issue.

In wireless broadband, testing is an integral part of the wireless network development. This includes both fixed and mobile broadband technologies. Other emerging wireless technologies such as EV-DO, HSDPA, and WiMax also require testing for interoperability and reliability of the networks.

There are several issues related to network testing. The following are some of the tests that can be performed to ensure better performance in a broadband network environment:

- ADSL modem interoperability test
- DSLAM (digital subscriber line access multiplexer) reliability tests
- point-to-point connection test over ATM and Ethernet

- remote area access network test
- developing ATM-based network and testing its different parameters
- VPN testing
- tests on interference, noise, signal, and the number of CPEs per base transceiver station (BTS)
- point-to-multipoint tests on broadband fixed wireless access networks
- wireless application tests

In this chapter, we will focus on ADSL modem interoperability and DSLAM reliability testing. The interoperability test is used to perform the reliability test. In this test, we use a single modem against different ports of DSLAM to check the performance of each port. This provides information on how reliable DSLAM ports are in the broadband network. We will also focus on wireless-related tests and the emerging WiMax technology. The rest of the chapter is designed as follows: Section 18.1 briefly describes the necessary tools for ADSL interoperability testing; Section 18.2 describes the test methods; Section 18.3 describes the telephone subscriber loop environment; Section 18.4 describes the point-to-multipoint broadband fixed wireless access networks; Section 18.5 describes the wireline broadband technology tests; Section 18.6 discusses fault tolerance issues; Section 18.7 discusses the PON testing issues; and finally, the summary of this chapter.

18.1 TESTING TOOLS FOR ADSL INTEROPERABILITY TESTS

Before testing any system we need to understand the architecture of that system. Depending on the architecture and services we need to choose the test tools and equipment. ADSL service is provided over traditional twisted telephone wires and is specifically designed to meet the demands of broadband communications for home users and small offices. The architecture of the ADSL system is shown below.

Figure 18.1 describes typical ADSL loop architecture. This shows the pot splitters connected at both ends (central office and customer premises) of the telephone line.

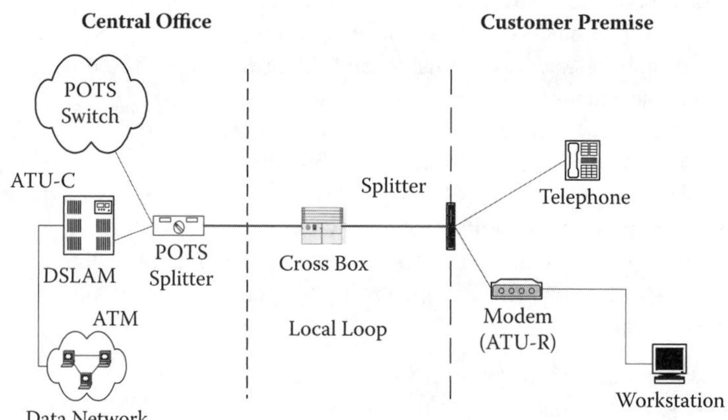

FIGURE 18.1 Typical ADSL loop architecture.

FIGURE 18.2 Stinger® DSL access multiplexer (DSLAM).

These splitters are low-pass filters. At the customer premises, the splitter passes a low-frequency signal to the telephone set and the high-frequency signals are passed to the ADSL modem. At the other side it follows the same procedure. At the central office the high-frequency signal or modem request connects to the central office's DSL modem pool. This is called digital subscriber line access multiplexer (DSLAM) and is shown in Figure 18.2. DSLAM is designed specially to handle broadband services and has connections to other core networks for broadband services. The Stinger DSLAM front and back panels are shown in Figure 18.3 and Figure 18.4, respectively.

This description shows an example of what we need for interoperability tests. We need to test various ADSL modems' interoperability with respect to DSLAM and the reliability test of DSLAM ports with respect to a single ADSL modem. If we bypass the cross-box, and splitters shown in Figure 18.1, then the important equipment we need to create the ADSL loop is DSLAM and the ADSL modem—the device under test (DUT). To connect these two pieces of equipment physically, we need a telephone wire. Since there are various scenarios of telephone loops, it is not feasible to use a live telephone loop for testing. That is why we need equipment that simulates various types of telephone loops. In the telecom industry, we call this equipment wireline simulator, and it is shown in Figure 18.5. To measure the throughput of different loop architectures with different parameters we also need a traffic generator, shown in Figure 18.6. Therefore, for complete ADSL interoperability testing we need DSLAM, DUT, wireline simulator (WLS), and traffic generator.

In our broadband laboratory at Prairie View A&M University (PVAMU), we used the Lucent Stinger for DSLAM, Sprirent's WLS, and Smartbit 200 for traffic generator. We describe each of these briefly in the following sections.

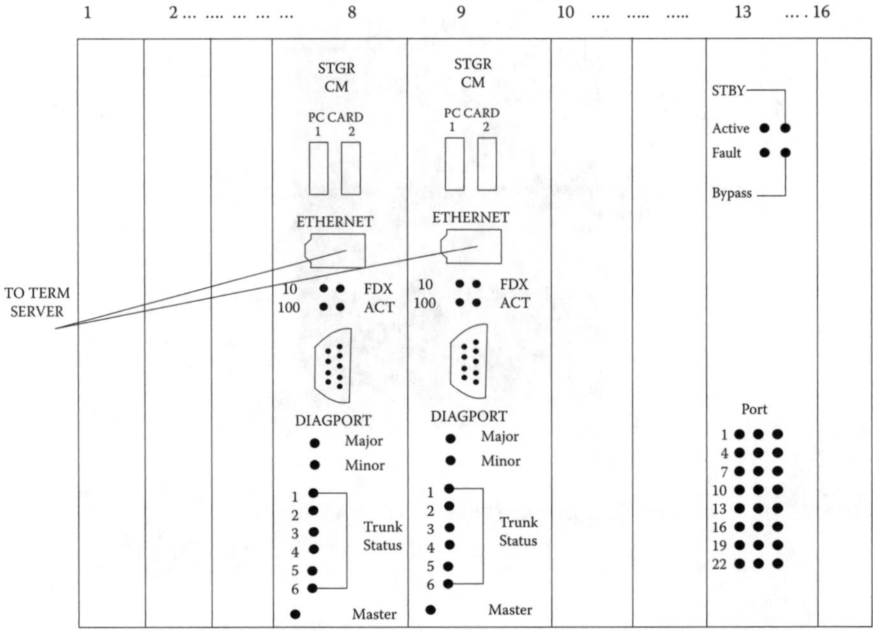

FIGURE 18.3 Stinger front panel.

18.1.1 Description of Test Equipment

DSLAM: Stinger

Vendor: Lucent

The Stinger architecture takes full advantage of ATM technologies, with integrated IP and flexible advanced queuing capabilities, transporting information at the highest possible speeds, with reliable Quality of Service (QoS), increased control over traffic flows, exceptional capacity, and reduced costs. Its modular design far exceeds any competitor in its ability to cut operating costs and allow rapid expansion—up to 31,000 connections via subtending. This modular approach also promotes cost-effective integration, enables complete system redundancy, and supports easy upgrades.

Wire Line Simulator: DL4-400IA

Vendor: Spirent Communications

A small variation in bit rate can mean a difference between passing and failing TR-048. Issues of noise floor, accuracy, and repeatability are key technical issues that must be considered. In response to this, Spirent has developed two wire line simulation solutions: the DLS 400IA for Cluster A (North America) and the DLS 400IAE for Cluster A & B (North America and Europe). These solutions offer all of

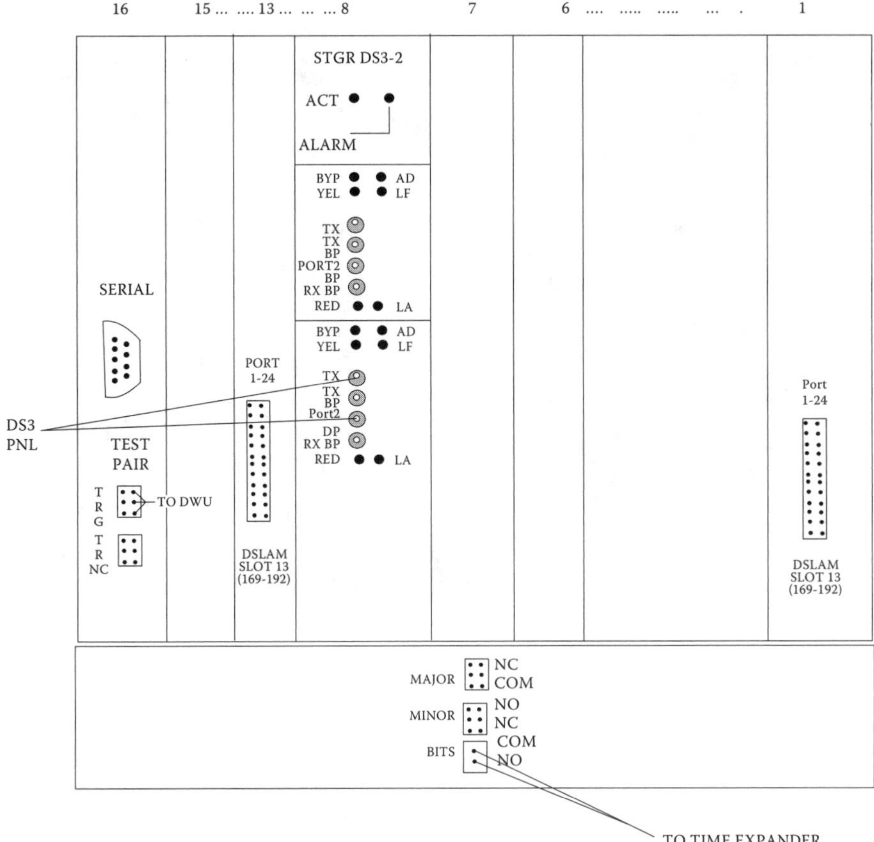

FIGURE 18.4 Stinger back panel.

the loops required for testing to TR-048 (proposed testing method by DSL Forum). Both the DLS 400IA and the DLS 400IAE include the latest technology in wire line simulation, both hardware and software. A key supporting product for the DLS 400IA/IAE is the DLS 1310 compensation and fine data rate adjustment engine, which ensures the simulator meets the accuracy, repeatability, and requirements specific to TR-048.

Traffic Generator: SMB-200 (Smartbit 200)

Vendor: Spirent Communications

The Smartbit 200 is a portable, lightweight, low-cost, four-port network performance analysis system. The SMB-200 is fully compatible with the larger SMB-2000 system, utilizing all of the same SmartCards and software applications. This generates/receives ATM traffic and Ethernet traffic. It comes with application software to measure network throughput and other parameters.

FIGURE 18.5 Wire line simulator (DL4-400IA).

18.2 TEST METHODS

The DSL Forum proposed test methods for testing ADSL interoperability tests. The test report discussed as an example is known as TR-048. The current test report is TR-069. We followed TR-048 in our laboratory at PVAMU to perform these tests. We performed a manual test at the beginning and found it very time consuming and we quickly came to the solution of performing the tests using an automated process. We developed an automated test process for this purpose using TCL scripts. The test plan 048 contains all the necessary scenarios of POT architecture and specific test requirements for each of the scenarios. It proposes different kinds of noise impairments such as white gaussian noise, HDSL noise, T1 line impairments, and other parameters that have influence on ADSL services.

TR-048 describes the physical layer tests, layer-2, and higher layer test cases. Asynchronous transfer mode (ATM), point-to-point protocol over ATM (PPPoA),

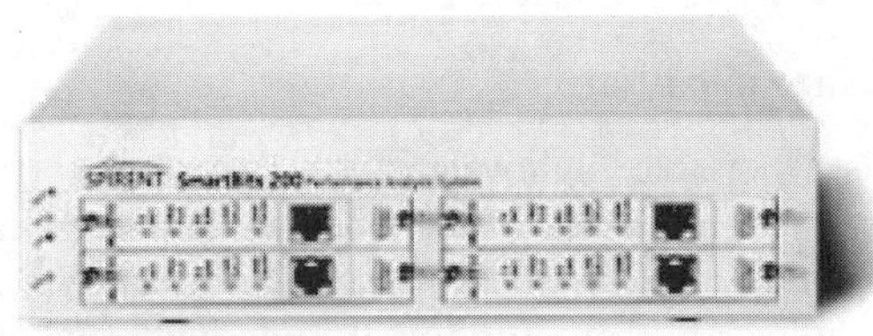

FIGURE 18.6 Traffic generator (SMB-200).

FIGURE 18.7 ADSL modems (DUT).

and point-to-point protocol over Ethernet (PPPoE) connectivity are tested for higher-layer test cases. Throughput and latency are measured to determine a modem's capability. These tests are significant because the performance of the ADSL line depends on the actual throughput. A user's activities such as Web browsing, downloading files, and multimedia applications using the ADSL connection depend on the throughput of the line.

We show the basic loops in Figure 18.8 that are defined by TR-048. Some loops are plain loops and some are with bridge taps. Bridge taps are nothing but unterminated loops. These cause reflections and distort data transmission. TR-048, which represents the test bed, is shown in Figure 18.9. We have used serial connections for controlling the wire line simulator, Telnet connection for DSLAM, and a traffic generator.

For the physical layer test, TR-048 proposes that a modem should synchronize with the DSLAM within 60 seconds. If it does not synchronize within this period, then the test for that instance fails. It also prescribes the sync rate for each instance of testing. If any test result does not meet the sync rate, then the result is considered a failure for that instance. This is followed for every physical layer test case. For the physical layer test, the specifications are as follows:

1. Loop tests with ports set for adaptive rate are done under the following conditions:
   - testing 0–18 kft with white noise only for adaptive sync rate
   - variable loop length test with 24 HDSL impairments and white gaussian noise
   - variable loop length test with 24 DSL impairments and white gaussian noise
   - variable loop length test with 5T1 adjacent binder impairment with white gaussian noise

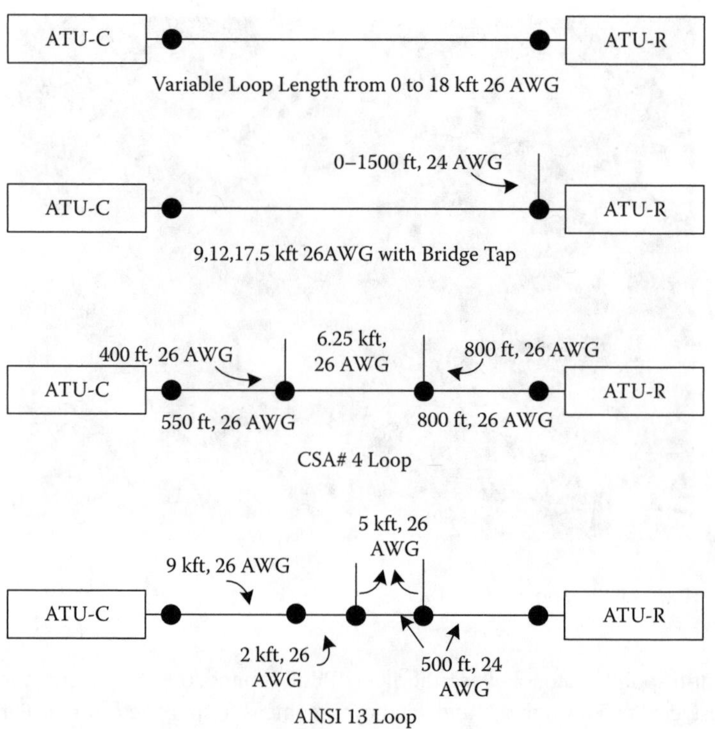

FIGURE 18.8 Variable loops, loops with bridge tabs, and standard loops CSA-4 and ANSI-13.

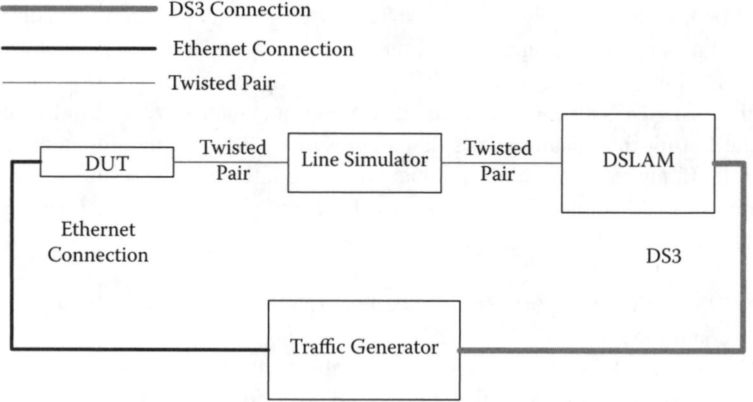

FIGURE 18.9 Basic testbed setup.

2. Loop tests with ports set for fixed rate (North American fixed rate tests):
   - testing 0–17.5 kft loops with white noise for fixed up and down sync rate
3. Full rate standard loop tests:
   - CSA #4 standard loop
   - ANSI 13 standard loop

   Noise impairments used for these tests are:
   - White noise (AWGN) impairment only
   - HDSL impairment
   - 24 DSL impairment
   - 5T1 adjacent binder impairment
4. Bridged tap (BT) tests:
   - 9, 12, and 17.5 kft loop with variable BT from 0–1500 ft with white noise.
5. ADSL functionality tests
6. DSL noise spikes/surges tests
7. Operation in the presence of impulse noise events

The higher layer test needs a traffic generator to measure the throughput of the modem with respect to the loop length, frame size, etc. The throughput test determines the maximum transmission rate at which a modem can forward traffic with no frame loss, or a user-defined acceptable frame loss. By increasing the transmission rate at specified levels, we can determine the modem's throughput. We configured the traffic generator to generate ATM traffic and the traffic is received via DSLAM and modem to the Ethernet port of the generator and vice versa.

Two types of tests are performed, one for variable loop length with adaptive rate with three frame sizes and one for fixed frame size for a standard loop.

For this test, we need to configure the test equipment specifically for the DSLAM before starting the tests. We could set up everything by automation or manually. We show in the next sections how to systematically configure the DSLAM. See also appendices A and B for more details.

18.2.1 DSLAM Configuration

We need to configure a virtual circuit (VC) between a DSLAM port and a modem port. This connection uses twisted telephone wire. To create the VC, we need to enable a DSLAM port (the DSLAM we use has 24 ports in one slot). Then we create the VC and name it Test_line (for example). We configure the various parameters of the VC according to the test plan. We show the process step by step in Figure 18.10. Appendices A and B show different configuration schemes that can be used for the testing loop.

Connect to Stinger

There are two ways to connect to Stinger DSLAM—via Telnet or via serial port. No matter which kind of connection is used, login information is required. We used a Telnet connection to connect to DSLAM.

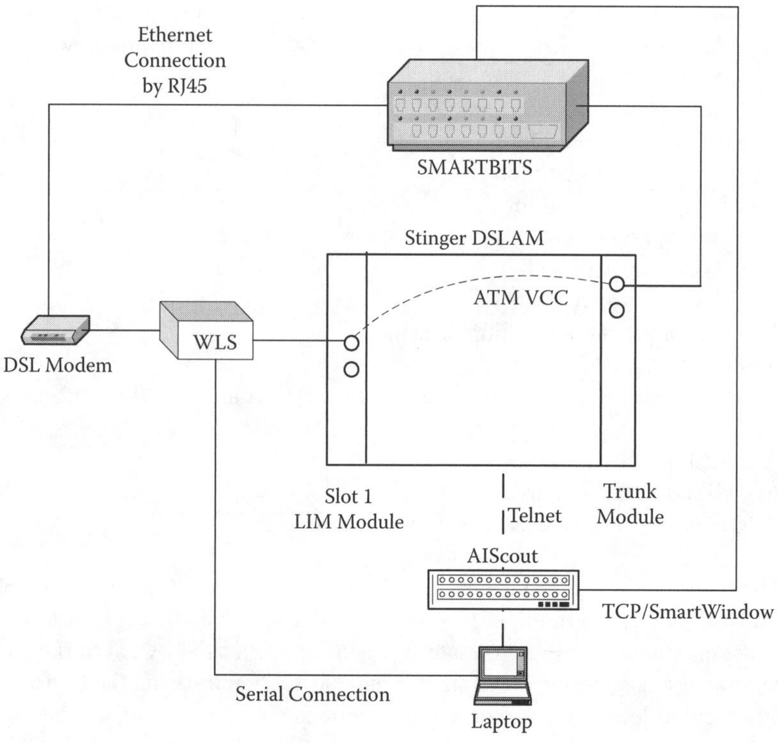

FIGURE 18.10 Detailed connection scheme.

From the command prompt we write: C:\>telnet 172.27.134.20 (172.27.134.20 is the IP address of the DSLAM we set before) as shown in Figure 18.11.

The Telnet process will ask for a user name and password. Input the user name as "admin" and the password as *** (not given here) to log in. Then the next screen

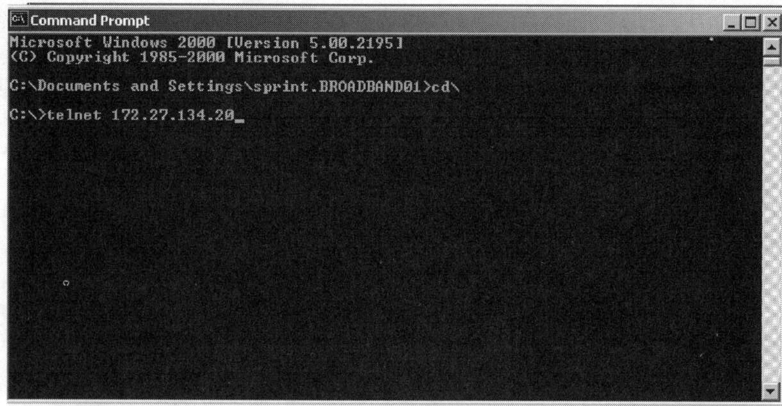

FIGURE 18.11 Telnet to DSLAM.

FIGURE 18.12 Access to DSLAM as admin user.

will appear (Figure 18.12). DSLAM uses "profile" for any setting. A profile is a group of configuration parameters related to a particular configurable entity. It can be considered as a template. Each configurable entity is an instance of the template. The index of the profile is used to distinguish between instances. In order to configure an entity, the profile with index should first be read into the workspace. The entire successive configuration is done on the working profile. A profile is described as {shelf -#, slot- #, port #}. For an example, for shelf 1, slot 1, and port 1 we write {shelf-1 slot-1 1} or simply {1:1:1}.

al-dmt is a profile of the ADSL line. To see all the ADSL lines (active or down), we check the ports by the command "dmtal –a." We see port 3 is down as shown in Figure 18.13. We can use port 3 to create a new ADSL line. The others are active ports.

18.2.2 LIM Port Configuration

A line interface module (LIM port) needs to be configured for physical connection to a CPE modem. The port needs to be enabled and other parameters like upstream and downstream data rate, rate adaptive/fixed rate, interleaved/fast mode, etc. need to be set. Read the profile in the workspace before reviewing or changing parameters in it. Then type "list" to show the parameters. We use "read al-dmt {1 1 3}" and then "list" for these operations as shown in Figure 18.14. This shows port 3 is not enabled. We enable it by the command "set." Then to save the configuration we must pass the command "write" or "wr" as shown in Figure 18.15.

In addition to parameters, a profile can also contain subprofiles. In the example shown in Figure 18.14, "name" and "enabled" are simple parameters. "line-config" and "fast-path-config" are subprofiles, which have more than one parameter in them. We see the content of a subprofile using the "list *subprofile_name*" command. In Figure 18.16, we showed the subprofile of "line-config" by the command "list line-config."

```
Command Prompt - telnet 172.27.134.20
admin> dmtal -a

All DMT ADSL lines:
                            <dvOp    dvUpSt  dvRq    sAdm    nailg>
  Line     <    1  1  1 >   <Up       Assign  UP      UP      00001>
  Line     <    1  1  2 >   <Up       Assign  UP      UP      00002>
  Line     <    1  1  3 >   <Down     Idle    DOWN    DOWN    00003>
  Line     <    1  1  4 >   <Up       Idle    UP      UP      00004>
  Line     <    1  1  5 >   <Up       Idle    UP      UP      00005>
  Line     <    1  1  6 >   <Up       Idle    UP      UP      00006>
  Line     <    1  1  7 >   <Up       Idle    UP      UP      00007>
  Line     <    1  1  8 >   <Up       Idle    UP      UP      00008>
  Line     <    1  1  9 >   <Up       Idle    UP      UP      00009>
  Line     <    1  1 10 >   <Up       Idle    UP      UP      00010>
  Line     <    1  1 11 >   <Up       Idle    UP      UP      00011>
  Line     <    1  1 12 >   <Up       Idle    UP      UP      00012>
  Line     <    1  1 13 >   <Up       Idle    UP      UP      00013>
  Line     <    1  1 14 >   <Up       Idle    UP      UP      00014>
  Line     <    1  1 15 >   <Up       Idle    UP      UP      00015>
  Line     <    1  1 16 >   <Up       Idle    UP      UP      00016>
  Line     <    1  1 17 >   <Up       Idle    UP      UP      00017>
  Line     <    1  1 18 >   <Up       Idle    UP      UP      00018>
  Line     <    1  1 19 >   <Up       Idle    UP      UP      00019>
  Line     <    1  1 20 >   <Up       Idle    UP      UP      00020>
  Line     <    1  1 21 >   <Up       Idle    UP      UP      00021>
[More? <ret>=next entry, <sp>=next page, <^C>=abort]_
```

FIGURE 18.13 Checking DMT ADSL line status.

```
Command Prompt - telnet 172.27.134.20
admin> read al-dmt {1 1 3}
AL-DMT/{ shelf-1 slot-1 3 } read
admin> list
[in AL-DMT/{ shelf-1 slot-1 3 }]
name = 1:1:3
physical-address* = { shelf-1 slot-1 3 }
enabled = no
sparing-mode = inactive
ignore-lineup = system-defined
line-config = { 3 15 automatic-at-startup automatic-at-startup 100 100 13 20 40+
fast-path-config = { 32 32 1000 8000 512 1000 }
interleave-path-config = { 32 32 1000 8000 512 1000 16 16 }
margin-config = { 6 6 6 6 31 31 0 0 0 0 0 0 0 0 no }
thresh-profile = default
bin-loading-profile = default

admin> _
```

FIGURE 18.14 Reading and listing profile of shelf-1, slot-1, port-3.

```
Command Prompt - telnet 172.27.134.20
admin> set enabled=yes
admin> wr
AL-DMT/{ shelf-1 slot-1 3 } written
admin>
```

FIGURE 18.15 Enabling port and saving configuration.

```
Command Prompt - telnet 172.27.134.20

admin> read al-dmt {1 1 3}
AL-DMT/{ shelf-1 slot-1 3 } read
admin> list line-config
[in AL-DMT/{ shelf-1 slot-1 3 }:line-config]
nailed-group = 3
vp-switching-vpi = 15
rate-adapt-mode-up = automatic-at-startup
rate-adapt-mode-down = automatic-at-startup
rate-adapt-ratio-up = 100
rate-adapt-ratio-down = 100
max-aggr-power-level-up = 13
max-aggr-power-level-down = 20
max-power-spectral-density = 40
line-code = auto-select
line-latency-down = fast
line-latency-up = fast
trellis-encoding = yes
gain-default = 20-db
upstream-start-bin = 6
upstream-end-bin = 31
downstream-start-bin = 32
downstream-end-bin = 255
loop-back = none
bit-swapping = no
fbm-dbm-mode = fbm
```

FIGURE 18.16 Listing subprofile of line configuration.

We need to change different parameters of "line-config," "fast-path-config," and "interleave-path-config" (see Figure 18.14) subprofiles according to the test plan. We use the basic command "set parameter name = value" to change the parameter value of those subprofiles. We must pass the command "wr" to change those permanently. We need to change the parameters based on the specific test to be performed.

These parameters are in general:

From subprofile "line-config":
- line-latency-down = fast or interleave
- line-latency-up = fast or interleave

From subprofile "fast-path-config" and "interleave-path-config":
- min-bitrate-up = value (for example 32, all bit rates are in bit)
- min-bitrate-down = value (e.g., 32)
- max-bitrate-up = value (e.g., 1000)
- max-bitrate-down = value (e.g., 8000)
- planned-bitrate-up = value (e.g., 256)
- planned-bitrate-down = value (e.g., 256)

From subprofile "margin-config":
- target-noise-margin-up = 6
- target-noise-margin-down = 6
- min-noise-margin-up = 6
- min-noise-margin-down = 6
- max-add-noise-margin-up = 31
- max-add-noise-margin-down = 31

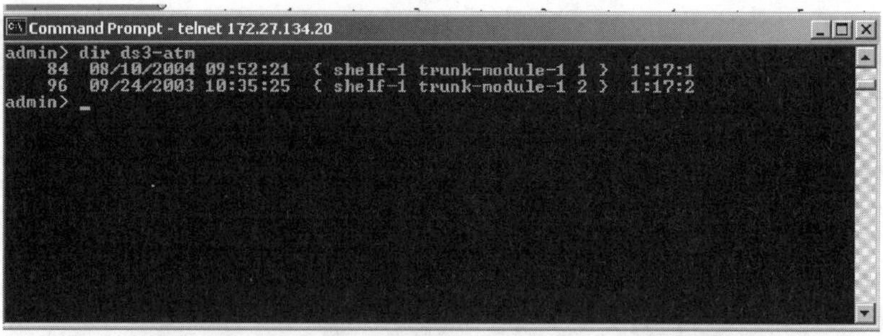

```
Command Prompt - telnet 172.27.134.20
admin> dir ds3-atm
    84  08/10/2004 09:52:21  { shelf-1 trunk-module-1 1 }  1:17:1
    96  09/24/2003 10:35:25  { shelf-1 trunk-module-1 2 }  1:17:2
admin> _
```

FIGURE 18.17 Checking index of trunk port.

18.2.3 Trunk Port Configurations

After doing these basic configurations, our LIM port 3 is ready for ADSL connection. Now we need to configure the trunk port. We have two DS3 trunk ports, which aggregate all the ADSL traffic and feed to the ATM switch in general. Check the index of the DS-3 trunk module by the command "dir ds3-atm" as shown in Figure 18.17. This shows the DS3 trunk module 1 at slot 17 with two ports, 1 and 2. We need to read the trunk ports. Let us assume we will use port 1. Then let us read port 1 by the command "read ds3-atm {1 17 1}." If this port is active then we do not need to activate it again, as shown in Figure 18.18.

By the command "set enabled=yes," we can enable the port and by the command "wr" we change the configuration permanently as shown in Figure 18.19. Now we need to connect trunk port 1 with LIM port 3 logically so that traffic can flow between these two ports. To do this, we need to see the nailed group of each individual port. To see this we use the command "which –n {shell# slot# port#}" as shown in Figure 18.20. Figure 18.20 shows that LIM port 3 is nailed with 3 and trunk port 1 is nailed with 801.

18.2.4 New Connection Setup

Now we will create a new connection for ADSL service. This is the last step. Create a new connection named "test_connection3" as shown in Figure 18.21.

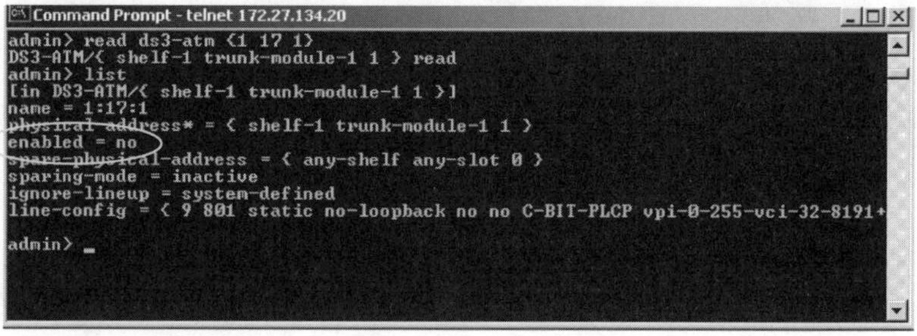

```
Command Prompt - telnet 172.27.134.20
admin> read ds3-atm {1 17 1}
DS3-ATM/{ shelf-1 trunk-module-1 1 } read
admin> list
[in DS3-ATM/{ shelf-1 trunk-module-1 1 }]
name = 1:17:1
physical-address* = { shelf-1 trunk-module-1 1 }
enabled = no
spare-physical-address = { any-shelf any-slot 0 }
sparing-mode = inactive
ignore-lineup = system-defined
line-config = { 9 801 static no-loopback no no C-BIT-PLCP vpi-0-255-vci-32-8191+

admin> _
```

FIGURE 18.18 Checking trunk port 1.

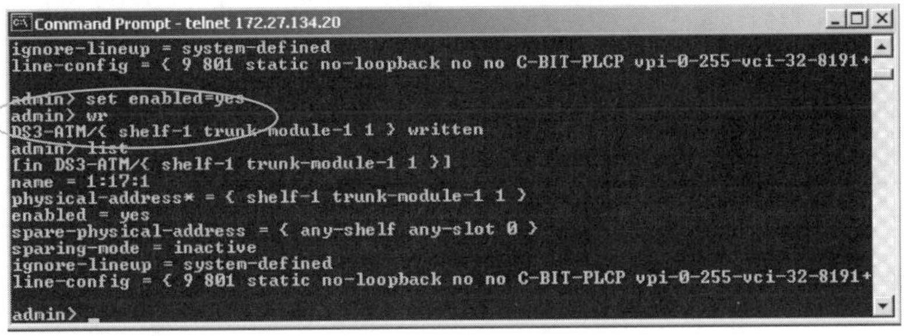

```
Command Prompt - telnet 172.27.134.20
ignore-lineup = system-defined
line-config = < 9 801 static no-loopback no no C-BIT-PLCP vpi-0-255-vci-32-8191+

admin> set enabled=yes
admin> wr
DS3-ATM/< shelf-1 trunk-module-1 1 > written
admin> list
[in DS3-ATM/< shelf-1 trunk-module-1 1 >]
name = 1:17:1
physical-address* = < shelf-1 trunk-module-1 1 >
enabled = yes
spare-physical-address = < any-shelf any-slot 0 >
sparing-mode = inactive
ignore-lineup = system-defined
line-config = < 9 801 static no-loopback no no C-BIT-PLCP vpi-0-255-vci-32-8191+

admin> _
```

FIGURE 18.19 Enabling trunk port 1.

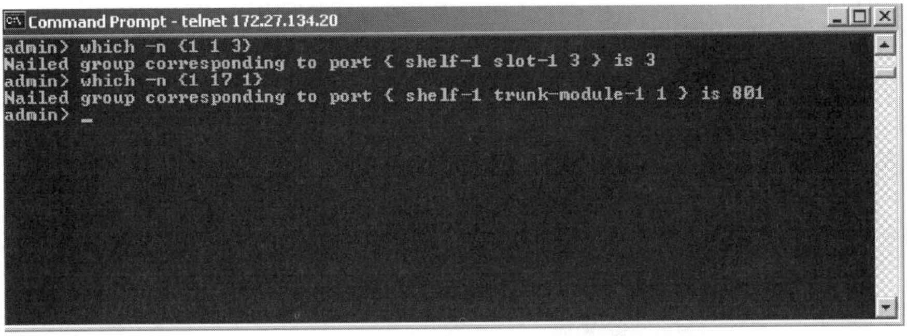

```
Command Prompt - telnet 172.27.134.20
admin> which -n (1 1 3)
Nailed group corresponding to port < shelf-1 slot-1 3 > is 3
admin> which -n (1 17 1)
Nailed group corresponding to port < shelf-1 trunk-module-1 1 > is 801
admin> _
```

FIGURE 18.20 Checking nailed group for individual port.

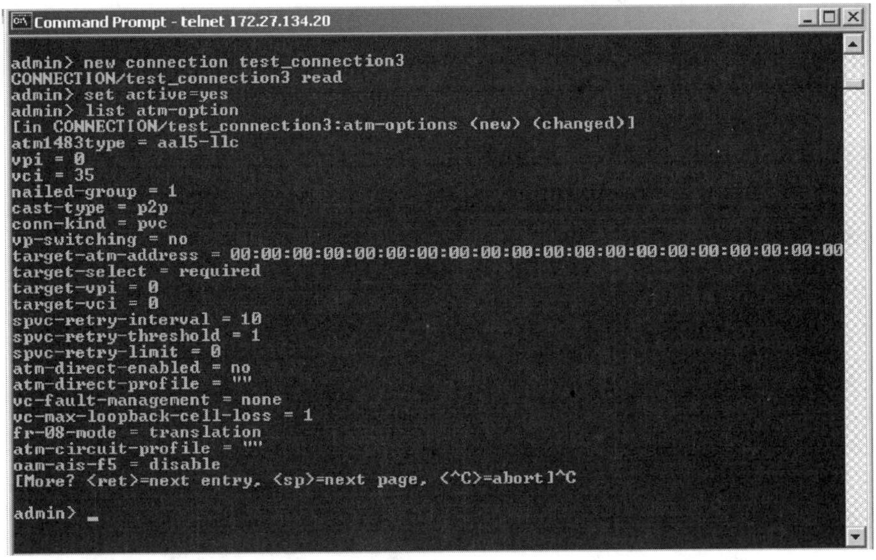

```
Command Prompt - telnet 172.27.134.20

admin> new connection test_connection3
CONNECTION/test_connection3 read
admin> set active=yes
admin> list atm-option
[in CONNECTION/test_connection3:atm-options (new) (changed)]
atm1483type = aal5-llc
vpi = 0
vci = 35
nailed-group = 1
cast-type = p2p
conn-kind = pvc
vp-switching = no
target-atm-address = 00:00:00:00:00:00:00:00:00:00:00:00:00:00:00:00:00:00:00:00
target-select = required
target-vpi = 0
target-vci = 0
spvc-retry-interval = 10
spvc-retry-threshold = 1
spvc-retry-limit = 0
atm-direct-enabled = no
atm-direct-profile = ""
vc-fault-management = none
vc-max-loopback-cell-loss = 1
fr-08-mode = translation
atm-circuit-profile = ""
oam-ais-f5 = disable
[More? <ret>=next entry, <sp>=next page, <^C>=abort]^C

admin> _
```

FIGURE 18.21 Creating new connection for ADSL service.

We need to change the different parameters as follows:

```
admin> set active=yes
admin> set atm-options nailed-group=3
admin> set atm-option vpi=0
admin> set atm-option vci=35
admin> set atm-connect-option nailed-group=801
admin> set atm-connect-option vpi=0
admin> set atm-connect-option vci=103
admin> wr
```

This ends our initial configuration of DSLAM. But we often need to change and view the different parameters of the subprofile when we run the tests. When we run the tests we need to check the sync rate (up and down) and noise level (up and down) for each test instance. To retrieve the data we pass the commands:

```
"read al-dmt-stat {1 1 1}\r"
"list physical-status up-stream-rate-fast\r"
"read al-dmt-stat {1 1 1}\r"
"list physical-status down-stream-rate-fast\r"
"read al-dmt-stat {1 1 1}\r"
"list physical-statistic noise-margin-up\r"
"read al-dmt-stat {1 1 1}\r"
"list physical-statistic noise-margin-down\r"
```

18.2.5 Configuration of Wire Line Simulator (WLS)

Wire Line Simulator (DLS 400HN):

We have a GUI for running the WLS, shown in Figure 18.22. This software allows you to change the different parameters required to test the ADSL interoperability. Everything is self-explanatory and hence students learn by playing with the software.

18.2.6 Traffic Generator Configuration (SMB 200)

SMB 200 has GUI. A handout is attached for handling the traffic generator SMB 200. To start the test you must systematically follow the test plan TR-048.

18.3 TELEPHONE SUBSCRIBER LOOP ENVIRONMENT

The term loop refers to the twisted-pair telephone line from the CO to the customer. Figure 18.23 shows a general telephone subscriber loop environment. Larger COs can serve up to 100,000 telephone lines. Feeder plant cables lead from the CO to the serving area interface (feeder distribution interface), which serves 1500 to 3000 lines. Distribution cables contain 25 to 1000 pairs. For residential and small business areas, the distribution cables lead to drop wires that serve each customer. The feeder and the distribution cables are bundled into binder groups of 25, 50, or 100 pairs. The pairs within a binder group remain adjacent to each other for the length of the cable.

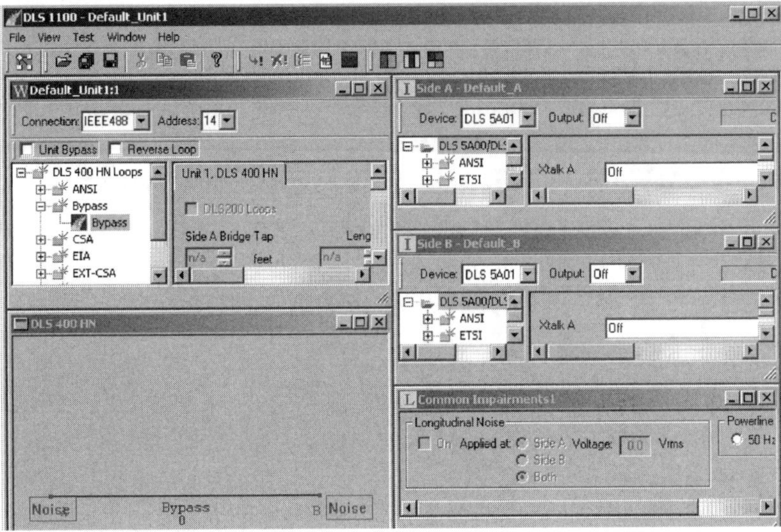

FIGURE 18.22 GUI for WLS.

As a result, the cross talk of pairs within a binder group is somewhat greater than cross talk between pairs in different binder groups.

A bridge tap is an unterminated wire and approximately 80% of loops in the United States have bridge taps. There are also series inductors in some loops, called loaded loops. A loop plant requires designing it a little bit differently for serving telephone service and DSL. Distance is a major factor for serving DSL. Shorter loop length achieves higher bit rate. Series inductors need to be removed because DSL does not operate on loaded loops. With the bridge taps in the loop, the service tends to be worse. In a binder group, there might be T1, ISDN, or HDSL disturbers and hence ADSL must cope with

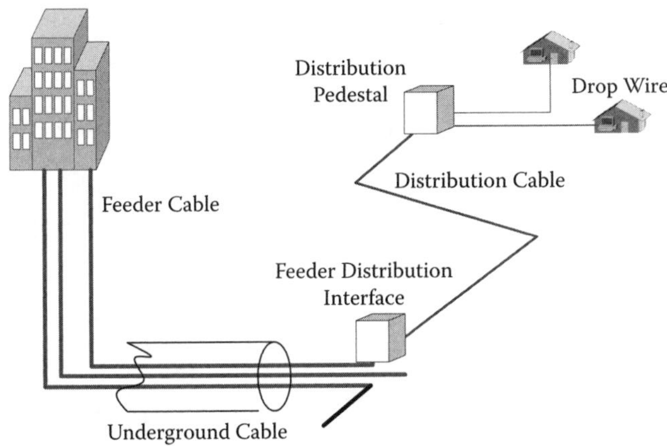

FIGURE 18.23 Telephone subscriber loop environment.

these noise impairments, length distribution, bridge tap, and other factors like noise spike, etc. To provide DSL services, a loop must be tested before deploying the service. The parameters that must be tested are open/short circuit, split pair, load coil, bridged tap, cable type/gauge, cable length, pulse velocity vector, blind spot, and impedance.

18.4 BROADBAND FIXED WIRELESS ACCESS NETWORKS

In this section, the testing of the broadband fixed wireless access networks is discussed.

18.4.1 Point-to-Multipoint Broadband Fixed Wireless Access Networks

Fixed wireless access network is seen as a strong business case for the last mile of broadband wireless access for homes and office buildings due to the increasing demand for Internet multimedia (voice, data, and video solutions). Currently, IEEE 802.11b is the most widely used indoor WLAN standard and has been extended to cover extended coverage areas by using point-to-multipoint access using suitable amplifier and antennas. Fixed wireless technology is promising because it allows the rapid deployment of new access networks for point to multipoint or for point to point and make it a feasible option for terrain-challenged and rural areas. This model provides high data rate and saves cost. Some broadband telecommunication corporations such as Sprint/Nextel have already initiated this service (Wi-Fi) in limited geographic regions but to spread widely, rigorous tests are required. There are issues with performance degradation due to noise, interference and fading effects, and finally throughput due to the distance of CPE to the base station. Figure 18.24 shows a typical network architecture for a point-to-multipoint fixed wireless access network.

Tests on interference, noise, signal, and the number of CPEs per BTS (base transceiver station) can be performed in a controlled RF (radio frequency) environment

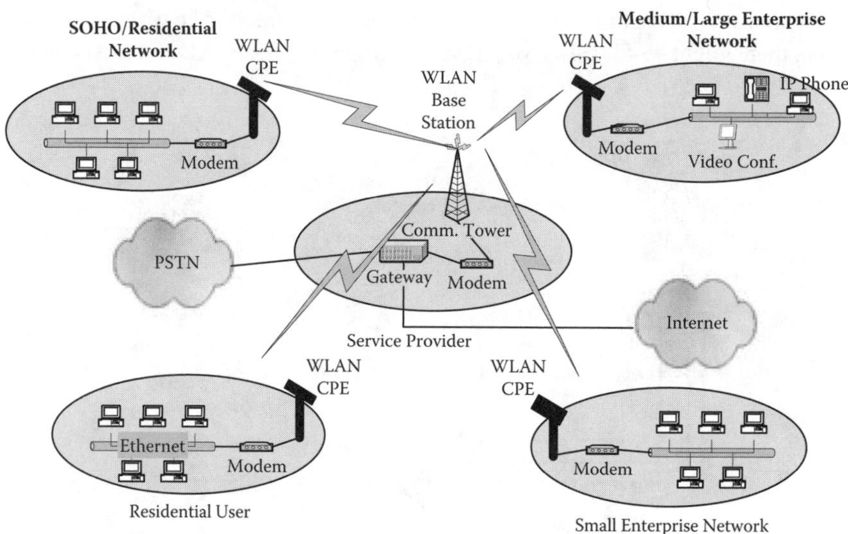

FIGURE 18.24 Network architecture using IEEE 802.11b WLAN access links.

to evaluate performance at various traffic loads when the SNIR (signal-to-noise and interference ratio) level is below a certain threshold. The findings can be used as a guideline in performing RF engineering for a service provider access network using WLANs. Figure 18.25 shows a general configuration for testing performance in the lab environment.

In this test procedure, the BTS and the CPEs are interconnected by coaxial cables and can be tested for a specific channel with specified frequency. A signal generator is used to create interference and noise. The effect of this interference and noise can be used to study throughput, packet loss, and key parameters of VoIP performance such as delay, jitter, and packet loss.

18.4.2 WiMAX Technology

WiMAX is an OFDM/IP-based wireless technology and allows higher data rates at lower cost. It has two standards: fixed WiMAX: 802.16d and mobile WiMAX: 802.16e. Table 18.1 shows a comparison for various wireless technologies.

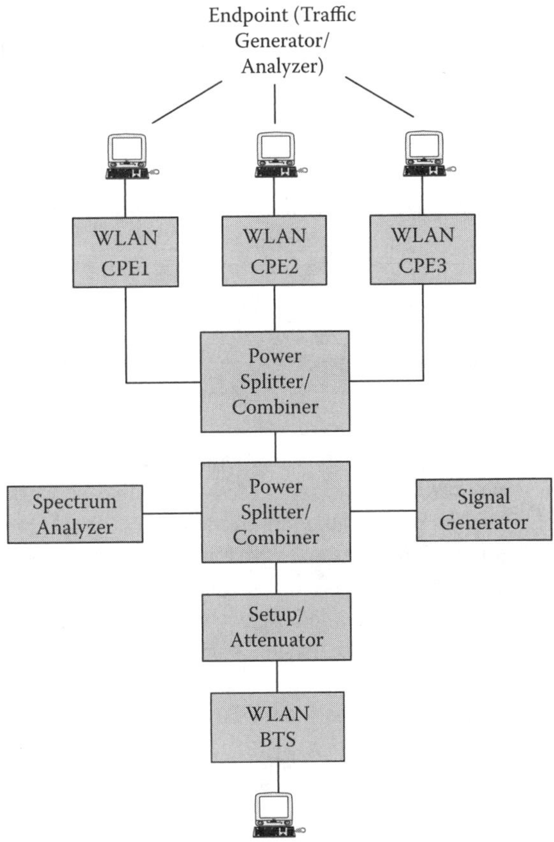

FIGURE 18.25 Laboratory test setup.

TABLE 18.1
Comparison of EV-DO, HSDPA, and WiMAX 802.16e Technologies

	EV-DO	HSDPA	WiMAX802.16e
Completed	In market	2006	Estimate 2nd Half 2006
Spectrum	CDMA 2000/1xEV-DO 400, 800, 1700, 1800, 1900, and 2100 MHz	WCDMA/UMTS 1800, 1900, and 2100 MHz	<11 GHz (<6 GHz practical) licensed (and unlicensed)
Channel Conditions	Non line of sight	Non line of sight	Non line of sight Grant-request-based MAC Enables QoS
Peak Raw Data Rate	Up to 2.5 Mbps CDMA 2000/1xEV-DO channel BW depends on technology	Up to 10 Mbps HSDPA Up to 2 Mbps WCDMA/UMTS	Up to 75 Mbps with 20 MHz BW channel. Full mobility: 80% performance of fixed usage model. Flexible channel BWs between 1.25 and 20 MHz.
Modulation	QPSK, 16 QAM, 64 QAM, others	QPSK, 16 QAM, 64 QAM, others	Scalable OFDMA (128-2k) QPSK, 16 QAM, 64 QAM
Mobility	Regional roaming	Regional roaming	Nomadic pedestrian; regional roaming
Typical Cell Radius	1–5 mi (7 km)	1–5 mi (7 km)	1–5 km (indoors) 2–7 km (outdoors)

Source: Mobile WiMax - A Complementary or Competitive Technology? Bechtel Corporation, 2006.

Sprint/Nextel announced in August 2006 that it is building a $3 billion nationwide mobile WiMAX data network and expects to start providing WiMaX service by late 2007. Sprint/Nextel calls it a "vision." It is essential that wireless devices are tested and pass defined criteria before service providers roll out those devices in the market. Obtaining certificates for the devices in the laboratory requires only a limited amount of testing, which is not sufficient for those used in the live network. Live networks operate differently from the perfect laboratory environment in which the devices have been developed. That is why field tests are performed vigorously, but it is costly and time consuming. To overcome these issues a wireless network simulation can provide alternative solutions to carry out detailed, repeatable tests in the laboratory. By creating or importing real-life network scenarios, some of these tests can be performed without the field trials.

For successful deployment, the following issues must be tested:

Performance characterization: capacity, reliability, susceptibility to various types of interferences, and sometimes comparisons to similar products to provide performance statistics due to various network scenarios.

Protocol conformance test (PCT): Provide preliminary assurance that the product will operate with specification-conforming products of other manufacturers.

Radio conformance test (RCT): A radio-interface conformance test as

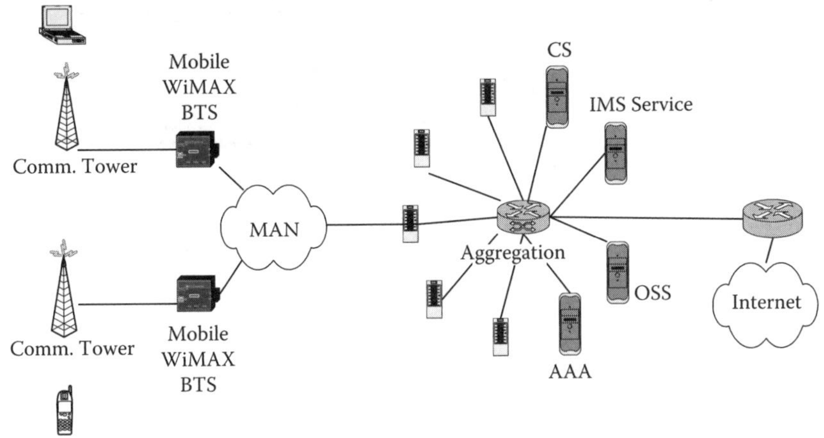

FIGURE 18.26 Mobile WiMAX network architecture.

specified by WiMAX RCT document that wireless MAN-OFDM and wireless HUMAN (-OFDM) devices must comply with.

Interoperability test (IOT): Confirms the ability of like products from different manufacturers to work together in a network and identifies incompatibilities. Need continuous improvement of the quality of interoperability testing. Equipment used for interoperability must receive a passing grade.

The WiMAX Forum continuously upgrades the certification processes. Figure 18.26 and Figure 18.27 show a general WiMAX network diagram while Figure 18.28 and Figure 18.29 show a typical lab test environment for transmitter and receiver.

18.4.3 Wireless Applications Test

Wireless applications are becoming increasingly popular. Service providers are tagging various wireless applications on top of their basic services. It is crucial that

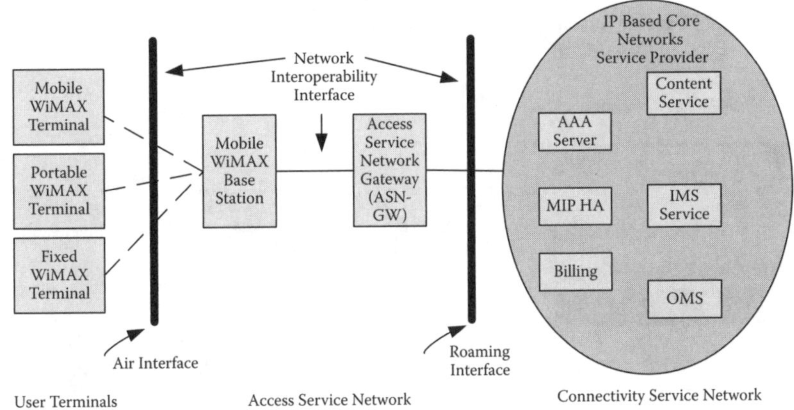

FIGURE 18.27 WiMAX network IP-based architecture.

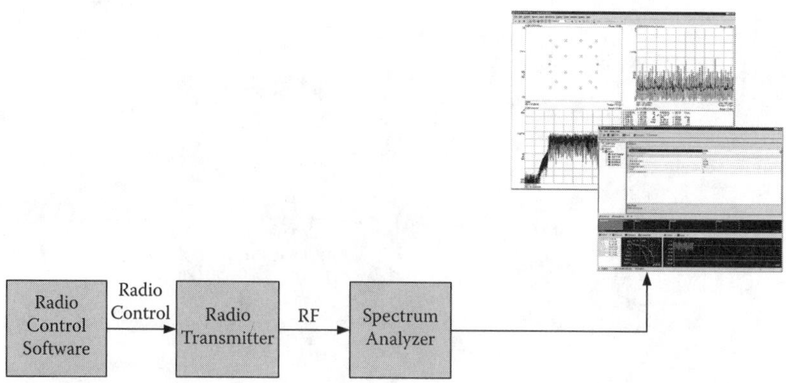

FIGURE 18.28 Transmitter measurement.

the applications work seamlessly without any interruption. Therefore, it is important that the applications for the various real network scenarios be tested. Wireless and cellular networks introduce a unique set of impairments that can dramatically affect an application's performance and functionality. These impairments include differences in bandwidth constraints as users move closer or further away from wireless base stations, connection problems as users pass from one cell to another, and performance issues that occur when users lose "line-of-sight" access to the base station. Wireless and cellular network impairments present a completely new set of challenges to developers, quality assurance professionals, and network designers. At the same time, overcoming these challenges has become increasingly critical as people become more dependent on wireless services to stay productive and meet business needs.

A virtual wireless environment can provide testing, planning, and troubleshooting solutions. This solution can save time and benefit costwise. With the simulated network, it is possible to replicate conditions on the wireless network—including mobility, inter-cell handoffs, and loss of line-of-sight—along with traditional network impairments such as latency, jitter, packet loss, and bandwidth limitations—over

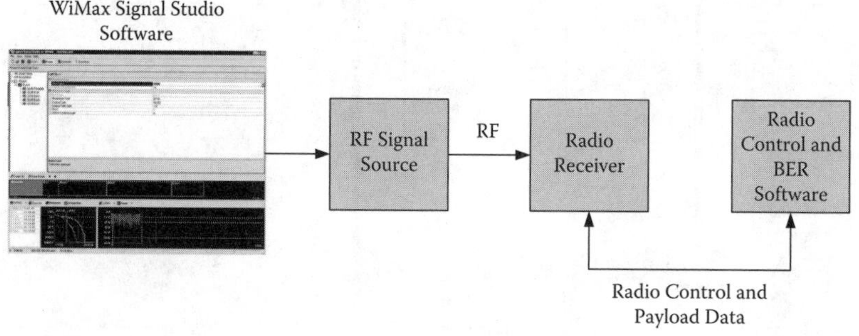

FIGURE 18.29 Receiver test.

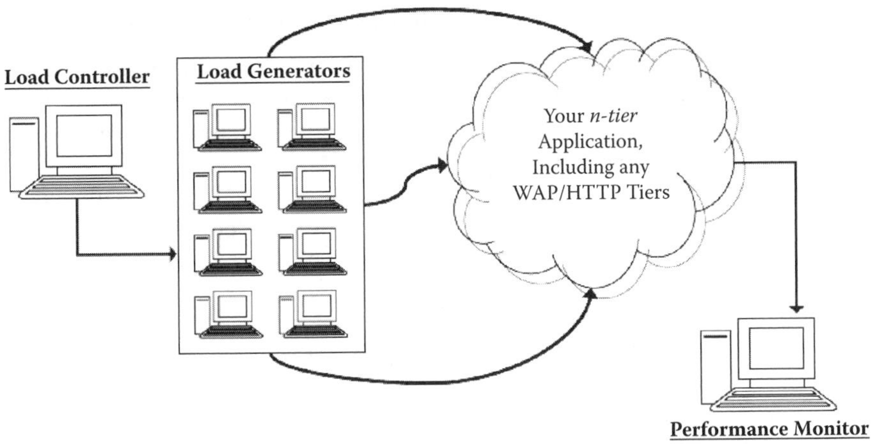

FIGURE 18.30 Performance testing example of back-end wireless application using simulated users via load generators.

an Ethernet LAN, enabling testers to use PC-based cellular emulators to validate applications with ease from the earliest stages of the development and testing lifecycle. An example configuration setup is shown in Figure 18.30.

18.5 WIRELINE BROADBAND TECHNOLOGY TESTBED

In this section, we will discuss the wireline broadband technology testbed that is capable of testing xDSLs.

18.5.1 Digital Subscriber Line (DSL)

All the different "flavors" of conventional telephone line DSL services, such as ADSL, ADSL2, VDSL, and HDSL, can be tested. All these services have limitations related to distance and throughput. The DSL Forum's TR-048 test plan had been executed in the past for interoperability and reliability for ADSL terminals. The TR-067 (Issue-2, September 2006) is the new standard for testing ADSL interoperability. The DSL Forum has standardized several other standards for testing various DSL terminals' interoperability and reliability.

One of the challenges of this technology is the distance to the customer premises and hence the throughput. The rival technology for this last mileage access is seen to be the wireless connection. This wireless technology for the last mileage access is in its infancy while DSL technology can evolve in its mature stages by testing various models quickly before the wireless technology hits the market. Service providers have invested in DSL technology, and for their interests, this technology needs to be matured. They must take steps to roll out new flavored xDSLs like ADSL2, VDSL2, etc. To be successful, these technologies must be tested. Figure 18.9 shows a general testbed for DSL interoperability and reliability tests.

18.5.2 Broadband Power Line Communications Testbed

A power line system is considered the third candidate for providing broadband services. The potential of this system to provide broadband Internet has been noted for years for its ubiquitous presence, ability to reach more homes than either cable systems or telephone lines, and its potential as a candidate for rural areas where DSL and other wireless services cannot be established. The Federal Communications Commission (FCC) in the United States approved nationwide deployment of new technologies that carry high-speed Internet connections over electric power lines in 2004.

The broadband power line (BPL) is a complicated communication media and the quality of the signal transmission is strongly polluted by noise, signal attenuation, multipath, and reflection. The critical issues for this technology are the interferences to other licensed radio services that occur due to the operation of this system in the so-called high-frequency (HF) band and the noise issues, signal attenuation, multipath, and reflection of the broadband power line (BPL) signals into the electric power line. The first issue with interference can be mitigated by design issues. The other issues can be solved by advanced signal processing methods.

Various IT businesses have started pilot projects with power companies, such as IBM with Houston's Center Point Energy (2005) in the United States. According to CNet News, more than 50 utility companies across the country are looking into it. Lots of work has to be done regarding various noise and interference issues for successful deployment of this service. It requires theoretical research as well as laboratory tests. Figure 18.31 shows a schematic testbed for analyzing BPL communications.

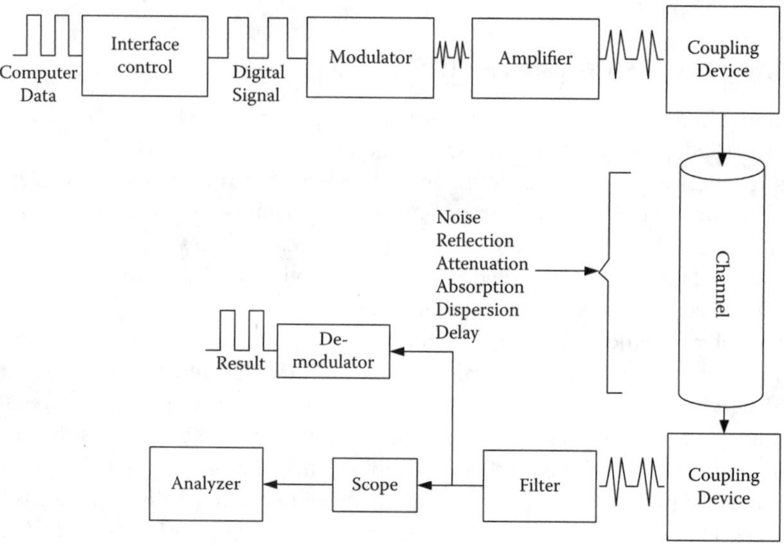

FIGURE 18.31 Broadband powerline communications testbed.

18.6 FAULT-TOLERANT OPERATIONS

In broadband networks, designing any system to tolerate faults requires that you first select a fault model. In addition, you select a set of all possible scenarios along with an understanding of frequency, duration, and impact of each scenario. The fault model lists the set of models to be considered. A decision based on a combination of expected frequency, impact on the system, and feasibility and cost of providing protection are all included in the fault model list. Most reliable broadband network designs tolerate multiple failures while others address the issues of single component failures.

18.6.1 Broadband LANs

In the LAN architectures used in broadband networks, the star topologies are typical. This involves having a switch, router, or other hub that is placed at the center of the topology where each node is connected to the hub. In terms of reliability issues, star topologies may have weaknesses. For example, a failure at the hub can entail failure of the broadband network. It should not be ruled out that other failures can occur even without failure of the hub. There are, however, many partial failures that can exist, such as

- port connection failures at the access nodes or at the hub
- amplifier failures
- cabling failures in the fiber itself
- transmitter failures at the access nodes because of laser failures
- receiver failures at the access nodes because of laser failures

In all of these failures, it will only be possible with the failure of one or more arms of the star topology. The solution to these kinds of failure is designing some form of redundancy into the network as shown in Figure 18.32. Additionally, two of the switches must be connected. Assume that the primary network interface cards (NICs) fail in server 1 in Figure 18.32. This server 1 will communicate through a secondary switch. It is not desirable to require other servers to communicate through the secondary switch first. It should be pointed out that irrespective of the fact that we have only shown two servers, this type of network normally has many servers connected to it. It is not that easy to reconfigure that many servers simultaneously. Second, due to initialization overheads, there is delay involved in creating new connections through switch 2.

The only way to avoid reconfiguration failures at all of the servers is to make all of the servers besides server 1 continue to communicate with the primary switch and the two switches to communicate with each other through the interswitch connection. Many enterprise networks and storage area networks (SANs) are built as star topologies. The enterprise networks are based on gigabit Ethernet (GigE), while SANs are based on fiber channel (FC).

18.6.2 Fault-Tolerant Designs for Wireless Broadband Networks

Wireless broadband networks have been deployed increasingly in recent years. There is therefore an expectation of high dependability in these enterprise broadband wireless networks. The problem of dependability also concerns broadband wireline networks;

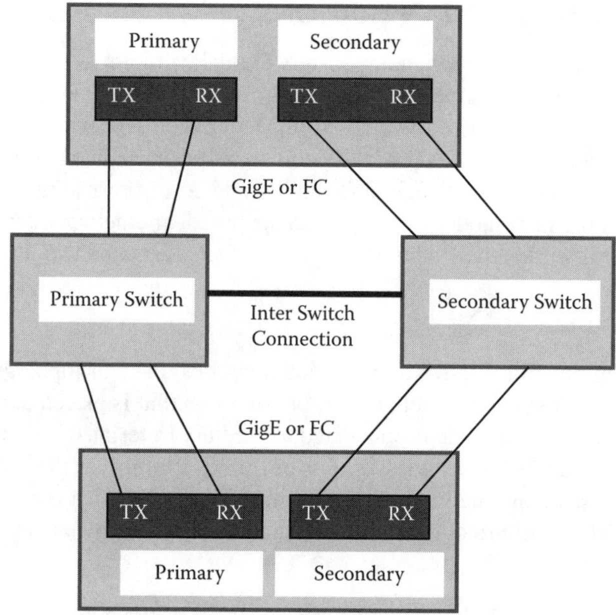

FIGURE 18.32 LAN redundant architecture in an enterprise network or using traditional star topologies.

however, user mobility, location and link dependence, and fault propagation make dependability a challenging task in wireless broadband networks. Redundancy is a key element in broadband wireless fault-tolerant designs. For example, at the block level, fault tolerance is provided by designing in block-level redundancy while at the network level fault tolerance is achieved by deploying a fault-tolerant SONET ring to interconnect multiple adaptable building blocks (ABBs) as shown in Figure 18.33.

Figure 18.33 shows an integrated fault-tolerant architecture for wireless networks with redundancy and replication techniques used for components and links in a block. The implementation cost, desired values of dependability attributes, and complexity determine the exact number of components, links, and levels required in a block.

Wireless Networks Fault Tolerance

As discussed in Section 18.6.2, one of the emerging solutions for wireless broadband dependability is the idea of designing a fault-tolerant multilevel network that can enhance the dependability of wireless broadband networks. The fault tolerance is added at component, link, block, and network levels. This technique seems to be effective in the following ways:

- Fault-tolerance is introduced by adding redundancy at component, link, block, and interconnection levels.

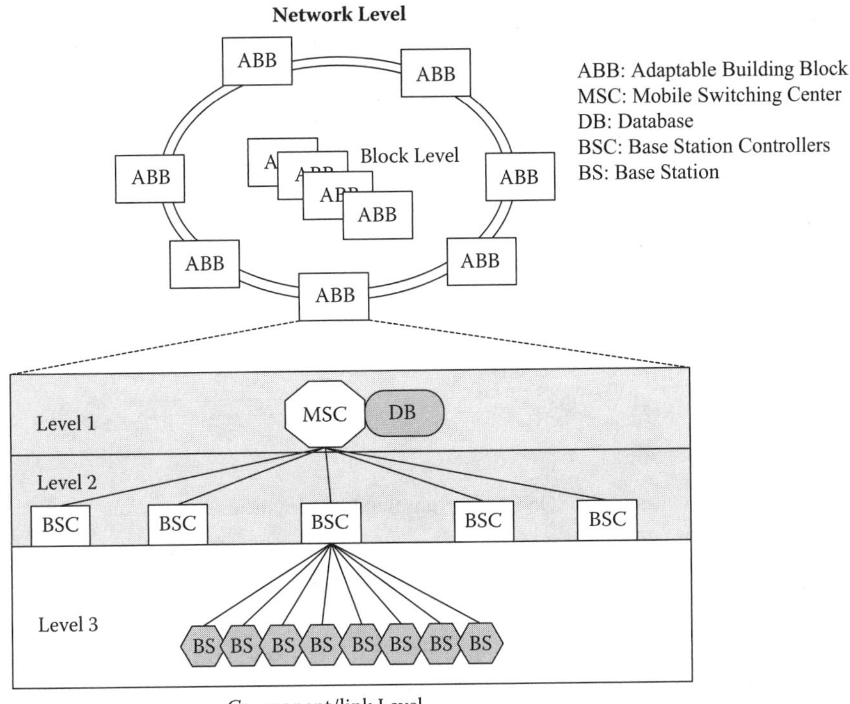

FIGURE 18.33 An integrated fault-tolerant architecture for wireless networks.

- There have been improvements in wireless dependability using multilevel fault tolerance.
- Selective redundancy could achieve a desired level of fault tolerance.
- "Macro-" level mobility is included to allow a system-level evaluation of fault-tolerant design.
- Wireless link performance is considered in evaluating the dependability of fault-tolerant broadband wireless architecture.
- Dependability performance of intercarrier and interconnected broadband wireless networks could be derived.
- The dependability evaluation of infrastructure-oriented broadband wireless networks of any size and number based on adaptable building blocks (ABBs) can be supported.

18.6.3 Fault-Tolerant Design Solutions for Broadband Telecommunication Systems

Fault-tolerant design solutions are techniques for providing survivability in broadband telecommunication systems. Fault-tolerance techniques are applied in transmission, signaling, switching, and power systems.

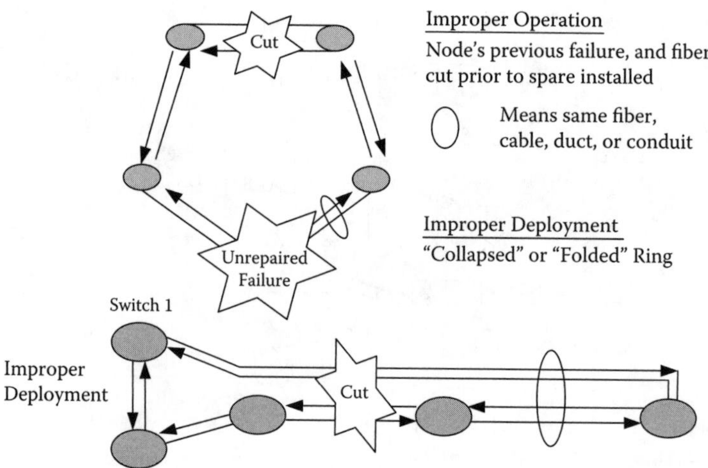

FIGURE 18.34 Fault-tolerant SONET ring: unreliable deployment or operation.

Transmission

Broadband telecommunication survivability entails providing transmission diversity via a protection channel. Digital cross-connect switches (DCS) and fault-tolerant SONET rings are two common techniques used. Within a millisecond period, once failure is sensed, the switching mechanism makes it possible to choose an alternative path for the system. However, cost is always a major problem in transmission diversity. Therefore, fault-tolerant rings are normally used instead. This is achieved with the aid of bidirectional rings. These bidirectional rings have primary channels in one channel and protection channels in another and all in the same cable. Figure 18.34 shows the two major modes of improper operation and deployment of SONET rings.

Signaling

Signaling System 7 (SS7), which was introduced in the 1980s, is a flexible and powerful broadband digital service applicable to both wireless and wireline services. It brings the telecommunications channels into a more integrated whole. Local telecommunications switches cannot process requests for new calls to or from any other building for users attached to a switch without connectivity to the SS7 network. Therefore, because of this problem, the ANSI Committee on T1 standards requires each local switch to have redundant access to the SS7 network. Figure 18.35 shows a common outage scenario involving SS7. There are two kinds of deployment, namely, proper and improper deployments. Through separate circuits, are the proper redundant deployment access links to different packet switch nodes [signal transfer points (STPs)] is shown where the access links not only are connected but the circuits are separate cables, fibers, conduit, paths, power supplies, timing sources, and electronic components. Improper deployment is shown in Figure 18.35 where access links

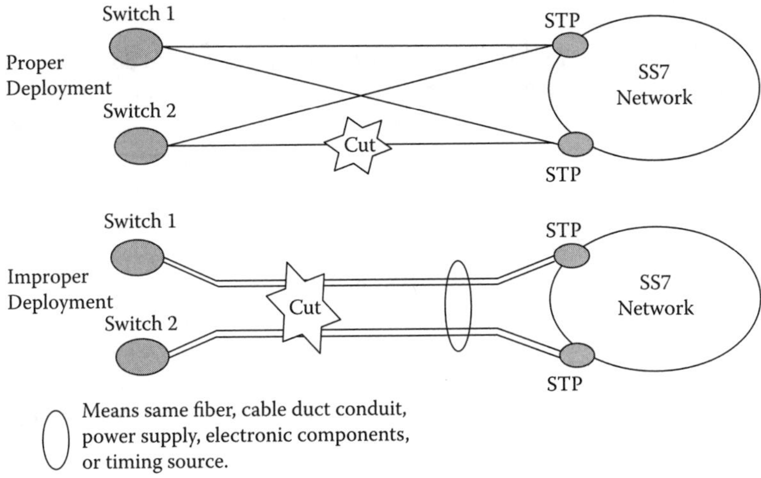

FIGURE 18.35 Fault-tolerant SS7 access links: proper and improper deployment.

are routed in the same path as fiber cable with both access links, isolating the local switches from the signaling network.

Power Systems

To make sure fault tolerance is in place, triple redundancy is implemented in most power systems. All broadband telecommunications equipment runs on DC voltage generated by AC-fed redundant rectifiers. The power companies supply the AC power. In the event of a power failure, there is a stand-by backup battery. At PVAMU, we have the same fault-tolerant architectural design in our Center of Excellence for Communication Systems Technology Research (CECSTR) broadband research laboratory. Figure 18.36 shows the fault-tolerant power system, depicting unreliable deployment or operation of the system.

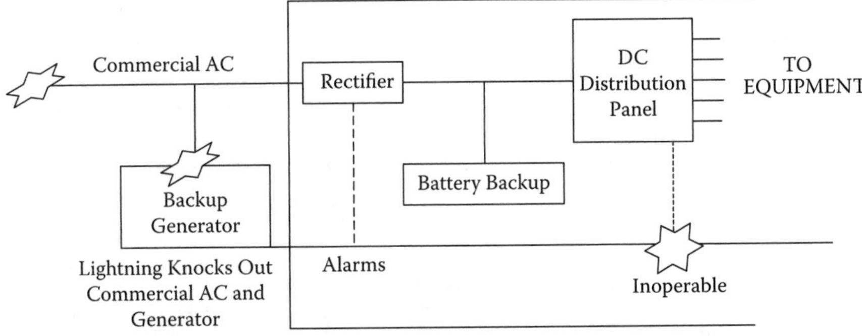

FIGURE 18.36 Fault-tolerant power: unreliable deployment or operation.

18.7 NEXT-GENERATION SONET TESTING

The never-ending need for higher data rates to support new and advanced services has resulted in increasing pressure on network operators to upgrade their networks. Fiber networks have been growing at a much faster pace compared to telephone and cable networks. Therefore, the strategic choice for many operators to deliver broadband services is to drive fiber deep in the access network using a fiber-to-the-premises (FTTP) network.

Current FTTP deployment utilizes 622 Mbps downstream data rate on 1490 nm wavelength and 155 Mbps upstream data rate on 1310 nm wavelength. The PON is shared among 32 users, and the video service is overlaid on a 1550-nm wavelength. In order to support IP television (IPTV) and other video applications (videoconferencing, video chatting, online gaming, video cams, peer-to-peer applications, etc.), the data rate will have to be at least four times higher. Hence, the need for a gigabit passive optical network (GPON).

Several parameters can be implemented to help monitor and troubleshoot the video overlay in GPON. Some of these parameters are

- Optical signal level: Used to monitor the optical level of the 1550-nm received signal.
- AGC setting: Used to adjust the gain of the video receiver.
- AGC mode: Used to specify whether the automatic gain control (AGC) is implemented through feedforward, narrowband feedback, or wideband feedback control loop.
- Video higher optical threshold: Used to alert when the 1550-nm optical power is higher than a predefined threshold.
- Video lower optical threshold: Used to alert when the 1550-nm optical input power drops below a predefined threshold.

These parameters along with others provide very valuable information in deploying, troubleshooting, and upgrading PON networks with video overlay.

Figure 18.37 shows the video test setup used to illustrate the expected gain in sensitivity when lowering the number of analog channels. The setup includes multi-frequency test equipment for the analog channels, and QAM generator and analyzer for the digital channels. In addition, the test setup includes the 1490-nm and 1310-nm digital data streams to make sure that any impact from the Raman crosstalk[1] is included. A fiber spool of 20 km is included to simulate the expected distance in field deployment. The typical channel lineup that is used by cable operators has 82 channels and 33 digital channels.

In PON systems, there are diagnostic systems that help to make sure that PON systems operate efficiently. Figure 18.38 shows the architecture of the PON and the

---

[1] Fiber nonlinearities can significantly limit the performance of WDM optical systems. Cross talk due to stimulated Raman scattering (SRS) is called Raman cross talk and it can potentially reduce carrier-to-interference ratios to unacceptable levels. The effects of SRS include cross talk that may produce interference at levels up to 30 dB below desired carriers. The reverse path is particularly susceptible to this impairment.

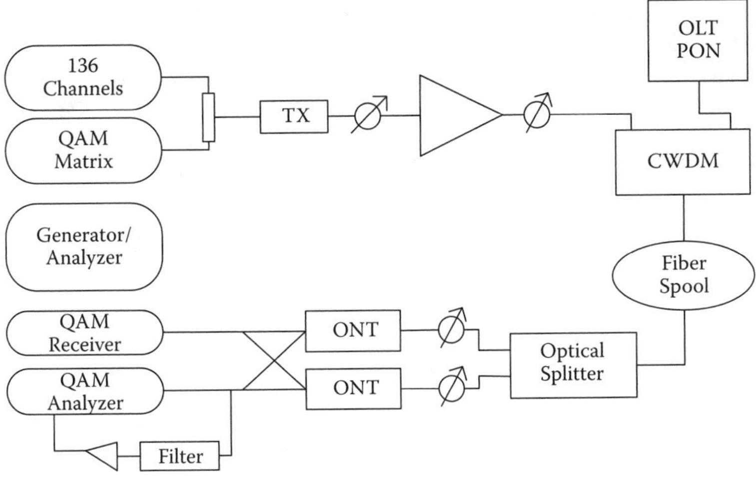

FIGURE 18.37 Video test setup.

functional partitioning of the OLT and ONU elements. The PON is based on optical passive filters to form the point-to-multipoint configuration. The network elements are partitioned to the following functions:

- optical distribution network (ODN)
- a physical (PHY) layer function that includes the PON transceiver
- medium access control (MAC) that aggregates both services and control cells
- service adaptation function (SAF) that translates specific services (example, data, and voice) into ATM cells
- management and control function

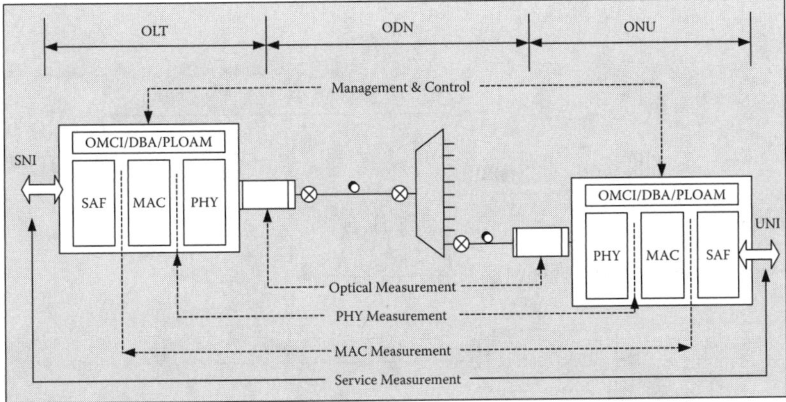

FIGURE 18.38 The diagnostic layers and interfaces in a PON architecture.

The diagnostics of the ODN is based on measurements of optical parameters using optical time-domain reflectometer (OTDR) and optical loss testers (OLTS). The system service layer is subject to service-specific operational testing (e.g., Ethernet LAN, voice, CATV, etc.). The specific functional testing of PON is related to the MAC and PHY functions, PON operations, and maintenance, including support for quality of service (QoS), dynamic bandwidth allocation (DBA), and ONU discovery and auto-ranging.

In the optical diagnostics, in the upstream direction, the OLT receiver sensitivity and bit-error rate (BER) differ from the regular point-to-point links. The conventional links can be measured regardless of the higher layer protocol (e.g., Ethernet, SONET/SDH, PON downstream direction) due to their continuous transmission mode. In the upstream direction of PON architectures, the bursty nature of the data and the distinguished distances of each ONU must be taken into account. Therefore, when the OLT receiver sensitivity is characterized, it has to be done as part of a PON in conjunction with several ONUs and not as a stand-alone link. Figure 18.39 shows the test setup for OLT receiver sensitivity. It has two ONUs on the near and far ends of the OLT. The OLT receiver has to capture the weak upstream burst that follows a strong one. The OLT sensitivity is measured in three scenarios:

- continuous mode measurement
- single ONU burst-mode measurement
- dual ONU burst-mode measurement

It is also possible that PON architectures can support an RF video overlay using additional wavelength. The optical isolation parameters between the two wavelengths of the ONU receiver in the downstream direction is emphasized. It is feasible to have a direct measurement. In running the test, a procedure that characterizes the BER and carrier-to-noise ratio (CNR) under crosstalk situation and then extracting the parameters after curve fitting of the results is used. Figure 18.40 shows the test procedure. The BER tester is used to characterize the digital baseband link while the network analyzer is used to generate the equivalent video signal. The power levels of both signals are controlled by optical attenuators and hence determine the crosstalk level.

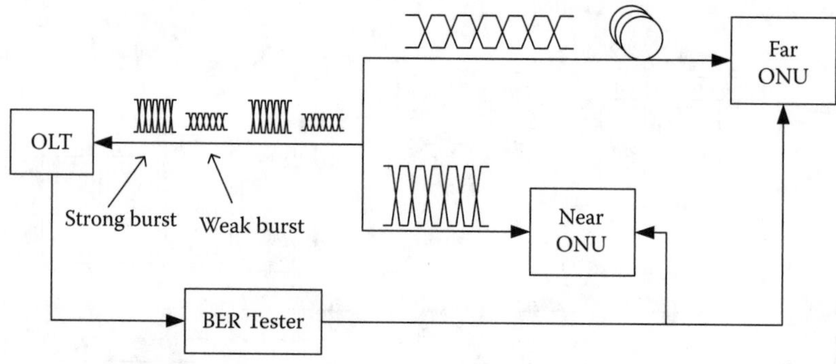

FIGURE 18.39 Test setup of OLT receiver sensitivity.

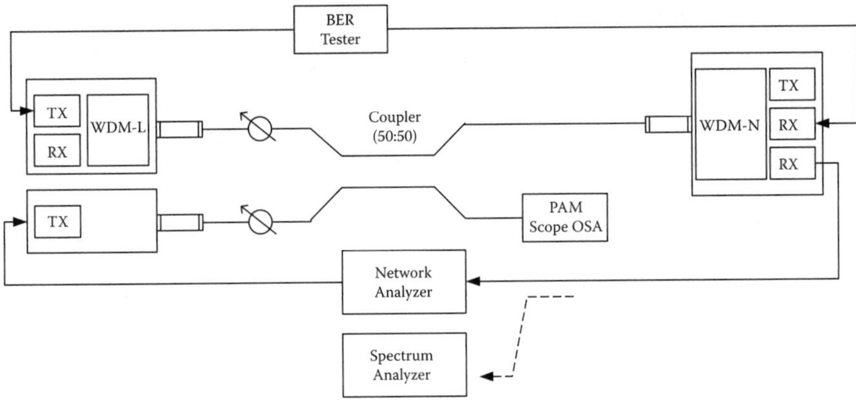

FIGURE 18.40 Test setup for ONU optical isolation.

SUMMARY

1. It is necessary to test broadband networks.
2. Interoperability and reliability tests of ADSL modems are required for the proper functioning of broadband networks.
3. The DSLAM must be properly configured before any test can proceed.
4. Manual tests can be done using the recommended test plan, but are time consuming and, in addition, are expensive.
5. The tests can easily be done by automation, as discussed in this chapter.
6. We can use TCL or another language for this automation.
7. Most of the network testing tools can be handled by simple TCL language.
8. In the industry, they also use the TCL language.
9. The general POT architecture has its limitation.
10. DSL services over traditional twisted copper wires seem to provide a much better service.
11. In wireless broadband, testing is an integral part of wireless network development.
12. Wireless broadband includes both fixed and mobile broadband technologies.
13. WiMAX is one of the emerging technologies in broadband wireless technologies.
14. The power line system is considered the third candidate for providing broadband services.
15. In broadband networks, designing any system to tolerate faults requires that you first select a fault model. In addition, you select a set of all possible scenarios along with an understanding of frequency, duration, and impact of each scenario.
16. Fault-tolerance techniques are applied in transmission, signaling, switching, and power systems.

REVIEW QUESTIONS

18.1 Today's broadband network is not heterogeneous in nature.
- a. True
- b. False

18.2 Interoperability is the key issue of internetworking for current service providers and vendors.
- a. True
- b. False

18.3 Every vendor confirms the interoperability of its product before it brings any product to market.
- a. True
- b. False

18.4 Although it takes months and even years to make a standard, this is a necessary part of product development, otherwise interoperability is jeopardized.
- a. True
- b. False

18.5 Before testing any broadband system, we do not need to understand the architecture of that system.
- a. True
- b. False

18.6 DSLAM is designed specially to handle broadband services.
- a. True
- b. False

18.7 DSLAM is:
- a. Digital subscriber line access multiplier
- b. Digital subtraction line access multiplexer
- c. Digital subscriber line access multiplexer
- d. None of the above

18.8 The device that is being tested is called:
- a. DOT
- b. DTE
- c. DWT
- d. DUT

18.9 Larger COs may serve up to 100,000 telephone lines.
- a. True
- b. False

18.10 Distribution cables contain 25 to 1000 pairs.
- a. True
- b. False

Answers: 18.1b, 18.2a, 18.3a, 18.4a, 18.5b, 18.6a, 18.7c, 18.8d, 18.9a, and 18.10a

PROBLEMS

18.1 List some of the tests we can perform to ensure better performance in a broadband network environment.

18.2 Why is it necessary to formulate standards in any product even when it takes months?
18.3 Is standard-making a necessary part of product development in broadband communications and why?
18.4 Draw and describe a typical ADSL loop architecture. Why is it important in a broadband network?
18.5 Describe the functions of DSLAM in the testing of broadband modems.
18.6 What are the processes required in the configuration of DSLAM in interoperability and reliability tests?
18.7 What do you understand as interoperability and reliability tests in a broadband network?
18.8 How are interoperability and reliability tests performed in a broadband network?
18.9 What is TR-48 test plan? Why is it important to have a test plan in the testing of broadband networks?
18.10 Describe the functions of a wire line simulator in the testing of ADSL modems.
18.11 What kind of equipment can we use in the generation of broadband traffic required for testing interoperability and reliability in a broadband network? Why is it important? What functions does it perform?
18.12 Describe the process for testing DSLAM using Smartbits.
18.13 Describe the configuration process for a LIM port.
18.14 How important is a telephone subscriber loop environment in the testing of ADSL systems?
18.15 What is the difference between interoperability and reliability tests?
18.16 What is the major function of the DSL Forum in testing of xDSL systems?
18.17 How is the work done by the DSL Forum important to the testing and development of xDSL systems?
18.18 Describe the effects different variable loop lengths may have in the testing of xDSL systems.
18.19 What are the different variable loop lengths that can be used in the testing of xDSL systems?
18.20 What is the major difference between TR-48 and TR-69 test plans?
18.21 How can SNIR be tested for in a point-to-point multipoint broadband fixed wireless access network to determine if the SNIR level is below a certain threshold?
18.22 Compare the differences in technologies among EV-DO, HSDIA, and WiMAX 802.16e.
18.23 What are the various issues that must be tested to ensure a successful deployment of WiMAX?
18.24 Why is a virtual wireless environment important in a broadband wireless application test?
18.25 Describe broadband powerline communication. How can it be tested?

Appendix A: Testing DSLAM Using SmartBits - SmartWindow

THE TESTING LOOP

The device under test (DUT) in this document is an ADSL modem connected in the testing loop shown in Figure A1.

The RJ-11 port of the modem connects to a WireLine Simulator, which can simulate different types of testing loops. The RJ-45 port of the modem connects to a ML-7710 10/100Mb Ethernet SmartCard in a SmartBits chassis. Connect the DS-3 cable from the Stinger trunk module to the AT-9045 45 Mb DS3 ATM SmartCard in the SmartBits chassis. The DSL line input of the Stinger comes from the WireLine Simulator.

The basic configuration parameters are shown in the following table.

Stinger Card	Type	Port	VPI/VCI
LIM	ADSL (DMT)	10	8/35
Trunk	DS3	1	0/300

Configuration of Stinger

See the "Stinger Manual.doc" for detailed configuration procedures.

SmartCards used in the test

The SmartCards we use are ML-7710 at slot 13 and AT-9045 at slot 19. The following figure shows the face panel of SmartBits 2000 (Figure A2).

Smart Window Testing

1. SmartBits 2000 provides the multi-user probability, so before using a SmartCard, it must be "reserved." Right click on both the ML-7710 card and AT-9045 card and select "Reserve Smart Card" (Figure A3).
2. Configure the ML-7710 card. Click on the card, and then select "Transmit Setup…" from the popup menu. Click the "This Port" button in the dialog box that appears. Make the configuration in the "Transmit Setup" dialog box as shown in Figure A4. The testing traffic is Ethernet frames, so we

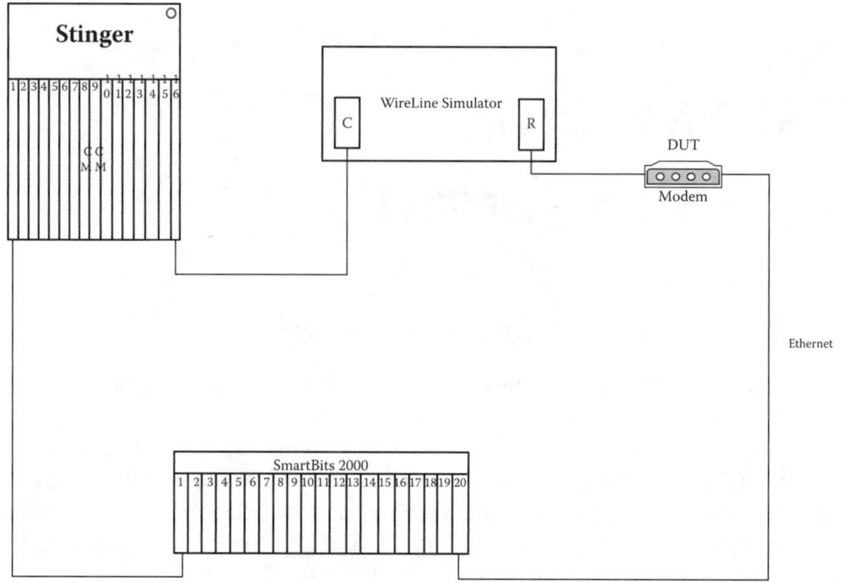

FIGURE A1

must define the static source and destination MAC address. The source MAC address for the traffic generated from the ML-7710 card is identified by its slot number (13) as "00 00 00 13 13 13." The traffic destination is the AT-9045 card at slot 19, so the destination MAC address is defined as "00 00 00 19 19 19." After the configuration click "OK" twice to close the dialog boxes (Figure A4).

3. Configure the AT-9045 ATM card. First, select "Line Interface Parameters…" from the popup menu. Be sure to select the "Framing Mode" as "DS3 PLCP." PLCP stands for Physical Layer Convergence Procedure/Protocol.

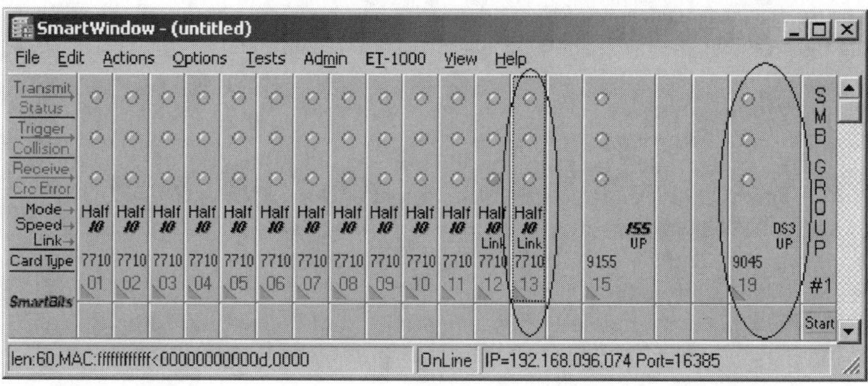

FIGURE A2

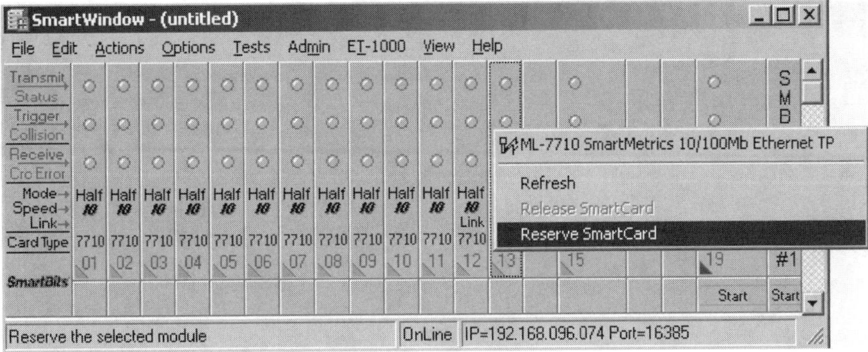

FIGURE A3

The parameter "Framing Mode" specifies what type of mapping is used to map ATM cells into the DS3 C-Bit Parity format. "DS3 Cell" maps the ATM cells directly into the DS3 frame; "DS3 PLCP" first maps the ATM cells into PLCP frames and then maps the PLCP frames into the DS3 frame (Figure A5).

4. Next, select "Transmit Setup…" from the AT-9045 ATM card popup menu (Figure A6).

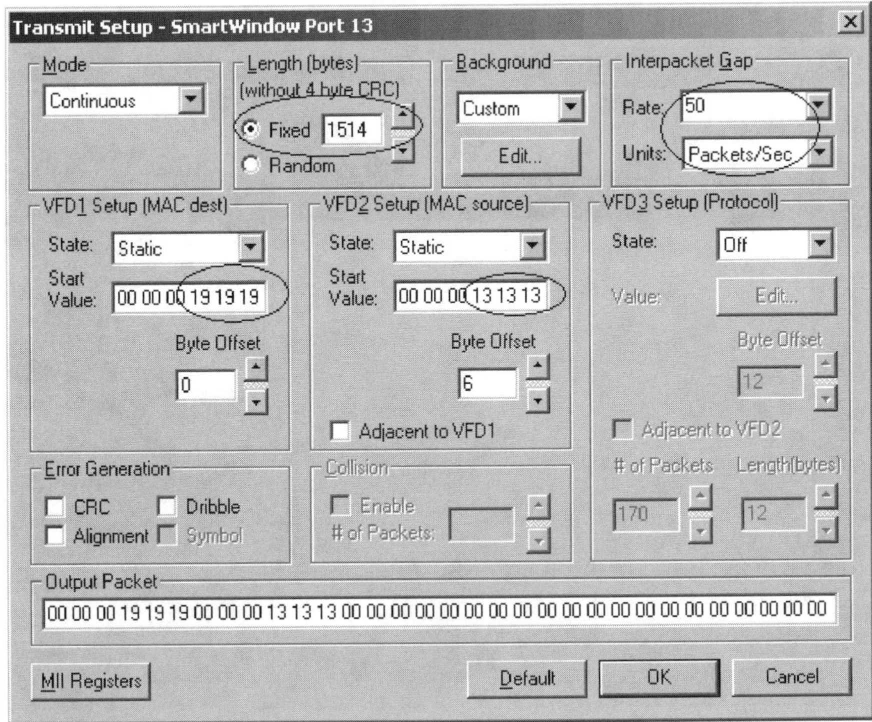

FIGURE A4

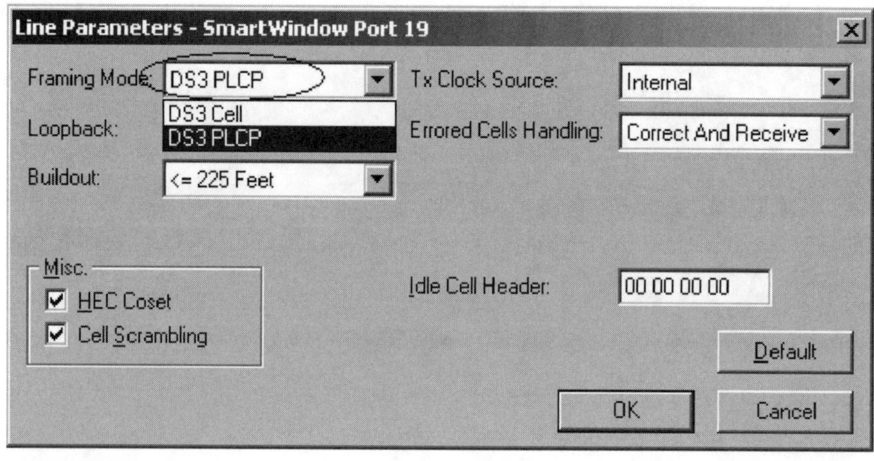

FIGURE A5

5. Click "Edit Stream…" Fill in the parameters as follows: The value of VPI/VCI matches the value defined at the Stinger trunk module port to which this AT-9045 ATM card is connected. In this example, it is 0/300 as shown in the table in "The Testing Loop" section. For the same reason as in step 2, we define the destination MAC address as "00 00 00 13 13 13"

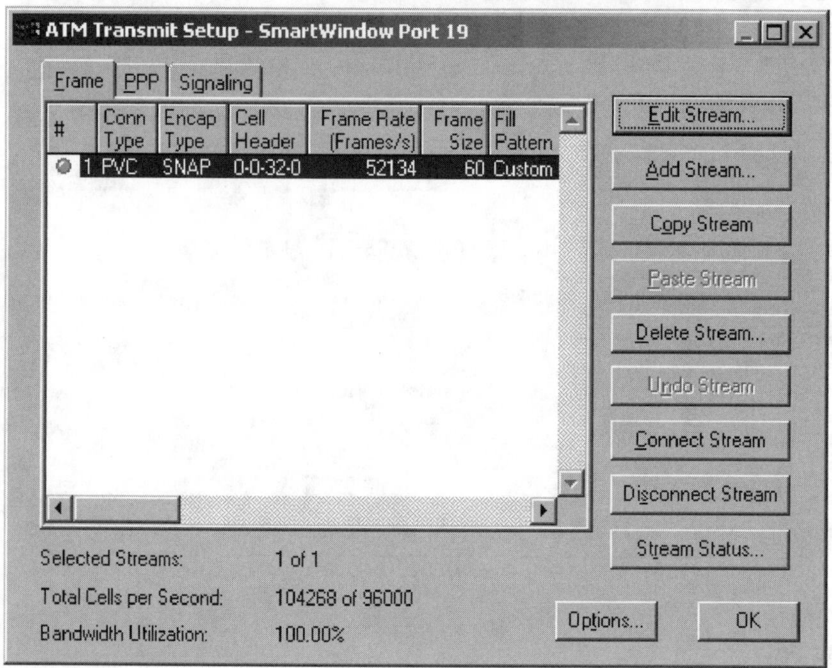

FIGURE A6

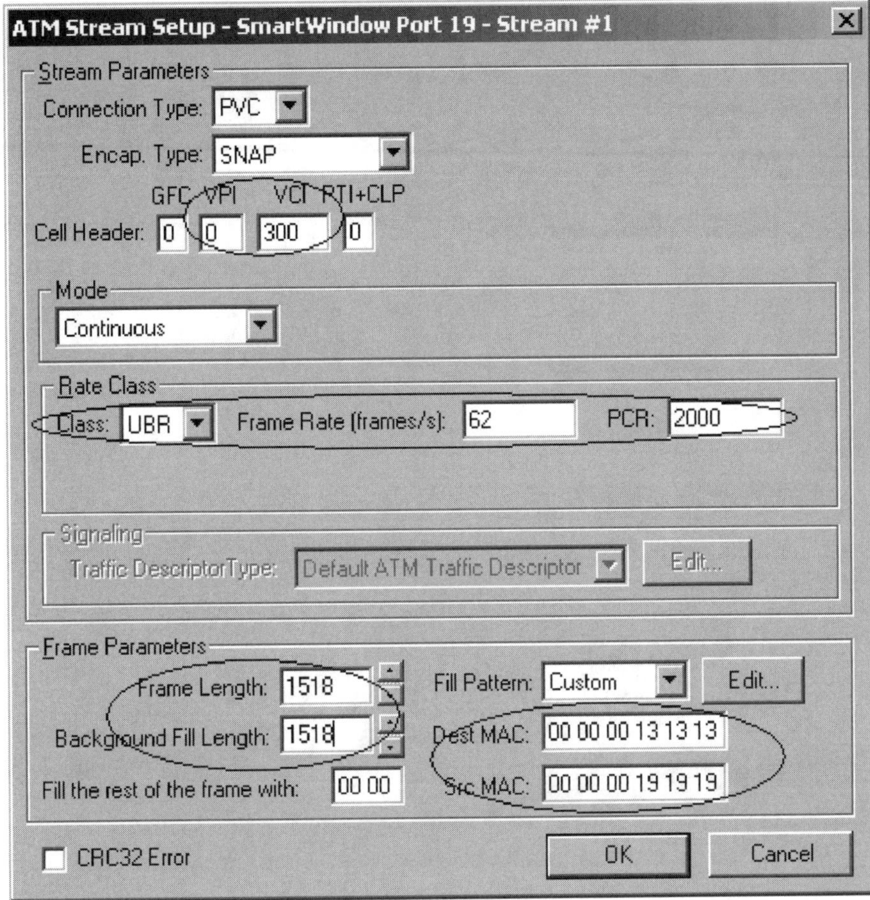

FIGURE A7

and the source MAC address as "00 00 00 19 19 19." Click "OK" twice to close the dialog boxes (Figure A7).

6. Select the menu "Actions | SmartCounters…" to open the counter window (Figure A8).
7. Click the left-most button "New" in the tool bar. The counter display interface will appear (Figure A9 and Figure A10).
8. Connect the ATM stream we defined in step 5 to the SmartCard. There are two methods to do this. One is to click the "Connect Stream" in the dialog box shown in step 4. This dialog box can also be used to disconnect a stream. The second method is to select the "Auto Connect" item in the AT-9045 card's popup menu. When the stream is connected, a plus sign will appear at the left of the "19 AT-9045B" port in the left panel of the window. Click on the plus sign to extend the tree. Then click on the item "1:0-300." The counters for the stream 0/300 appear in the right panel (Figure A11).

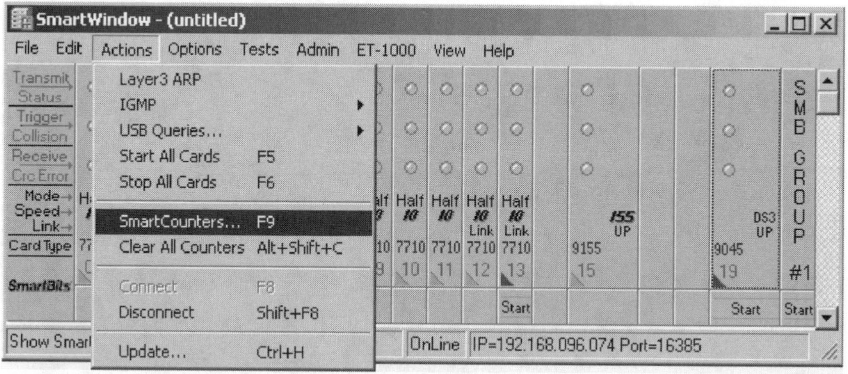

FIGURE A8

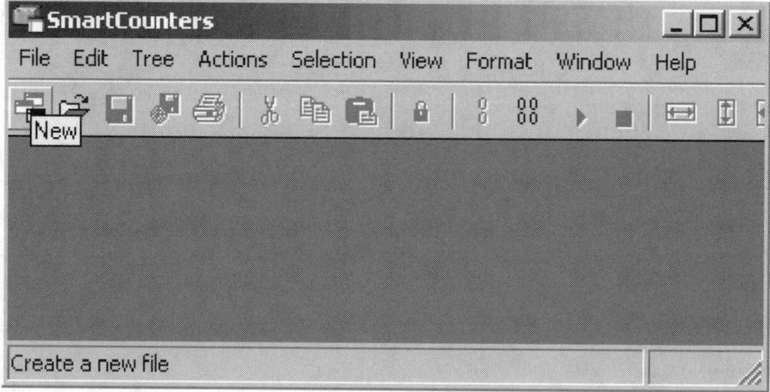

FIGURE A9

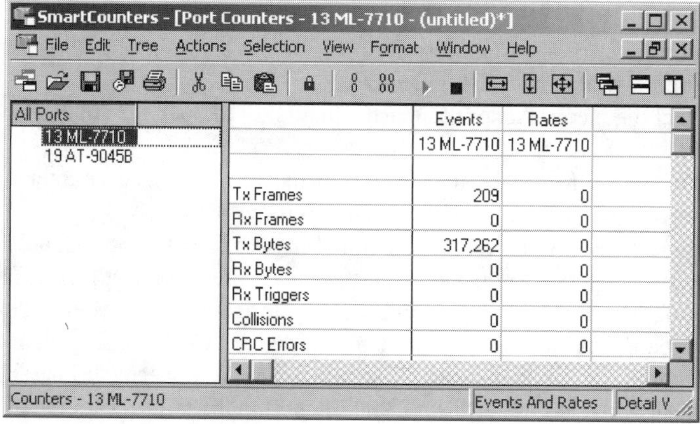

FIGURE A10

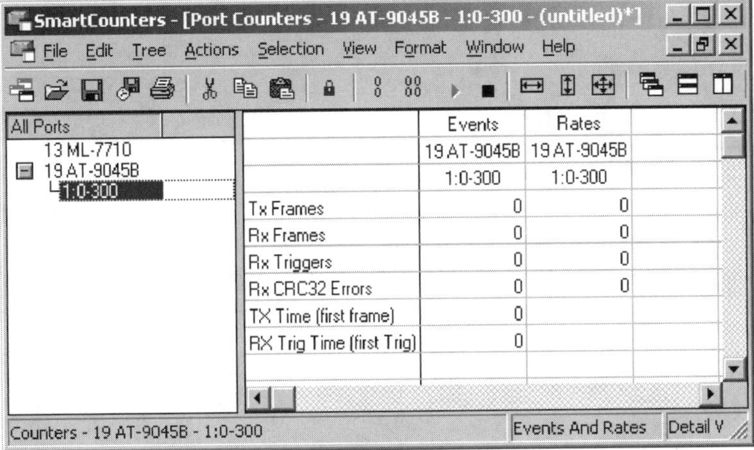

FIGURE A11

9. Select the item "13 ML-7710" to display the counters for the Ethernet port, then hold the control key and click on the item "1:0-300." Notice that two sets of counters appear in the right panel at the same time (Figure A12).
10. Click the start button under ML-7710 card to start generating Ethernet traffic. Observe that the counter of "Tx-Frames" for ML-7710 and the counter of "Rx Frames" for AT 9045B are increasing. This shows that the upstream traffic is running correctly (Figure A13).
11. Stop generating Ethernet traffic by clicking the "Stop" button under ML-7710 card. In the SmartCounters window click on the tool button "Clear All SmartCards" to reset all the counters (Figure A14).

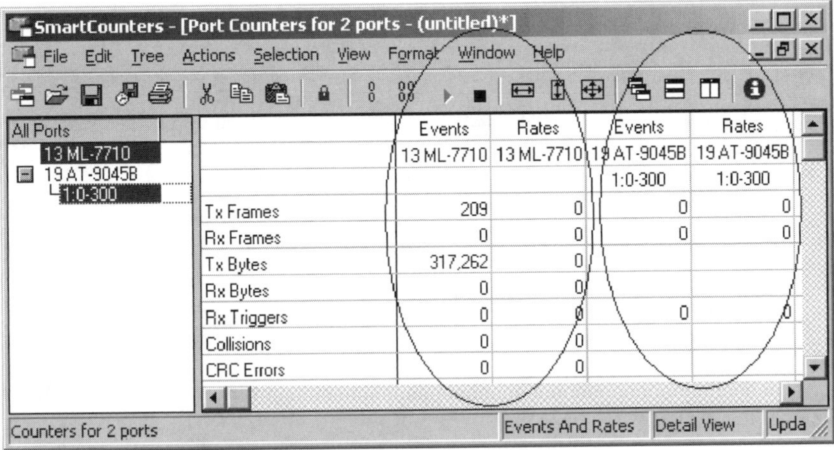

FIGURE A12

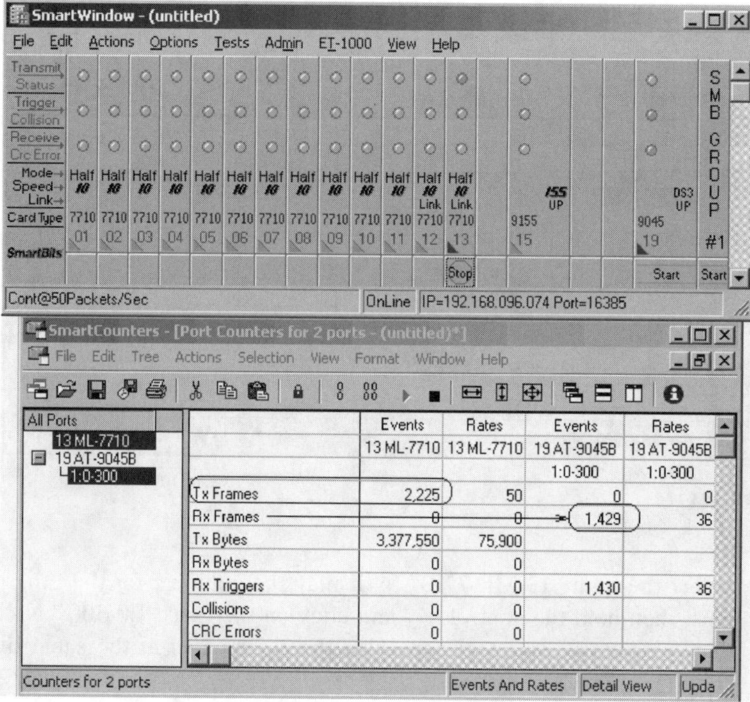

FIGURE A13

12. Then click the "start" button under AT-9045B card to generate ATM traffic. Observe that the counter of "Tx-Frames" for AT 9045B and the counter of "Rx Frames" for ML-7710 are increasing. This shows that the downstream traffic is running correctly (Figure A15).
13. Generate Ethernet traffic again to observe the bi-directional traffic at the same time (Figure A16).

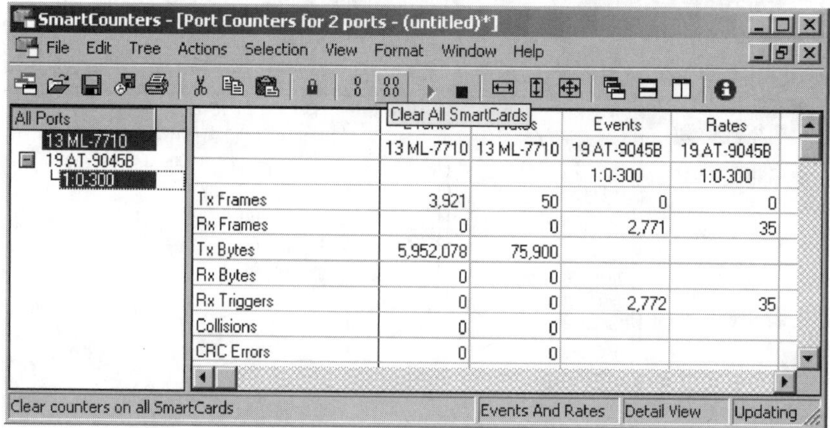

FIGURE A14

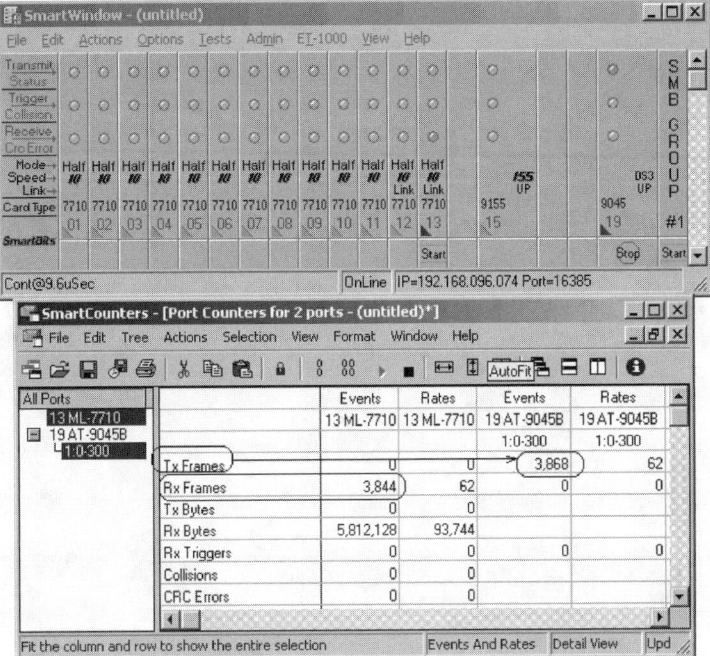

FIGURE A15

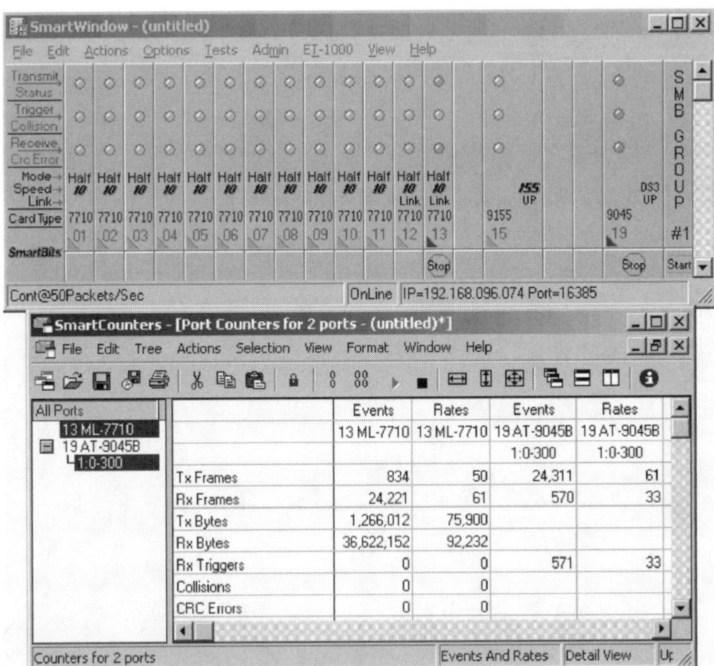

FIGURE A16

Appendix B: Testing DSLAM Using SmartBits - SmartxDSL

THE TESTING LOOP

The device under test (DUT) in this document is an ADSL modem connected in the testing loop shown in Figure B1.

The RJ-11 port of the modem connects to the WireLine Simulator, which can simulate different types of testing loops. The RJ-45 port of the modem connects to an ML-7710 10/100Mb Ethernet SmartCard in SmartBits chassis. Connect the DS-3 cable from the Stinger trunk module to the AT-9045 45 Mb DS3 ATM SmartCard in SmartBits chassis. The DSL line input of the Stinger comes from the Wire Line Simulator.

The basic configuration parameters are shown in the following table.

Stinger Card	Type	Port	VPI/VCI
LIM	ADSL (DMT)	10	8/35
Trunk	DS3	1	0/300

Configuration of Stinger

See the "Stinger Manual.doc" for detailed configuration procedures.

Smart Cards Used in the Test

The Smart Cards we use are ML-7710 at slot 13 and AT-9045 at slot 19. The following figure shows the face panel of SmartBits 2000 (Figure B2).

SmartxDSL Testing

SmartxDSL can be used to do the frame tests, such as frame loss test, frame latency test, and throughput test. According to the online manual of SmartxDSL, the throughput test "determines the maximum transmission rate at which the DUT/SUT can forward traffic with no frame loss, or at a user-specified acceptable frame loss. By increasing the transmission rate at specified levels, you can determine the DUT/SUT's capacity." The flow chart of the throughput test is shown in Figure B3.

1. The first preparation for the test is to set up the SmartBits connection. Click on the "Setup SmartBits Connections" tool button on the tool bar or select from the menu "Setup | SmartBits Connections…" (Figure B4).
2. Input the IP address of the SmartBits that we will use in the test and click the "Add" button (Figure B5).

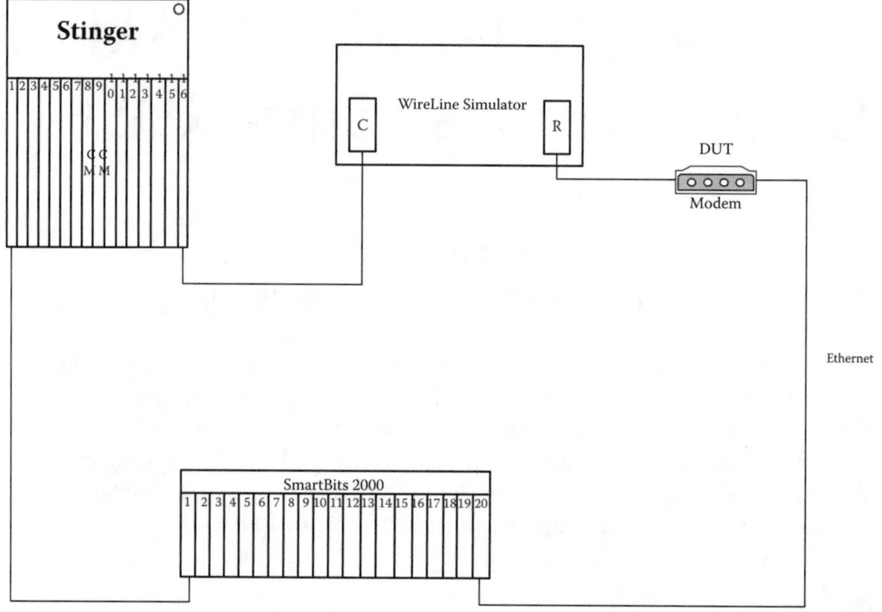

FIGURE B1

3. In order to avoid a conflict, close all the other SmartBits applications so that SmartxDSL can get full control over all the cards in the specified SmartBits chassis.
4. From the "Action" menu select "Connect" to connect to the SmartBits defined above. When the connection to SmartBits is created, the "Test Layout" panel in the main interface shows all the available Smart Cards at both the Network Side and Access Side.

The "Network Side" refers to the ATM SmartCard that connects to the trunk module of DSLAM such as AT-9045B DS3 45Mb SmartCard.

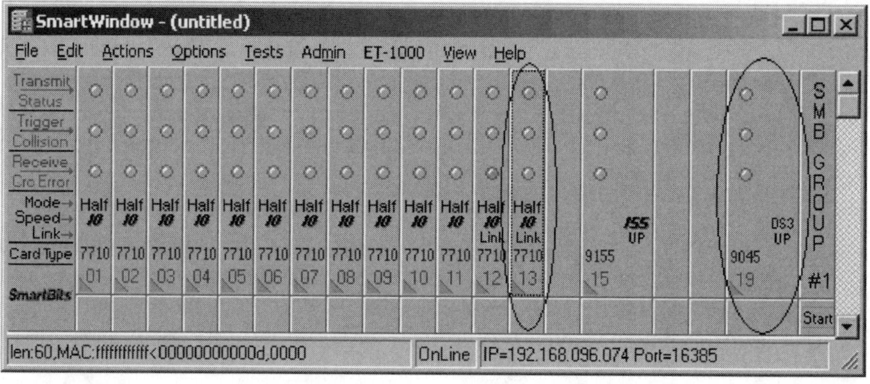

FIGURE B2

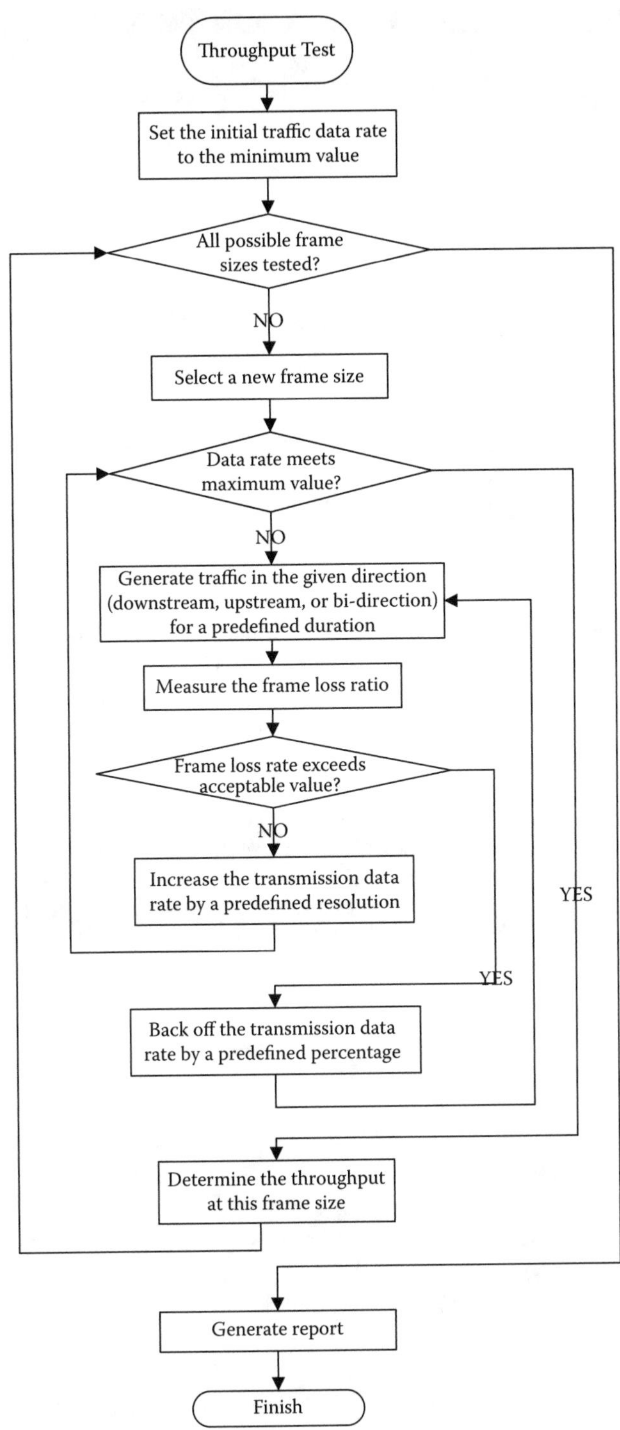
Throughput Test
Set the initial traffic data rate to the minimum value
All possible frame sizes tested?
NO
Select a new frame size
Data rate meets maximum value?
NO
Generate traffic in the given direction (downstream, upstream, or bi-direction) for a predefined duration
Measure the frame loss ratio
Frame loss rate exceeds acceptable value?
NO
Increase the transmission data rate by a predefined resolution
YES
YES
Back off the transmission data rate by a predefined percentage
Determine the throughput at this frame size
Generate report
Finish

FIGURE B3

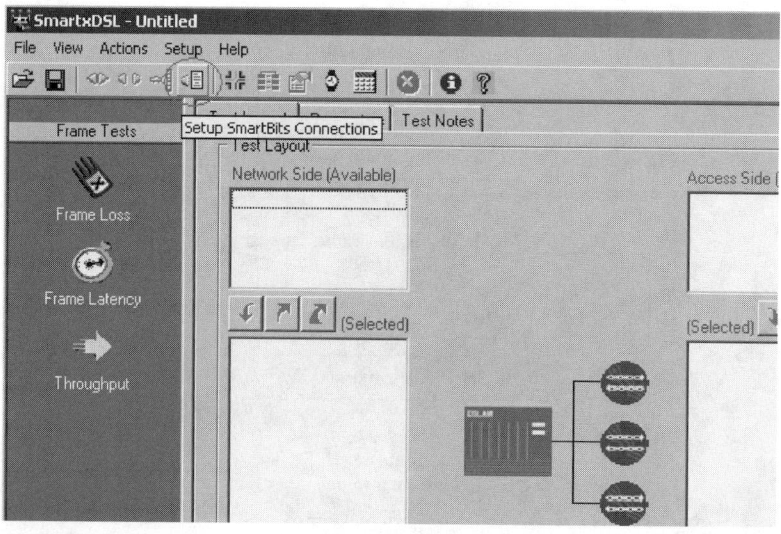

FIGURE B4

The "Access Side" refers to the SmartCard that connects to the modem such as ML-7710 Ethernet SmartCard (Figure B6).

5. Select the required ATM SmartCard "(1,19,1): AT-9045B" and press the down arrow button to move this card to the "Selected" list in the bottom. Perform the same operation for the Ethernet card (Figure B7).

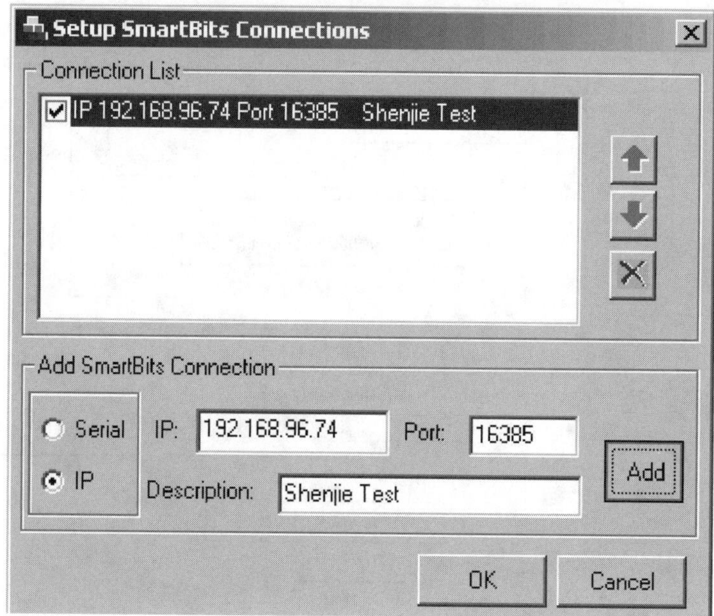

FIGURE B5

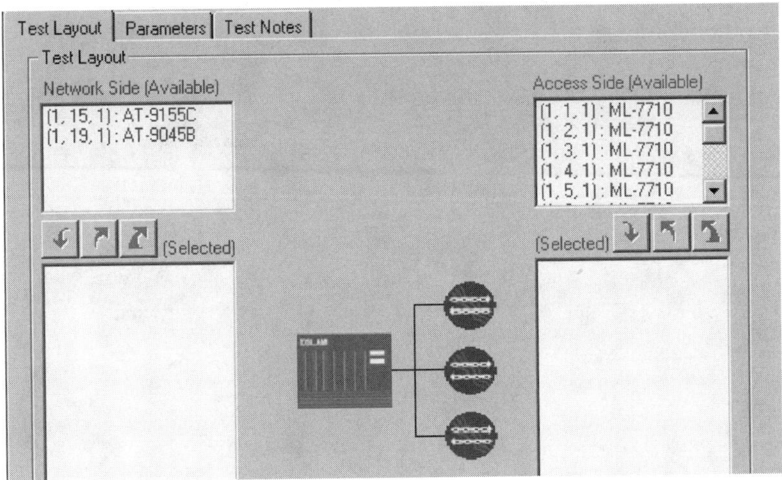

FIGURE B6

6. Figure B8 shows how the screen will look after the cards are selected at both sides.
7. Click the "Stream Definition…" button under the "Test Layout" panel to define the stream parameters.
8. Click the "Network" or "Access" button at the far bottom-left of the window to switch over between the configuration of ATM parameters and Ethernet parameters (Figure B9).
9. In the network stream parameters configuration, change the "MAC address" to "00 00 00 19 19 19" (which indicates the slot number 19 of the AT-9045 card used in this test), and change the "VCI" to "300" according

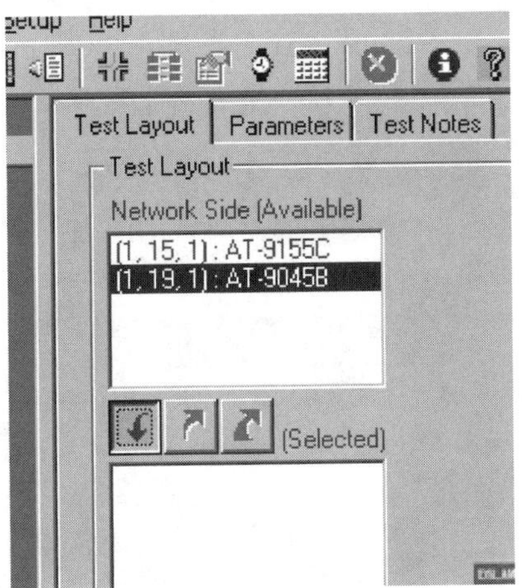

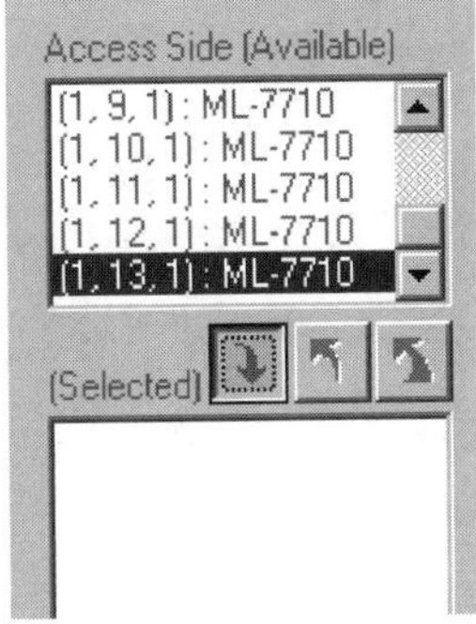

FIGURE B7

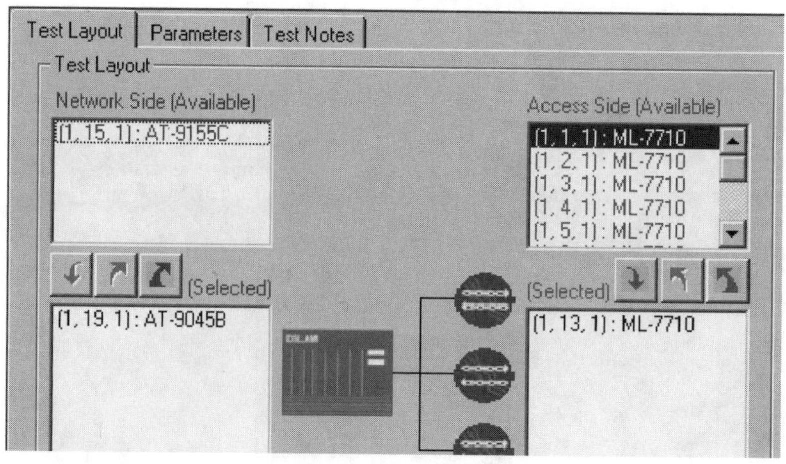

FIGURE B8

to the Stinger trunk module port parameters as listed in the table at the beginning of this document.

10. In the access stream parameters configuration, change the "MAC address" to "00 00 00 13 13 13" (which indicates the slot number 13 of the ML-7710 card used in this test).

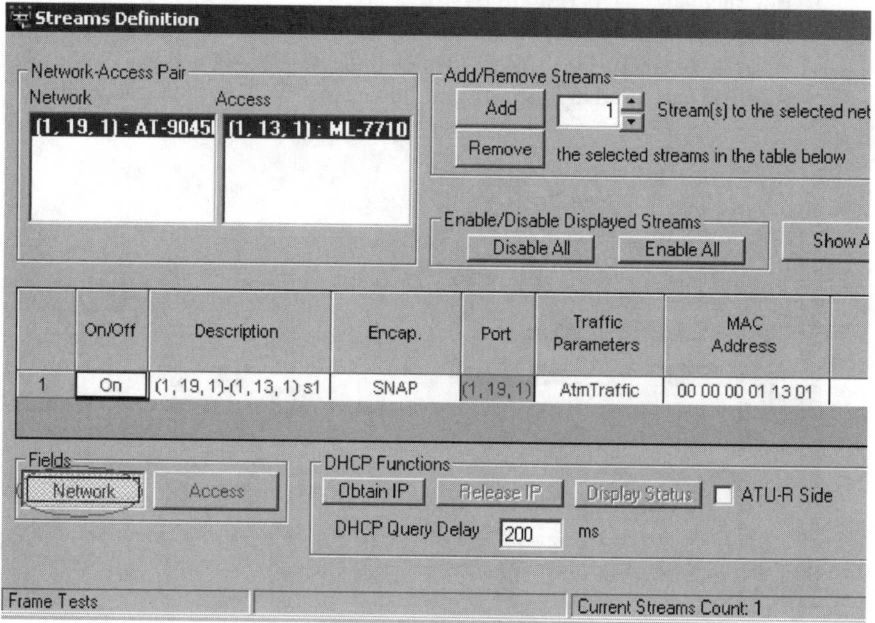

FIGURE B9

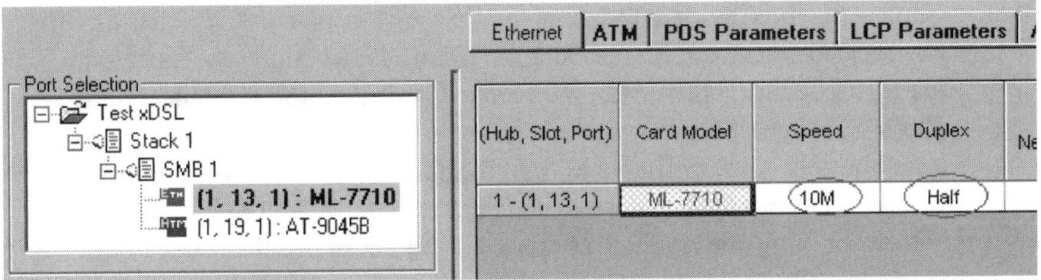

FIGURE B10

11. Click the "Setup Port Parameters" tool button in the toolbar.
12. Select the "ML-7710" card in the left panel. In the "Ethernet" page of the right panel, change the "Duplex" parameter to "Full." Change the "Speed" parameter according to the type of modem. In this example, the modem does not support 100 Mbps, so keep the default value as 10 M (Figure B10).
13. Select the "AT-9045B" card in the left panel. Select the "ATM" page in the right panel. Right click the parameter under column header "Framing Mode," and select "ATM DS3 PLCP" from the popup menu. Click "OK" to close the dialog box (Figure B11).
14. In the right panel of the main window, select the "Parameters" page. Configure the throughput testing parameters in this interface. The "Load Parameters" values are in percentages. The reference value depends on the traffic source. For example, if we select the "Traffic Direction" as "Down-Stream," the traffic source is the DSLAM trunk card, so the reference

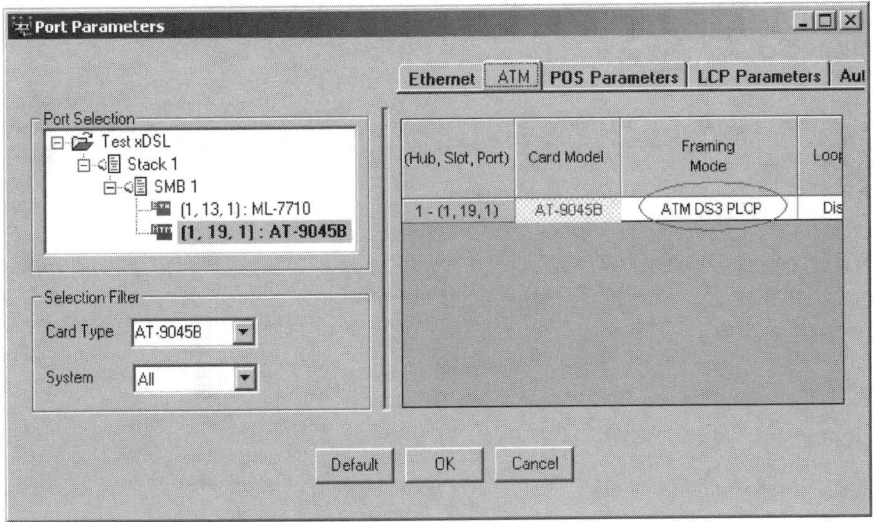

FIGURE B11

value is the maximum data rate of the DSLAM trunk card, which is 45 Mb for the DS3 interface. If we select the "Traffic Direction" as "UpStream," the traffic source is the ML-7710 SmartCard, thus the reference value is 100 Mb. Maybe because of a bug in the software, the "Select Subscriber Rate" button is not available in this interface. But we can get around it by going to the "Frame Loss" page and clicking the "Select Subscriber Rate" button, making the configurations, and then copying the values back to the "Throughput" page.

The "Frame Size" parameters define the starting, ending, and step of the frame size during the test. The "Time Duration per Step" parameter defines how long each step will last. The "Backoff Ratio" is defined as "The rate, measured in percentage, that the transmission rate decreases after a failure at the current rate" (Figure B12).

15. In order to get the load parameters in percentage units, we go to the "Frame Loss" page and click on the "Select Subscriber Rate" button. In this example, the reference data rate is 45 Mbps, and we define the start data rate as 128 kbps (128/1024 = 0.125 Mbps), the step as also 128 kbps, and the max data rate as 6 Mbps. The percentage value is calculated automatically and is shown in the interface (Figure B13).
16. Copy the percentage values to the "Throughput" test parameters (Figure B14).
17. We usually select "Custom" for the "Frame Size" parameter. Click the "Size" button to display the customized frame sizes. These seven packet sizes are the most commonly used sizes. The first one (74) is an exception.

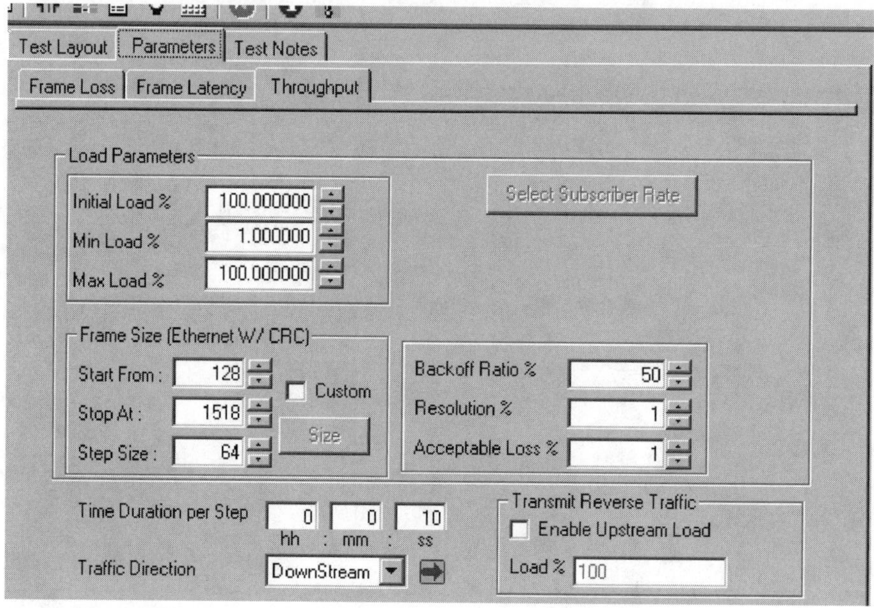

FIGURE B12

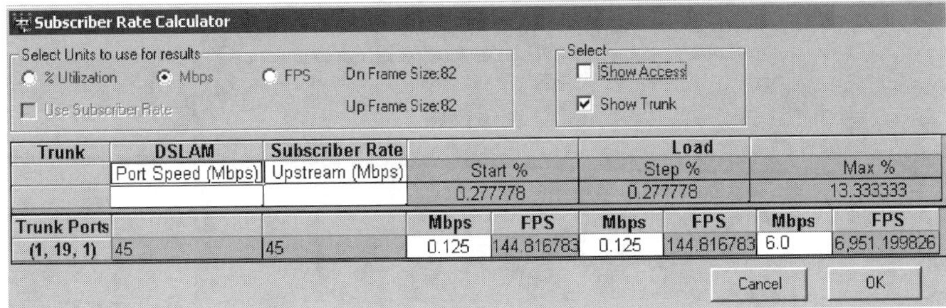

FIGURE B13

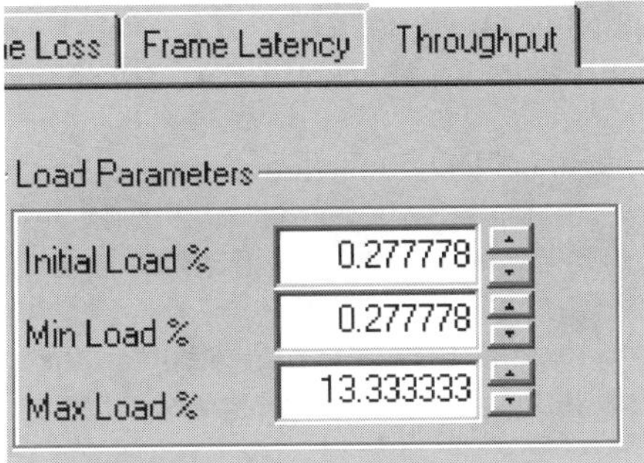

FIGURE B14

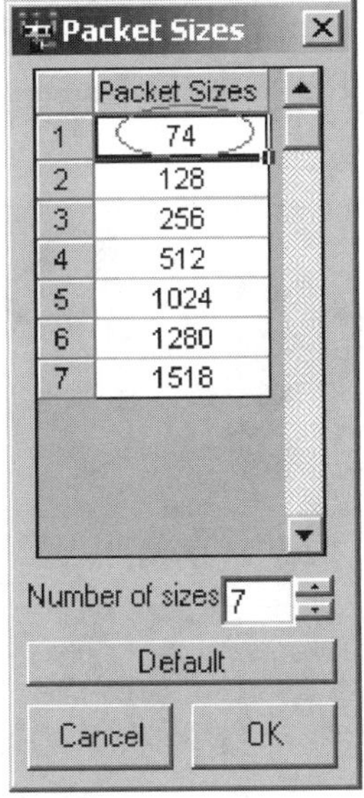

FIGURE B15

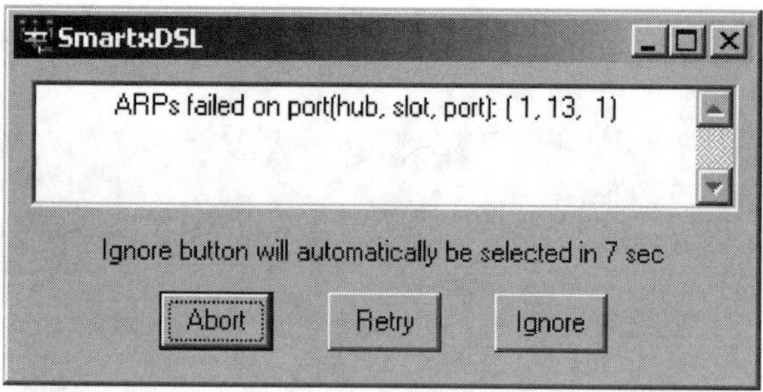

FIGURE B16

The standard frame size should be 64, but the smallest frame size that SmartBits supports is 70 (Figure B15).

18. Keep the default values for "Backoff ratio," "Resolution," and "Acceptable Loss."
19. Click the "Throughput" icon in the left panel to start the test. After a few seconds, a popup dialog box shows that ARP fails. Just click "Ignore" or take no action. The "Ignore" button will be automatically selected in 10 seconds (Figure B16).
20. As the test is running, an Excel file is created and the test result is stored in this file. The test may last for several hours.

TEST RESULTS

The test results automatically appear in an Excel spreadsheet format after the test is initiated. The spreadsheet shows the test data in real time as the test progresses.

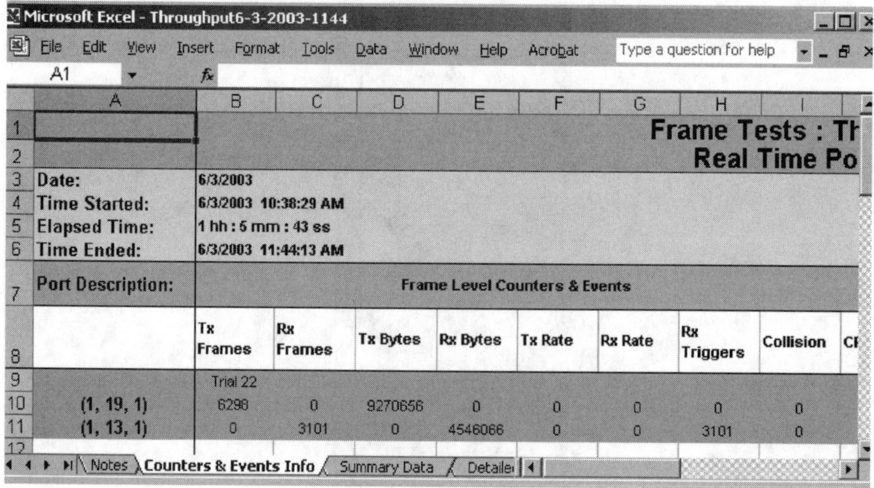

FIGURE B17

Index

D

E

F

G

H

I

J

K

L

M

N

O

P

Q

R

S

T

U

V

W

X

Z